Migration, Environment and Climate Change:

ASSESSING THE EVIDENCE

Edited by
Frank Laczko and Christine Aghazarm

in collaboration with

IOM International Organization for Migration

UNITED NATIONS UNIVERSITY
UNU-EHS
Institute for Environment and Human Security

ccema

with the financial support of

THE ROCKEFELLER FOUNDATION

Contents

Foreword by William Lacy Swing

The consequences of climate change on migration present humanity with an unprecedented challenge. The numbers of storms, droughts and floods have increased threefold over the last 30 years with devastating affects on vulnerable communities, particularly in the developing world. In 2008, 20 million persons have been displaced by extreme weather events, compared to 4.6 million internally displaced by conflict and violence over the same period. How many people will be affected by climate change by 2050? Forecasts vary from 25 million to 1 billion people with a figure of 200 million being the most widely cited estimate.

Extreme environmental events such as cyclones, hurricanes, tsunamis and tornadoes tend to capture the media headlines, but it is gradual changes in the environment that are likely to have a much greater impact on the movement people in the future. For example, over the last 30 years, twice as many people have been affected by droughts as by storms (1.6 billion compared with approximately 718 million).

It is important, however, not to view migration as simply the failure of communities to adapt to climate change. Migration has always been one of the ways in which people have chosen to adapt to changing environments. Migration can also help those left behind in environmentally degraded areas. Studies in Côte d'Ivoire, for example, have shown that migrants who moved from Burkina Faso regularly sent home remittances which were invested in schools and hospitals and in water and irrigation systems. Moreover, migrants are often the first to provide assistance when natural disasters occur. Research in countries such as El Salvador, Jamaica , Botswana and the Philippines has shown that migrant remittances increase significantly when disasters occur providing essential relief assistance to affected communities.

As the world's leading migration agency, the International Organization for Migration (IOM) endeavours to stay abreast of trends and issues that impact the more than 212 million migrants worldwide. Since the early 1990s, IOM has been active in the area of migration, climate change and the environment, and has carried out programmes in more than 40 countries from the Pacific Islands, to Latin America and on the Asian and African continents. In many of these areas, we have assisted those affected by hurricanes, severe flooding and drought.

IOM's programmatic action has constructed a solid foundation of first-hand experiences and lessons learned that have energized the Organization's policy and research. We have sought to enhance our knowledge base through research and publications that examine the complex relationship between migration, climate change and the environment. In doing so, we have been able to identify emerging trends, raise awareness, and work towards innovative solutions that are sensitive to specific local conditions.

The main purpose of this new book is to suggest concrete ways in which the international community can begin to address the huge gaps in our knowledge relating to the likely impact of climate change on migration. The book does this by taking stock of the existing evidence on the effects of climate change and environmental degradation on migration, providing a comprehensive overview of the findings of recent research studies. Throughout, our focus is centred on how research can best inform policy and provide the evidence which decision-makers will need in the future to plan for and respond to environmentally induced migration.

Addressing the unprecedented challenge before us requires unprecedented partnership - collaboration among international organizations, civil society, the private sector, the academic world, and governments. In preparing this new book , we hope to share our expertise with our partners and contribute to global dialogue and efforts within the United Nations Framework Convention on Climate Change and beyond.

William Lacy Swing
Director General

Introduction and Overview: Enhancing the knowledge base

Frank Laczko, Christine Aghazarm[1]

[1] Research and Publications Division, IOM, Geneva

1. Introduction

Over the last few years there has been an upsurge of interest in the likely impact of climate change on population movements. Estimates have suggested that between 25 million to one billion people could be displaced by climate change over the next 40 years. For the most part these figures represent the number of people exposed to the risk of climate change in certain parts of the world and do not take account of the measures that could be taken to adapt to these changes. Although experts have dismissed such figures as, at best, "guesswork" these statistics have helped to focus policy makers' attention on the likely implications of climate change on migration.

Despite the lack of precise figures, there is now little doubt that parts of the earth are becoming less habitable due to factors such as climate change, deterioration of agricultural lands, desertification, and water pollution. The number of natural disasters has more than doubled over the last two decades, and more than 20 million people were displaced by sudden-onset climate-related natural disasters in 2008 (OCHA-IDMC, 2009). Further climate change, with global temperatures expected to rise between 2 and 5 degrees centigrade by the end of this century, could have a major impact on the movement of people. Policy makers are therefore asking the research community and other experts to provide them with guidance in regards to a number of key questions.

First, there is a call for better data to answer questions relating to the likely scale and pattern of movement such as, how many will migrate due to environmental/climate change? Who will migrate? When and where will they migrate; will new destinations have to be found? Will migration be temporary or permanent, internal or international? What will be the consequences of migration for the people who move, for those left behind and for the places of destination? There is also a concern to understand better the here and now – how is environmental change affecting migration today and can we already identify especially vulnerable populations or regions?

A second set of questions concerns policy responses, and how policy makers in countries of origin and destination should address these challenges. Partly the response will depend on how we choose to frame the policy challenge from the outset. There is a tendency at present to frame the policy challenges in fairly negative terms with media headlines often suggesting that millions of people will be uprooted and forced to seek protection in Europe and North America. There is also a tendency to focus on the inadequacy of policies and legal frameworks to assist those displaced due to extreme environmental events, with much less discussion of how migration could help some countries adapt to climate change.

"The literature on climate change and migration is generally very pessimistic about mobility arising from climate change. This creates a starting point bias in thinking about policy responses, eschewing the development of policies that seek to harness migration as a strategy to promote adaptation to climate change..."

(Barnett and Webber, 2009, p.19).

There has been an overwhelming tendency to focus on the negative consequences of migration for the environment, with fewer studies exploring how migration can be a coping or adaptation strategy or how migration can relieve pressure on environmentally degraded areas. The World Bank in a new report on Climate Change and Development (2010) warns that there are risks in presenting the policy challenges linked to environmental migration in too negative terms:

"The negative portrayal of migration can foster policies that seek to reduce and control its incidence and do little to address the needs of those who migrate, when migration may be the only option for those affected by climate hazards. Indeed, policies designed to restrict migration rarely succeed, are often self-defeating, and increase the costs to migrants and to communities of origin and destination."

(World Bank, 2010, p. 25).

Developing an agenda for research which will help the international community understand these complex questions and issues is going to be extremely challenging. In the first instance, it is useful to try to take stock of the existing evidence base and that is the key purpose of this book. The book assesses the existing body of evidence relating to the likely impact of environmental and climate change on migration, and proposes several concrete ways in which to enhance the current knowledge base. In April 2008, IOM together with the UN University Institute for the Environment and Human Security (UNU-EHS), the Munich Re Foundation and the UN Environment Programme (UNEP), organized an expert workshop, "Research Workshop on Migration and the Environment: Developing a Global Agenda for Research", in Munich with financial support from the Rockefeller Foundation to discuss how best to develop a global agenda for research on migration, climate change and the environment. Experts at this workshop all recommended that a systematic review of existing research should be conducted prior to launching any major new studies. In particular, it was felt that such an assessment could help to frame the future research agenda by highlighting gaps in current knowledge, outline conceptual issues and highlight policy challenges. This book and the themes which have been selected are the result of the recommendations from the Munich workshop.

This book focuses on seven key areas of research relating to the topic of migration, the environment and climate change, covering issues such as data challenges, research methods, sudden environmental and slow-onset events, and policy responses. The focus is not limited to climate change as much of the research literature tends to focus on migration and the wider concept of environmental change. The book is mainly focused on the impact of environmental/climate change on migration given the current policy interest in this issue, but it is recognized that there is a considerable body of literature on the impact of migration and refugee movements, on the environment (see Bilsborrow Chapter 3 in this volume).

This book offers a selective review of key research to date on the topic of migration, the environment and climate change within the aforementioned themes. It examines the existing evidence with respect to the ways in which changes in the environment and climate change are affecting the movement of people and the types of policy responses and protection gaps which potentially exist. Furthermore, it offers an overview of innovative approaches to measuring and collecting data on the migration and environment nexus.

The main aims of this first chapter are to contextualize the migration, environment and climate change debate and to provide a summary of the main findings, knowledge gaps and key messages of the seven chapters commissioned for this study. At the end of the chapter we recommend a number of steps that could be taken to enhance understanding of the linkages between changes in the environment and the movement of people.

2. Contextualizing the migration, environment and climate change debate

The migration, environment and climate change nexus is a complex one. By way of background and in order to contextualize the debate, the following section provides a brief overview of the issue in terms of its "re-discovery", the impacts of environmental and climate change on human mobility, its development implications and how the issue links to wider migration and demographic trends.

A 're-discovered' issue

The movement of people as a result of changes in the environment is not a new phenomenon. People have been moving in response to changes in their environment, often seasonally, for centuries. For nomadic peoples and pastoralists such movement is part of their livelihood. However, it is only in the last 20 years or so that the international community has begun to slowly recognize the wider linkages and implications that a changing climate and environment has on human mobility.

As early as 1990, the Intergovernmental Panel on Climate Change (IPCC, 1990:20) warned that "the greatest single impact of climate change could be on human migration" – with millions of people displaced by shoreline erosion, coastal flooding and severe drought. In addition, in 1992 IOM together with the Refugee Policy Group published a report on "Migration and Environment" in which it is stated:

"Large numbers of people are moving as a result of environmental degradation that has increased dramatically in recent years. The number of such migrants could rise substantially as larger areas of the earth become uninhabitable as a result of climate change."

(IOM, 1992)

At the time the issue was framed within a wider security debate, but the momentum did not last. Though it was a first attempt to explicitly link migration with environmental change, the topic of migration and the environment and its inter-linkages was largely ignored by migration experts and policy makers until recently. Indeed, in the 2005 report of the Global Commission on International Migration, there is barely a mention of the topic. Part of the reason for the neglect may be due to the fact that there has been little consensus over the years among researchers about whether or not environmental migration is a distinct form of migration worthy of special study. There has been little consensus between researchers about the relationship between environmental change and migration. As Shurke, 1993 points out, the research literature on environmental migration has tended to fall into two broad categories (1) work done by "minimalists" who suggest that the environment is only a contextual factor in migration decisions and (2) work done by "maximalists" who claim that the environment directly causes people to be forced to move.

Due to several high level conferences, expert meetings, and new research[2] published over the last few years, the issue has re-surfaced. Experts have increasingly raised awareness of the linkages between the environment and human mobility and the importance of unifying these issues at all levels of policy dialogue and cooperation – local, regional, national and global.

Impacts of environmental and climate change on human mobility

Climate change, on its own, does not directly displace people or cause them to move but it produces environmental effects and exacerbates current vulnerabilities that make it difficult for people to survive where they are. Climate change is expected to make the world hotter, rainfall more intense, and result in more extreme weather events such as droughts, storms and floods. These changes, in turn, will likely result in further population movements. According to the UN International Strategy for Disaster Reduction (UNISDR) – storms, floods and droughts – have increased threefold over the past 30 years.

Extreme environmental events such as cyclones, hurricanes, tsunamis and tornadoes tend to capture the media headlines, but gradual

[2] UNFPA-IOM Expert Seminar on Migration and the Environment, Bangkok 2007, IOM, UNU-EHS, UNEP and Munich Re Foundation Research Workshop on Migration and the Environment, Munich 2008 and 2009; Environment, Forced Migration and Social Vulnerability (EFMSV) conference, Bonn 2008

changes in the environment may have a much greater impact on the movement of people in the future. Gradual environmental changes, such as desertification, coastal and soil erosion, tend to be less dramatic and therefore attract less attention than natural disasters. However, gradual and slow-onset changes in the environment tend to affect a larger number of people and will continue to do so in the long term. For example, during the period 1979-2008, 718 million people were affected by storms compared to 1.6 billion people affected by droughts (International Emergencies Disaster Database (EM-DAT), 2009).

As Susan Martin explains in Chapter 7 in this volume, climate change could affect the movement of people in at least four different ways:

1. the intensification of natural disasters;
2. increased warming and drought that affects agricultural production and access to clean water;
3. rising sea levels make coastal areas uninhabitable and increase the number of sinking island states. (44% of the world's population lives within 150 kilometers of the coast);
4. competition over natural resources may lead to conflict and in turn displacement.

Some environmental changes, such as hurricanes and earthquakes, occur with little or no warning and require that people move quickly. Others develop more slowly and may provide time for people to assess their options, leave in an orderly manner and even bring resources with them. However, certain thresholds or "tipping points" may be reached, where there is little choice left but to move. The extent to which the environment, including climate change, is the primary driver of migration remains debatable for several reasons as discussed in Section 3 below. Migration can also affect the environment in terms of additional stress on already degraded lands and competition for scarce resources in both rural and urban settings.

All regions are likely to experience some adverse effects of climate change, but less developed regions are especially vulnerable because a large share of their economies depend on climate-sensitive sectors and their adaptive capacity is low due to low levels of human, financial and natural resources, as well as limited institutional and technological capability (IOM, 2008). Certain "hotspots" – regions or countries already facing environmental, migration and population pressures - are expected to worsen in the coming years such as the sinking small island states, like Tuvalu and the Maldives and in delta regions (*inter-alia* the Mekong, Inner Niger Delta and the Ganges Delta), and regions already facing severe drought and desertification such as the Sahel region.

Development implications: migration as adaptation

The potential benefits of migration for development are now widely recognized at the highest levels beginning with the High Level Dialogue on Migration and Development in 2006, and continuing with the recognition of the impact of remittances on countries of origin and on individual migrants and their families, the role of Diasporas and the work being carried out to integrate migration into poverty reduction strategy papers (PRSPs).

However, as mentioned earlier, discussions of migration triggered by environmental changes, usually see migration as the result of a failure to adapt to the environment, rather than as a possible way of enhancing adaptation to climate change. Migration may itself be one of several adaptation strategies and a coping strategy, for example in the Sahel region in times of drought or as a response to regular flooding in the floodplains of India. The IOM Colombian Temporary and Circular Labour Migration programme, for example, provides an opportunity for families affected by natural disasters to find temporary work abroad (IOM 2009).

The concept of “adaptation” has gained international attention within the discourse on climate change, notably with the context of the Copenhagen Process and beyond. There is increasing recognition that migration has a role to play within this discourse (see Martin Chapter 7 in this volume; IOM 2007, 2008; World Bank 2010; UNDP 2009). Migration when a planned and voluntary coping mechanism can serve as a social safety net for loss of income for example through the sending of remittances, and could potentially serve to alleviate pressure on already degraded lands.

Therefore, bringing together migration, development, climate change and the environment policy perspectives is a priority and challenge for policy makers if the issue is to be addressed holistically.

3. The state of current knowledge and gaps: A summary of key findings

The complexity of the migration and environment nexus as described above, requires not only contextualizing the debate, but also a coherent framing of the issues which surround it. The following sub-sections, while providing a summary of the key findings and knowledge gaps identified throughout the book, also serve to do just that – identify and frame the main issues of relevance for policy makers and researchers alike. They are grouped as follows: (i) conceptualizing the relationship between climate change, the environment and migration, (ii) data and methodological challenges and approaches, (iii) current migration trends in response to sudden and slow-onset disasters, and (iv) policy responses and legal frameworks.

(i) Conceptualizing the relationship between climate change, the environment and migration

> The multi-causal nature of migration represents a challenge in identifying environmental factors as the primary driver of migration. The voluntary/forced migration debate also raises additional issues where "Population mobility is probably best viewed as being arranged along a continuum ranging from totally voluntary migration,...to totally forced migration" where reality is somewhere in between (Hugo, 1996). How those who migrate for environmental reasons are defined has implications for both legal frameworks and research purposes.

Although many experts accept that environmental degradation and climate change are factors which can impact on the decision to migrate, the conceptualization of these factors as a primary cause of migration or forced displacement has been questioned (Black, 2001). Given the multi-causal nature of migration, which can result from a combination of various "push" and "pull" factors that can be inter-alia economic, social, political, establishing a direct causal link is a challenge. The key is assessing the extent to which the environment or climate change is the primary driver or simply one of many drivers of migration (see Kniveton Chapter 2, Warner et al. Chapter 4 in this volume).

Whether movements in relation to environmental or climate change are forced or voluntary is also the subject of much debate. In some situations, such as natural disasters, people may have little choice but to move. However, migration patterns beyond the immediate move may shift between forced and voluntary. In other situations where environmental change is gradual, movement is more likely to be voluntary and linked to other economic, social and political factors.

Drawing a clear line between voluntary and forced movements is not always straightforward. Therefore, it is perhaps more useful, instead, to think in terms of a continuum:

"Population mobility is probably best viewed as being arranged along a continuum ranging from totally voluntary migration,... to totally forced migration, very few decisions are entirely forced or voluntary."

(Hugo, 1996)

(i) Definitions

Definitions are crucial in two ways 1) in guiding the policies of governments and international agencies in regards to how to respond to population movements; and 2) in the generation of statistics which also depends on how those who migrate for environmental reasons are defined (see ii.a. below).

Terms, such as "climate change refugee" or "environmental refugee" are widely used in the media but these terms are a misnomer under international law and risk undermining the very precise legal definition of a refugee and the protection regime which exists (see Zetter Chaper 8 in this volume). Moreover, as the bulk of environmental migration tends to occur within countries rather than between countries it makes more sense to talk of internally displaced persons than refugees, a term which is defined in relation to cross-border movement. Others have argued that "the term 'environmental refugee' is simplistic, one-sided and misleading. It implies a mono-causality which very rarely exists in practice" (Castles, 2002, p.8).

The term "environmental refugee" was first popularized by Lester Brown of WorldWatch Institute in the 1970s and further by El-Hinnawi in the early 1990's. The term "environmental refugees" has been used to describe the whole category of people who migrate because of environmental factors. This broad definition, while evoking an image that has brought public attention to the issue, is not sufficiently precise to describe all the various types of movements which may be linked to environmental factors.

The terminology and definitional issues have been taken up by various experts and international agencies. Notably, the Inter-Agency Standing Committee (IASC) Working Group on Migration/Displacement and Climate Change has devised a typology which summarizes the nature of movement, the affected persons and the protection framework under which those who are moving may be granted protection (2008).

In the absence of an internationally agreed definition, IOM developed a working definition in 2007 which defines "environmental migrants" as follows:

"Environmental migrants are persons or groups of persons who, for compelling reasons of sudden or progressive change in the environment that adversely affects their lives or living conditions, are obliged to leave their habitual homes, or choose to do so, either temporarily or permanently, and who move either within their country or abroad."

The purpose of this definition is to try to encompass population movement or displacement, whether it be temporary or permanent, internal or cross-border, and regardless of whether it is voluntary or forced, or due to sudden or gradual environmental change.

(ii) Research on migration, environment and climate change: data and methodological challenges and approaches

A persistent lack of data is one of the primary challenges to measuring the migration and environment nexus, while data collection on migration and the environment represents a challenge in itself. A range of methodologies exist which can be utilized and enhanced for research purposes.

The topic of environmental migration has often been studied within separate fields in the natural and social sciences or within a sub-set of the larger field of environment studies. How the subject has been framed impacts on both data collection as well as the methodological approaches used. Though migration theory does historically take into account environmental indicators, it is only recently that it has received renewed attention.

(ii.a) Data collection challenges

As mentioned earlier, there are many estimates of the total number of people who will be displaced (incidentally figures usually do not refer to numbers expected to migrate) due to climate change. These figures are often called into question (see Kniveton et al. Chapter 2 in this volume) because they rely on crude population estimates and they assume that populations will permanently leave areas affected by climate change. Probably the most reliable data are the statistics relating to the impact of natural disasters on population movements. The findings of a joint report by the United Nations Office for the Coordination of Humanitarian Affairs (OCHA), the Internal Displacement Monitoring Centre (IDMC) and the Norwegian Refugee Council (NRC) show that at least 36 million people were displaced by sudden onset natural disasters that occurred in 2008. More than half of the displacement – 20 million persons – was estimated to be climate-related.

There are several factors which make it difficult to measure current levels of environmental migration and to predict the likely scale of future follows. It is extremely difficult to predict the impact of climate change and climate modelling techniques to date have not yet begun to account adequately for the impact of individual choice, the potential for international action and the variability of future emissions and meteorological scenarios (see Kniveton et al. Chapter 2, Bilsborrow Chapter 3 and Leighton Chapter 6 in this volume; Brown, 2008).

The absence of an adequate definition to cover migrants affected by climate and environmental change, discussed above, also presents a challenge for statistics gathering. Disaggregating the role of climate or other environmental change from other economic, political and social factors which drive migration, while also taking into account migrants' perception and behaviour in relation to such change, is a difficult task. Furthermore, challenges exist as well in how terms and concepts such as 'environment', 'disasters', 'adaptation', 'vulnerability', etc. are defined and utilized, thereby adding to the complexity of data collection (see Warner et. al Chapter 4, Naik Chapter 5 in this volume).

There is a basic lack of migration data available, especially in developing countries which are likely to be most vulnerable to climate change. Many countries still do not include basic questions about migration in their censuses. Even in the current 2010 census round, many important countries still do not ask where people were born – including Japan, Mexico, Korea, the Philippines, and Egypt. Roughly a third of countries do not ask about previous residence in another country (Center for

Global Development (CGD), 2009). There is a lack of reliable surveys, in addition to the census, for many key parts of the world. Though small scale local case studies provide a valuable context specific analysis of a certain area and/or environmental phenomenon, they need to be expanded and "scaled up."

Moreover, few social scientists who focus on migration (relying on data from censuses and household surveys) have been engaged in data collection or research on the environment. Similarly, relatively few of those who focus on the environment, whether social or natural scientists, work on migration. However, as the interest in climate change and its impacts on population movement has increased at the global level, more interdisciplinary cooperation has begun.

Finally, lack of data collection and research capacity, especially in developing countries most vulnerable to environmental changes, remains a serious issue. In its recent research strategy paper, the United Kingdom's Department for International Development (DFID), highlights the current lack of research capacity in developing countries (2008). There are only 48 researchers for every million Africans living south of the Sahara compared with nearly 3,000 for every million people in OECD countries Therefore, building the capacities of researchers in these regions is a crucial component of addressing the research and data challenge.

(ii.b) Methodological approaches

Data collection challenges are not easily separated from methodological ones and inherently impact upon the results of the approach utilized. What data is being collected, both migration (ie. length of residence, characteristics of migrants and non-migrants, household composition, remittances) and environmental (i.e. land use, energy use, land degradation, drought and flooding) are crucial in order to understand the linkages between the two.

Several methodological approaches do exist which are promising and would benefit from further enhancement and inter-disciplinary cooperation. These range in aim and scope from *inter-alia* assessments such as impact, vulnerability and adaptation assessments, climate modelling such as agent based modelling, the use of remote sensing data to measure environmental variables and household and community based surveys (for more detail on methods see Kniveton et al. Chapter 2, Bilsborrow Chapter 3 and Warner, et al. Chapter 4 in this volume).

Additional methods include modelling techniques such as agent-based modelling, not often used in climate studies, which aims to capture the decision making behaviour of individuals while also being able to predict such behaviour as well as the interplay between these decisions and larger scale outcomes. Such multi-agent approaches offer a different perspective to traditional economic models, which consider individuals as rational actors making decisions to maximize their well-being with unlimited cognitive resources. Instead agent-based models underline individuals do not necessarily make decisions in isolation, have social interactions and their perceptions may be biased or incomplete. Other modelling techniques created to integrate and correlate socio-economic data and biophysical datasets could potentially be used within a migration and environment context bearing in mind the caveats, in particular to migration data, referred to previously.

Surveys, both household and community, present yet another set of approaches to measuring migration in relation to the environment. For household surveys, specific modules can be integrated which include questions to identify migrants from the household, work activity of the migrant prior to migration, reasons for migration, work at the last destination and remittances received. For the environment these would cover issues related to land use, production, degradation, change in weather, etc. Remote sensing from a series of satellite images can also be used to analyse land changes over time and scale. At the community level, surveys are most relevant in rural areas where boundaries are most easily defined. Bilsborrow specifically highlights two case studies using these methods in Chapter 3: Guatemala, where questions were added to a Demographic and Health Survey (DHS) on migration and the environment highlighting the value of adding questions to existing surveys. The second case in Ecuador used an innovative multiple source and method approach integrating samples of households and communities through surveys, together with satellite imagery over a period of eight years.

The EACH-FOR project, as analysed by Warner, et al. is another example of a project which attempted to isolate the environment as the primary variable or driver in migration through a series of small scale surveys in 23 countries. Though the samples were not representative on a national or regional scale, they did identify several research challenges such as defining "the environment" issues of temporal and geographic scales as well as migrants' perceptions of their environment and risk which are all important considerations in undertaking future work in this field. The EACH-FOR project helps to lay the groundwork for further larger scale studies.

(iii) Current migration trends in response to sudden and slow-onset disasters

Migration which does occur in response to both sudden and slow on-set environmental events exhibits varied characteristics which can be organized according to the following spectrum: voluntary – forced; temporary – permanent; internal – international; vulnerability – resilience (see Naik Chapter 5 in this volume). Vulnerability and adaptive capacity are further influenced by several factors such as inter-alia gender, age and ethnic background which may in turn inform decisions to migrate at the societal, household and/or individual level.

Migration related to both sudden and slow-onset events are likely to be predominantly internal, with movements being rural-rural and rural-urban, and in terms of international migration generally to the nearest border – the Pacific small island developing States represent a particular case where "statelessness" could be an issue. Longer distance international migration requires financial resources and social networks which facilitate such a move. Such movements which do occur do so mainly along already existing migration patterns.

With regard to sudden onset disasters, a clear finding is that migration post-disaster takes on various patterns based on the nature, intensity and duration of the disaster, the group affected and the location (see Naik Chapter 5 in this volume for a more detailed discussion of migration post sudden onset natural disasters; see also Warner et al. Chapter 4 in this volume for additional case examples). Though data does exist on the incidence of natural disasters and the overall numbers of those affected, data is lacking on the pattern and scale of migration after disasters occur.

Most of the population movement following disasters tends to be short-distance and temporary in nature. Though the initial movement may be on the forced side of the spectrum, subsequent movements may also be more voluntary. Though the numbers of natural disasters is increasing, there has not been a significant increase in international migration for the reasons explained previously. Even in the case of the Asian Tsunami in 2004, there was little out-migration to neighbouring countries in Asia; most of those affected by the Tsunami were displaced within their own countries (see Naik Chapter 5 in this volume).

Exposure to risk and adaptive capacity are also critical factors which influence migration decision making processes. While a disaster may influence movement, it is important to note that migration does not always occur, as the most vulnerable may lack the means to migrate. Economic factors play a decisive role in both the risk and response to natural disasters, where developing countries are most vulnerable due to lack of resources to prevent, respond and cope with their effects. Under certain circumstances, disaster affected locations may even draw in migrants, such as in the aftermath of Hurricane Katrina (Donato, 2006).

There are numerous similarities between sudden and slow-onset disasters as described above primarily in terms of vulnerability and adaptive capacity and their intervening socio-economic factors as well as the predominantly internal nature of most movements (for a detailed discussion on migration in relation to slow-onset events see Leighton Chapter 6 in this volume; see also Bilsborrow Chapter 3 and Warner Chapter 4 et al. for additional case examples).

Slow-onset events and gradual degradation such as droughts and desertification, can also directly threaten rural household income sources, especially those from agriculture, leading many rural agricultural families to diversify their income which may entail migration. Slow-onset events may provide more time with regard to the decision making process, therefore making the voluntary/forced spectrum unclear in certain instances. Seasonal or circular migration patterns have been identified as a coping strategy, for example, at the end of the growing season in Mali, Senegal, Ethiopia, Argentina and India among others. Such movements can also be correlated with periods of extended drought and loss of agricultural production which in some circumstances may lead to permanent migration as a survival strategy. International migration has also been documented during periods of drought, though primarily within the same region such as the Sahel, but also from Mexico to the United States. However, it has also been suggested that drought may actually inhibit and lead to a decline in longer distance international migration, for example from Mali to France, during drought years, as the financial resources needed to undertake long distance travel are affected.

Again as case study examples are context specific it is difficult to draw overarching conclusions or make generalizations with regard to migration patterns and trends. What also remains unknown is the extent to which climate change may exacerbate existing environmental and socio-economic problems in the future and potentially shift migration patterns from temporary to more permanent migration and/or to other destinations, in particular leading to more international migration.

(iv) Policy responses and legal frameworks

> More efforts are needed to identify, test and implement new programmes, policies, frameworks to manage future movements of people linked to environmental and climate change. The capacities of governments to implement existing ones need to be enhanced. Migration can be among several adaptation strategies as exemplified in several National Adaptation Programmes of Action (NAPAs). Countries of destination lack coherent policies to address potential future flows. Measures specifically targeting migration in relation to the environment are ad hoc at best.

Discussions of mechanisms to manage environmental migration are still in their early stages. There is a widespread perception that climate change may lead to a substantial increase in migration to the Global North but as illustrated in sub-section (iii) movements are most likely to be internal or to the closest international border within a region. International migration beyond the nearest border is likely to largely follow existing migration patterns established by migrant social networks.

As Susan Martin explains in her chapter in this volume, due to the complexity of the migration, environment and climate change nexus, policy makers have been slow in identifying and formulating policy responses. In order to manage the movement of people more effectively due to environmental or climatic change, policy makers need to take into account the full spectrum of responses at various stages of movement from prevention, mitigation and adaptation to migration (voluntary and forced) to return or resettlement and finally integration in the final destination area (for a detailed discussion of policy responses and legal frameworks see Martin Chapter 7 and Zetter Chapter 8 in this volume).

(iv.a) National policy responses

Adaptation in least developed countries

Migration may be considered as one of several adaptation strategies in response to changes in the environment. Adaptation may take on various forms, including - the institutional level, technological developments, community development, education and training initiatives. Remittances play an important role in development and post-disaster recovery, thereby also contributing to adaptation (see Naik

Chapter 5 and Leighton Chapter 6 in this volume). In terms of national policy responses and adaptation, National Adaptation Programmes of Action (NAPA), have been developed for least developed countries (LDCs) to respond and adapt to climate change. However, criticisms of the NAPA process include that they have not been integrated within other processes which seek to promote development and reduce vulnerability such as poverty reduction strategies (UNDP, 2008).

Nonetheless, NAPAs are among the few policy instruments that exist which specifically refer to migration in relation to adaptation policies, even if the references are often to the need to reduce migratory pressures or limit the negative impact of migration on urban areas. There are cases where migration in itself is cited as an adaptation strategy either as a way to reduce pressure on eco-systems or in terms of planned resettlement where trends point to this inevitable need. The inclusion of migration overall, however, remains limited. More broadly as highlighted by Susan Martin, few of the major middle income developing countries which are the main source countries of international migrants, such as Mexico, India and China, have included any reference to migration in their climate change adaptation plans.

Policies in countries of destination

Few destination countries have elaborated specific policy measures to respond to the international movements of environmental migrants. None currently have a pro-active policy to resettle those affected by environmental disasters. Most migrants would fall under already established labour migration or family reunification schemes. Other measures are generally on an ad-hoc basis usually in response to natural disasters focused on exceptions to deportation for affected persons already on the territory. For example, after the Asian Tsunami in 2004 several countries such as Canada, Switzerland the United Kingdom suspended deportations of those from affected areas. Further still, some countries have special policies allowing individuals to remain temporarily without fear of deportation in the event of natural disasters, armed conflict and other severe situations as is the case in the United States with the designation of temporary protected status (TPS).

(iv.b.) Legal frameworks and protection gaps

"There is a general consensus that there is little opportunity to create an entirely new set of legal instruments to address environmental migration, but rather the possibility of revising those legal instruments that already exist in order to include this conceptually "new" category of migrants" (Zetter Chapter 8 in this volume). Through an analysis of: human rights law, environmental law, subsidiary and complementary protection, regional Instruments, ad hoc disaster protection regulations and the 1998 Guiding Principles on Internally Displaced Persons, Zetter asserts that the issue of environmental migrants or environmentally displaced persons is and could be already included within existing frameworks. In particular the experience of the 1998 Guiding Principles on Internally Displaced Persons, may provide an interesting starting point. However, the lack of capacity of States most vulnerable to the effects of climate change to implement existing frameworks is a critical weakness as is a detailed empirical understanding of the availability, scope and efficiency of current rights-based frameworks.

Furthermore, as Zetter explains in detail in his chapter, important protection gaps do remain especially with regard to "trans-border" movement and in cases where people can become "stateless" as their country disappears due to climate change. Although our focus in this book is on migrants and migration, he also reminds us that it is equally important to ensure that norms and legal instruments protect the rights of persons who do not migrate and who may be among the most vulnerable groups affected by climate change.

4. Recommendations for further policy-oriented research

Though important steps have been taken in recent years to enhance the state of knowledge on the interactions between migration, the environment and climate change as highlighted by the main findings of this book in Section 3, wide-ranging knowledge gaps continue to exist. Below we suggest some priority areas for further research and policy analysis.

(i) Shifting the research agenda and the framing of key issues

Research on migration tends to focus primarily on the study of international migration, rather than internal migration, and most studies tend to be conducted by scholars in the North. To better understand the likely impact of climate change will likely require a shift in the migration research agenda towards a stronger focus on internal migration, and more emphasis on understanding South-South migration where most migration is likely to take place. From a policy perspective we need to keep in mind that most migrants move within developing countries or between developing countries, and it is these countries which will be most affected by climate change. Therefore we will need a research agenda which gives greater priority to informing policy in the South if we are to develop adequate policy responses to environmental migration.

Furthermore, future research needs to be more balanced to include a greater focus on slow-onset changes in the environment. Though extreme environmental events receive a great deal of media and policy attention, gradual changes in the environment may have a much greater impact on the movement of people. The future research agenda will also need to explore in much more detail how migration can contribute to efforts to adapt to climate change, in order to understand better the potential benefits of mobility.

Additionally, as all the chapters have highlighted, data collection and research methodologies would benefit from broader and larger scale inter-disciplinary work. Though slowly changing, the two fields, migration and the environment, remain separate both in terms of data collection, methods and at the level of policy responses and analysis.

(ii) Towards better data and methods

It is hard to persuade policy makers of the likely importance of climate change related migration without reliable figures. Several measures could be taken to improve data on environmental migration, beginning with the establishment of a special commission consisting of data experts from the migration, environment, development and humanitarian fields who would be tasked with developing new practical ways of making better use of existing data sources as well as developing new indicators of environmental migration.

Establishing a Commission on Migration and Environment Data (CMED)

One way to address the data and methods challenges discussed earlier is to establish a Commission on Migration and Environment Data (CMED). A first task of such a commission could be to develop different definitions of population movement linked to environmental and climate change. Clearer definitions of what is being measured are needed to more accurately assess the scale of environmental migration. The data commission would bring together experts and representatives from agencies which collect data on migration and the environment to work together on developing a practical set of recommendations, identifying in the first instance some low-cost measures which could be taken in the short-term to advance understanding of the nexus between migration, the environment and climate change. For example, it would be useful to investigate the extent to which migration or environment questions could be added to existing databases, administrative registrars and surveys. For example, the authors of this chapter recently met with researchers working in Brussels compiling data for the EM-DAT database and discussed the possibility of adding migration related questions to the existing template which is used to collect data on natural disasters. More and more people are being affected by disasters but it is not clear how displacement impacts on migration trends within and between countries affected by disasters.

Other measures which could be taken by such a data commission, could be to promote the systematic sharing of existing data through the creation of an online clearinghouse, and to advocate for specialized surveys to produce new data. The commission could work towards promoting the sharing of existing data and innovative new research findings in a more systematic fashion through the creation of an online Clearinghouse on Migration and the Environment. While it is important to make better use of existing data sources, there is nonetheless a pressing need for new larger-scale specialized surveys to help fill some of the major data gaps. It is evident from the earlier analysis (see especially Bilsborrow Chapter 3 and Leighton Chapter 6 in this volume) that there has been relatively little large-scale empirical work carried out recently to investigate the migration-environment nexus. For example, Bilsborrow notes that in a review of literature over the last 50 years, of 321 publications, including 153 articles in peer-reviewed journals and 29 books, only two articles were found which investigate the effects of environmental factors on out-migration based on quantitative multivariate methods.

More specialized household surveys which include both migration and environmental questions are therefore urgently needed. The recent and extremely valuable EACH-FOR project based on 23 small-scale case studies around the world, which is presented in this volume, needs to be complemented by larger-scale more nationally representative studies. Larger-scale studies are also needed to map, in a more systematic fashion, the populations around the world which are likely to be most vulnerable to increased pressure to migrate due to climate change. While not all people exposed to environmental stresses are able to or perceive the need to move, vulnerability assessments allow the identification of regions where pressure to migrate might be expected to increase. A number of methods exist to assess vulnerability including the formation of expert groups and statistical analyses to select and combine indicators of vulnerability to climate impacts (see Kniveton et al. Chapter 2 in this volume).

There is also a need for better data to help develop more robust forecasts of likely future trends and scenarios. Integrated assessments, agent based models, GIS mapping could be developed that are able to simulate complex emergent social phenomena and these should be used to develop scenarios of future migration flows linked to future climate and environmental change.

Furthermore, there is a need for comparative research to help us understand better why in some circumstances environmental pressures contribute to migration and in others they do not. Comparative research, comparing countries or regions facing similar environmental pressures but which result in sometimes quite different migration trends is required.

Improved adaptation assessments

Adaptation assessments are able to incorporate the range of social-economic, cultural and psychological connections and contexts within which decisions to migrate or not are made and carried out.According to Kniveton in his chapter, at national level expert groups and statistical analyses should be carried out to select indicators of migration drivers and combine these to identify populations where migration as a response to increased environmental pressure might be expected to increase. Furthermore, current and past conceptualizations of the migration decision making process provide knowledge of the drivers of the migration decision, while information on migration flows allows for the assessment of the validity of the indicators chosen. However, the authors in this volume caution against trying to focus new research on whether and where climate change is the one predominant indicator of migration, but instead recommend that new studies should focus on what role climatic changes play among the other determinants of migration.

Identification, evaluation and monitoring of policy responses

Though the emphasis and focus has largely been on the need for better data and research to inform policy, there is also a need to document, monitor and analyse policy responses to environmental migration from prevention, adaptation, resettlement to managing future flows. Much of the emphasis of current policy in regards to climate change and migration tends to be on extreme environmental events. This policy discussion tends to focus on questions such as how best to provide emergency assistance to those who are displaced, how to reduce disaster risk and how to improve the legal and normative framework for the protection of the displaced. However, it is equally important to consider movements due to slow-onset events and frame the issue both in terms of displacement and more voluntary movement.

In short, the chapters in this volume suggest that studies of current policy responses could fall under the following four headings 1) analysis of appropriate legal and policy frameworks (in terms of protection, migration management) which would be of value to policy makers with particular attention to known possible protection gaps such as statelessness and cross-border movement; 2) identification of adaptation strategies which would allow people to remain or to migrate as a voluntary coping strategy; 3) identification of resettlement strategies which protect a communities' rights and livelihood through a systemic evaluation of past resettlement experience and lessons learned so that future resettlement respects the right of those resettled; and 4) an analysis of the response mechanisms of humanitarian organizations with regard to disaster to better understand not only the response, but the gaps, needs for information and the perspective of those affected.

(iii) Capacity Building

As mentioned earlier, it is likely that most environmental migration is likely to occur within and between developing countries where research capacities are likely to be very limited. It is essential therefore that new programmes for research on migration, environment and climate change include a research capacity-building component which will enable developing countries to build up their own knowledge base. It will also be necessary to invest in new data collection systems and to build capacities to make better use of existing data sources in some of the poorest countries of the world which will be most affected by climate change.

Managing environmental migration will also present new challenges for policy makers in developing countries where governance structures are already very weak. Therefore, capacity building should extend beyond the research community to the policy realm as well.

One concrete way of beginning to link the research and policy domains in terms of capacity building would be to evaluate the current capacities for national governments to implement existing and/or new frameworks or policies. As outlined above, there is considerable scope for adapting or building upon existing norms, instruments and programmes to develop a framework, guidelines or policies to protect those displaced for environmental reasons. Therefore, new research would be especially needed to explore how far governments and civil society at the national level have the capacity to implement both protection policies and adaptation and resilience strategies.

5. Concluding remarks

Fears that millions of people from some of the poorest countries in the world could be forced to migrate to richer parts of the world due to climate change, have led to a renewed interest in research on migration and the environment. In advancing the research agenda on this topic, however, it is important not to define the future agenda for research on migration and the environment too narrowly. It is essential to start from the position that migration is not always the problem, but can in certain circumstances, where migration contributes to adaptation, be part of the solution. In short, migration linked to climate change will create both risks and opportunities.

It is also necessary to keep in mind that the relationship between the environment and migration is complex, which will require much more nuanced studies in the future. It is far too simplistic to assume that environmental and climate change will in all circumstances automatically result in the increased movement of people. To take but one example, mentioned earlier, the number of persons affected by natural disasters has more than doubled in recent years but we have not seen a major increase in international migration in many of the disaster affected regions.

Finally, in addition to investing in new research along the lines mentioned in the preceding section, it is important to ensure that existing data sources and research findings are utilized by policy makers, especially those in developing countries who may have poor access to such information. This will likely require special efforts to foster closer dialogue and cooperation between researchers and policy makers in the future.

6. References

Barnett, J. and Webber, M.
2009 "Accommodating Migration to Promote Adaptation to Climate Change", Swedish Commission on Climate Change and Development, Stockholm, www.ccdcommission.org.

Black, R.
2001 "Environmental Refugees: Myth or reality?" UNHCR working paper 34, UNHCR, Geneva.

Black, R. et al.
Demographics and climate change: future trends and their policy implications for migration. Report prepared for the Department for International Development (DFID) by the Development Research Centre on Migration, Globalisation and Poverty, University of Sussex.

Brown, O.
2008 "Migration and climate change", International Organization for Migration Research Series, IOM, Geneva. Available at: http://publications.iom.int/bookstore/index.php?main_page=product_info&products_id=96.

Castles, S.
2002 Environmental change and forced migration: making sense of the debate, Working paper No. 70. UNHCR, Refugee Study Centre, Oxford University, Oxford.

Center for Global Development (CGD)
2009 "Migrants Count: Five Steps Toward Better Data," Report of the Commission on International Migration Data for Development Research and Policy, Washington DC.

Donato, K. & Hakimzadeh, S.
2006 "The Changing Face of the Gulf Coast: Immigration to Louisiana, Mississippi and Alabama", *Migration Information Source*. Available at: http://www.migrationinformation.org/Feature/display.cfm?ID=368

El-Hinnawi, E.
1985 Environmental refugees, Nairobi, United Nations Environment Programme.

EM-DAT (International Emergencies Disaster Database)
2009 Available at: http://www.emdat.be/index.html

Hugo, G.
1996 "Environmental concerns and international migration," *International Migration Review*, 30(1): 105–131.

Hugo, G.
2009 "Migration, development and environment", International Organization for Migration Research Series, IOM, Geneva.

Inter-Agency Standing Committee (IASC), informal group on Migration/Displacement and Climate Change
2008 "Climate Change, Migration and Displacement: Who will be affected?" Working paper, Geneva, 31 October 2008.

International Organization for Migration (IOM)
1992 *Migration and the Environment*, IOM and the Refugee Policy Group (RPG), Geneva.

2007 Discussion Note: Migration and the Environment, MC/INF/288, International Organization for Migration (IOM). Available at: http://www.iom.int/jahia/webdav/site/myjahiasite/shared/shared/mainsite/microsites/IDM/workshops/evolving_global_economy_2728112007/MC_INF_288_EN.pdf

2007 "Expert Seminar: Migration and the Environment," International Dialogue on Migration no. 10, IOM, Geneva.

2008 *World Migration 2008, Managing Labour Mobility in the Evolving Economy,*IOM, Geneva.

2009 "Migration, Climate Change and the Environment", IOM Policy Brief. Available at: http://www.iom.int/jahia/webdav/shared/shared/mainsite/policy_and_research/policy_documents/policy_brief_envmig.pdf

2009 *Compendium of IOM's Activities on Migration, Climate Change and the Environment*, Geneva.

Intergovernmental Panel on Climate Change (IPCC)
1990 *First Assessment Report*, Cambridge University Press, Cambridge.

2007 *Climate Change 2007: Impacts, Adaptation and Vulnerability*. Contribution of Working Group II to the Fourth Assessment Report of the Intergovernmental Panel on Climate Change, M.L. et al. (Eds), Cambridge University Press, Cambridge.

Kniveton, D., et al.
2008 "Climate change and migration: improving methodologies to estimate flows", International Organization for Migration Research Series, IOM, Geneva. Available at: www.iom.int/jahia/webdav/site/myjahiasite/shared/shared/mainsite/published_docs/serial_publications/MRS-33.pdf.

Naik, A. et al.
2007 "Migration, development and natural disasters: Insights from the Indian Ocean tsunami", International Organization for Migration Research Series, IOM, Geneva. Available at:http://www.iom.int/jahia/webdav/site/myjahiasite/shared/shared/mainsite/published_docs/serial_publications/MRS30.pdf

OCHA/IDMC
2009 *Monitoring disaster displacement in the context of climate change - Findings of a study by the United Nations Office for the Coordination of Humanitarian Affairs and the Internal Displacement Monitoring Centre*, http://www.internal-displacement.org/8025708F004BE3B1/(httpInfoFiles)/12E8C7224C2A6A9EC125763900315AD4/$file/monitoring-disaster-displacement.pdf

Suhrke, A.
1993 "Pressure Points: Environmental Degradation, Migration and Conflict," Paper prepared for the workshop on "Environmental Change, Population Displacement, and Acute Conflict," held at the Institute for Research on Public Policy, Ottawa, June 1991. Available at: http://www.cmi.no/publications/1993%5Cpressure_points.pdf

UK Department for International Development (DFID)
2008 DFID Research Strategy 2008-2013," Working Paper Series: Climate Change, DFID, London.

UN International Strategy for Disaster Risk Reduction (UN ISDR)
2007 "World experts unite to confront growing risks of disasters," Pres release UN/ISR 2007/8, Geneva.

United Nations Department of Economic and Social Affairs, Population Division
2006 The world population in 2300. Available at: www.un.org/esa/population/publications/longrange2/WorldPop2300final.pdf.

United Nations Department of Economic and Social Affairs, Population Division
2009 Trends in global migration stock: the 2008 revision; New York

2006 Tends in global migration stock: the 2005 revision, New York. Available at: www.un.org/esa/population/publications/migration/UN_Migrant_Stock_Documentation_2005.pdf.

United Nations Development Programme (UNDP)
2007 *Human Development Report 2007/2008, Fighting Climate Change: Human Solidarity in a Divided World*, UNDP, New York. Available at: http://hdr.undp.org/en/media/HDR_20072008_EN_Complete.pdf

2009 *Human Development Report 2009, Overcoming Barriers: Human Mobility and Development,* UNDP, New York. Available at: http://hdr.undp.org/en/media/HDR_2009_EN_Complete.pdf

World Bank
2010 *World Development Report 2010, Development and Climate Change,* The International Bank for Reconstruction and Development / The World Bank, Washington DC.

Challenges and approaches to measuring the migration–environment nexus

Dr. Dominic Kniveton, Christopher Smith, Richard Black, Kerstin Schmidt-Verkerk[1]

[1] School of Global Studies, University of Sussex; E-mail: d.r.kniveton@sussex.ac.uk

1. Setting the scene

Introduction

The impact of environmental change on global society is a matter of increasing concern for policy makers and the wider public as awareness of human-induced climate change increases. Rising sea levels, deforestation and dryland degradation, as well as natural disasters, pose challenges in terms of their effect on development and livelihoods, settlement options, food production and health. These environmental events and processes have been predicted to lead to the large-scale displacement of people – both internally and internationally – with estimates of some 200 million to 1 billion migrants resulting from climate change alone, by 2050 (Jacobson, 1988; Myers, 1997, 2002; Stern et al., 2006).

The variation in migratory responses to climate change has been shown by a number of events. At one extreme, the experience of Hurricane Katrina in 2005 showed that a single climatic event can cause considerable population displacement. In contrast, studies of the Sahel have shown that, rather than encouraging migration, decreases in rainfall and subsequent bad harvests tend to limit household abilities to invest in long-distance movement (Findley, 1994; Henry et al., 2004). It has thus been argued that accurately predicting climate change-induced migration is almost impossible (Black, 2001), due to uncertainties in the extent and magnitude of the climatic signals responsible for pushing and pulling migrants, and to the variation in the contexts and perception of climate threats and, thus, the behaviour of the people upon whom they have an impact.

While climate change dominates debates about the environment and migration, the nexus of the two also extends to the impacts of natural hazards, such as earthquakes and volcanic eruptions, and technological hazards, including nuclear accidents and chemical spills. Climate change and natural and technological hazards share problems of predictability; however, climate change is unique in being global in its extent and attributable to universal human activity (albeit largely due to lifestyles in developed countries). Strictly speaking, the climate system incorporates natural and technological hazards, given its definition as an interactive system consisting of five major components: the atmosphere, the hydrosphere, the cryosphere, the land surface and the biosphere. This chapter will refer to both environment and climate as interacting with migration, with the process of anthropogenically induced climate change forming a subset of environmental phenomena.

Before embarking on a discussion of the challenges and approaches to measuring the migration and environment nexus, it is worth considering the aims and objectives to be achieved by this measurement. Within the context of climate change, decision makers are increasingly calling upon the research community to provide (Carter et al., 2007):

- good-quality information on what impacts are occurring now, their location and the groups or systems most affected;
- reliable estimates of the impacts to be expected under projected climate change;
- early warning of potentially alarming or irreversible impacts;
- estimation of different risks and opportunities associated with a changing climate;
- effective approaches for identifying and evaluating both existing and prospective adaptation measures and strategies;
- credible methods of costing different outcomes and response measures;
- an adequate basis for comparing and prioritizing alternative response measures, including both adaptation and mitigation.

In terms of migration, the objectives of the research are to:

a) document current levels of migration that involve environmental factors in shaping migration decisions;
b) assess the degree to which existing migration flows – both 'voluntary' and 'involuntary' – are sensitive to future climate change and variability, providing a basis for estimation of future flows that is set in the context of existing migration patterns and drivers;
c) establish models of migration processes in order to understand and identify the thresholds and ranges of climate conditions that are likely to significantly increase migration, or fundamentally change its nature;

d) identify appropriate policy responses to migration and assessing their relative impact on future flows of migration and the underlying social, ecological and economic conditions of the populations affected.

In order to achieve these objectives, we need to first examine the spatial and temporal characteristics of migration and environmental processes to determine the scales of their interaction before reviewing conceptual models of migration and society–environment interactions upon which to base a strategy of measurement and model development.

The term 'model' is used varyingly in different disciplines to describe conceptualizations of processes or theories; attempts to artificially simulate (often in a laboratory setting) the processes being studied, called hardware modelling; and to numerically reproduce fundamental processes and feedback in the system being analysed. Due to the global nature of climate change, hardware modelling is often not applicable to impact studies of climate change, leaving much of the work on simulating impacts of future climate to numerical modelling. Importantly, the development of numerical models allows for the testing of hypotheses, testing of the sensitivity of processes to changes in one or more of the parameters of the model, theory-testing and building, future scenario simulation and policy formation, and testing for conditions outside of those experienced in the past. This latter ability is particularly apt for a changing climate whereby a unique set of conditions may fall outside those experienced previously by society. In this chapter, the term 'model' will be confined to discussions of methodologies relating to numerical or theoretical type models.

Time and space

The environment varies on a range of timescales – from the minutes it takes for an earthquake to occur, to the millennia of glacial cycles and ice ages. It also varies in terms of its spatial impact – from a few kilometres, in the case of a tornado, to climate change at the global level. In terms of migration, we are interested in timescales of up to a century long and all spatial scales. Migration also varies temporally – from temporary migration, to seasonal, annual, decadal and permanent – and spatially, ranging from local to regional, rural to rural, rural to urban, and international. Like environmental processes, migration also varies in terms of magnitude – from displacement of individuals to mass migrations. A number of studies have attempted to integrate the time and space scales of environmental processes and migration to gain insight into the different types of environmental migrants (Renaud et al., 2007; Raleigh, 2008). The goal here is not to create different typologies of migrants but, rather, to show that there is a wide variety of scales of interaction between migration and the environment, each with possibly different characteristics (see table 1). For instance, a much-cited study by Henry et al. (2004) found that short-term rainfall deficits tend to increase the likelihood of short-term migration for Sahelian households to rural areas, yet decrease the likelihood of long-term moves to distant destinations. Calculating the number of possible combinations of interactions between environmental processes and migration responses (see table 1) gives an indication of the size of the task needed to further the understanding of the global environment–migration nexus. The resulting 120 possible interactions need to be researched and explored.

Table 1: Range of spatial and temporal scales of environment and migration processes

Environmental processes		Migration response	
Time	Space	Time	Space
Gradual Sudden	Local National Global	Temporary Seasonal Yearly Permanent	Local Rural to rural Rural to urban Urban to rural International

Environmental change context

"Human alteration of Earth is substantial and growing. Between one-third and one-half of the land surface has been transformed by human action; the carbon dioxide concentration in the atmosphere has increased by nearly 30 per cent since the beginning of the Industrial Revolution; more atmospheric nitrogen is fixed by humanity than by all natural terrestrial sources combined; more than half of all accessible surface fresh water is put to use by humanity; and about one-quarter of the bird species on Earth have been driven to extinction. By these and other standards, it is clear that we live on a human-dominated planet."

(Vitousek et al., 1997: 494)

Environmental variability and change result from both natural processes and human activities. Undoubtedly, in the next few centuries, changes will be dominated by human alteration, with the impact of anthropogenically induced climate change chief among the forcings. Predictions of future climate changes are inherently unreliable, with questions about how emissions of climate-altering gases and particles might change in the future; the degree of natural variability in the climate system; and our limited scientific understanding of climate system processes and feedbacks. Despite these uncertainties, there exists strong scientific consensus that changes in the composition of the atmosphere from human activities are likely to produce: 1) a warmer future climate, with warming greatest over land and at most high northern latitudes, and least over the Southern Ocean and parts of the North Atlantic Ocean; 2) a contraction of snow cover area, increases in thaw depth over most permafrost regions and a decrease in sea ice extent; 3) Arctic late-summer sea ice disappearing almost entirely by the latter part of the 21st century; 4) an increase in the frequency of hot extremes, heat waves and heavy precipitation, and an increase in tropical cyclone intensity; 5) a pole-ward shift of extra-tropical storm tracks, with consequent changes in wind, precipitation and temperature patterns; and 6) precipitation increases in high latitudes and likely decreases in most subtropical land regions (IPCC, 2007a). The impacts of these changes on water availability have been predicted to include an increase in annual river runoff at high latitudes (and in some tropical wet areas) and decreases in some dry regions in the mid-latitudes and tropics. There is also high confidence that many semi-arid areas (e.g., the Mediterranean Basin, western United States of America (USA), southern Africa and north-eastern Brazil) will suffer a decrease in water resources due to climate change (IPCC, 2007b).

Particular systems and sectors most likely to be affected by climate change include: the tundra, boreal forest and mountain regions, because of their sensitivity to warming; Mediterranean-type ecosystems and tropical rainforests, because of reductions in rainfall; mangroves, coral reefs and salt marshes, due to multiple stresses; the sea ice biome, because of its sensitivity to warming; water resources in some dry regions at mid-latitudes and in the dry tropics, due to changes in rainfall and evapotranspiration, and in areas dependent on snow and ice melt; agriculture in low latitudes, due to reduced water availability; and low-lying coastal systems, due to the threat of sea-level rise (IPCC, 2007b). Regions most likely to be affected include the Arctic, because of the impacts of high rates of projected warming on natural systems and human communities; Africa, because of low adaptive capacity and projected climate change impacts; small islands, where there is high exposure of population and infrastructure to projected climate change impacts; and Asian and African mega-deltas, due to large populations and high exposure to sea-level rise, storm surges and river flooding

(IPCC, 2007b). Considering climate change alone, it can be seen that the livelihoods of many people around the world are likely to be under increasing pressure in this century from environmental change. In the next section, the nature of the interaction between the environment and migration will be outlined.

Environment and migration

The impact of the environment on human migration has been illustrated throughout the Anthropocene, with evidence of the movement of both humans and their remote ancestors, hominins, in response to large-scale climatic change, stretching back 2 million years (Dennell, 2008). Yet a review of the literature reveals that the nexus between climate change, environmental degradation and migration in contemporary society has not been explored empirically in a way that generates conclusive results (Black et al., 2008). A large number of empirical studies have attempted to understand the causes of migration, employing methodologies and concepts borrowed from economics, sociology, geography and political science. It is generally agreed that the key drivers of migration are:

(a) factors related to the region or country of origin, including political instability and conflict, lack of economic opportunities, and lack of access to resources ('push' factors);

(b) factors related to the region or country of destination, including the availability of employment and demand for workers, higher wages, political stability or access to resources ('pull' factors);

(c) intervening factors that facilitate or restrict migration, including ease of transportation, family or social networks, government immigration or emigration policies, economic ties such as trade and investment linkages, or social and cultural exchanges.

However, what is not so clear is how these different factors interact with each other to inform migration behaviour or, more importantly, for the purpose of this chapter, the extent and magnitude of the role that the environment plays in these decisions, with respect to the other socio-economic drivers. Studies of climate-induced migration in the past have commonly calculated the numbers of 'environmental refugees' by projecting physical climate changes, such as sea-level rise or rainfall decline, on an exposed population (TERI, 1996; Nicholls and Tol, 2006; Warren et al., 2006). These studies generally make simplistic assumptions about populations' ability to cope with variations in climate, and often assume that migration reflects a failure to cope. Yet, in reality, migration responses are the result of much more complex behavioural decisions. Migration already involves many hundreds of millions of people worldwide; it ranges from seasonal to temporary and permanent moves; and it may be local, national or international in scope. Crucially, migration may represent a planned adaptation to climate change or uncertainty, or it may be a last resort – a decision made in distress when other options have failed. It is within this varied and complex context that this chapter aims to review methods for measuring the environment–migration nexus.

Migration–environment studies form a subset of a larger study subject of environment–society interactions. The following sections explore different approaches used to measure these interactions and assess their applicability to migration studies. The next section looks at the methods used in studies of climate change impact, adaptation and vulnerability (CCIAV). The standard climate scenario-driven approach (impact assessment) is used in most climate change-based assessments of climate–society interactions, but other approaches are increasingly being used. These include assessments of current and future adaptations to climate, adaptive capacity, social vulnerability, multiple stresses, adaptation in the context of sustainable development and risk-management frameworks.

In conceptual terms, classical migration frameworks offer potential for the specific inclusion of the environment as a contextual characteristic. Studies related to modelling migration have been based around statistical analyses and methodologies taken largely from economics, geography and sociology. Section 3 will explore how migration conceptualizations and approaches have been and can be incorporated into the measuring and modelling the nexus of migration and the environment. Lastly, we examine the uncertainty and data requirements for measuring the migration–environment nexus. The concluding section of this chapter brings together the findings of the previous sections to present a seven-point action plan of research needed for measuring the migration and environment nexus.

2. Climate change impact, adaptation and vulnerability (CCIAV)

The last report of the Intergovernmental Panel on Climate Change (IPCC) in 2007 concluded that estimates of the number of people who may become environmental migrants were, at best, guesswork (Wilbanks et al., 2007) because:

(a) migration in areas affected by climate change were not all necessarily one-way and permanent, as assumed by previous attempts to estimate them, but multidirectional and often temporary or episodic;

(b) there are often multiple and complex reasons for people migrating and these reasons do not relate straightforwardly to climate variability and change;

(c) while migration can be a longstanding response to seasonal variability in environmental conditions, it also represents a wealth accumulation strategy or a route out of poverty that benefits receiving and original countries or regions;

(d) of a lack of reliable censuses or surveys for many key parts of the world, on which to base such estimates;

(e) of a lack of agreement on the definition of an environmental migrant (Unruh et al., 2004; Eakin, 2006).

These criticisms point to a lack of data about people's behaviour and a complexity in migration decisions that are not represented by current analyses of the environment–migration nexus. However, a number of methods for measuring and modelling climate–society interactions in other areas of societal behaviour do exist. These include the standard climate scenario-driven (impact assessment) approach, adaptation- and vulnerability-based approaches, and integrated assessment. The various characteristics of the different approaches are given in Table 2.1.

Impact assessment

The impact approach is the most widely used of the CCIAV techniques and focuses on evaluating the likely impacts of climate change under given scenarios (Carter et al., 2007). Used most often to model impacts on physical systems, this approach suffers from the large uncertainty and variability in future climate change scenarios, which causes the impact assessments to become invalid as new scenarios are produced, and a top-down logic that promotes the propagation of pressure–response relationships between climate and the system being assessed for impact. More recent attempts to use this approach have emphasized the importance of providing a socio-economic and technological context for characterizing future climatic conditions, the presentation of impacts, in terms of probabilities, and the use of control runs for comparisons.

Table 2.1: Characteristics of different approaches to CCIAV assessment

	Approach			
	Impact	**Vulnerability**	**Adaptation**	**Integrated**
Scientific objectives	Impacts and risks under future climate	Processes affecting vulnerability to climate change	Processes affecting adaptation and adaptive capacity	Interactions and feedbacks between multiple drivers and impacts
Practical aims	Actions to reduce risks	Actions to reduce vulnerability	Actions to improve adaptation	Global policy options and costs
Research methods	Standard approach to CCIAV Drivers-pressure-state-impact-response (DPSIR) methods Hazard-driven risk assessment	Vulnerability indicators and profiles Past and present climate risks Livelihood analysis Agent-based methods Narrative methods Risk perception including critical thresholds Development/sustainability policy performance Relationship of adaptive capacity to sustainable development		Integrated assessment modelling Cross-sectoral interactions Integration of climate with other drivers Stakeholder discussions Linking models across types and scales Combining assessment approaches/methods
Spatial domains	Top-down Global -› Local	Bottom-up Local -› Regional (macro-economic approaches are top-down)		Linking scales Commonly global/ regional Often grid-based
Scenario types	Exploratory scenarios of climate and other factors (e.g., SRES) Normative scenarios (e.g., stabilization)	Socio-economic conditions Scenarios or inverse methods	Baseline adaptation Adaptation analogues from history, other locations, other activities	Exploratory scenarios: exogenous and often endogenous (including feedbacks) Normative pathways
Motivation	Research-driven	Research-/ stakeholder-driven	Stakeholder-/ research-driven	Research-/stakeholder-driven

Source: Carter et al., 2007

A number of studies have used an impact approach to explore the impact of raised sea levels on current coastal populations (Fankhauser, 1995; Nicholls and Tol, 2006; Olsthoorn et al., 2008; Nicholls et al., 2008). In figure 2.1, the population land area and GDP exposed to mean sea-level rise as a function of sea level is shown from Nicholls et al., (2008). In table 2.2, the minimum impact of a one- and five-metre sea-level rise on area, population and total income are shown, assuming no coastal adaptation. Unlike those in figure 2.1, these data are based on change relative to high water. Importantly, these data are calculated assuming that all assets and population in this area are lost or forced to move.

Table 2.2 (based on 2000 data)

	1-m rise	5-m rise
Population (millions)	131	410
Land area (thousand Km^2)	2,463	4,107
GDP, N, MER (billions US $)	1,015	2,425
GDP, R, MER (billions US $)	1,009	2,482
GDP, N, PPP (billions US $)	1,132	2,959
GDP, R, PPP (billions US $)	1,239	3,342

Source: Nicholls et al., 2008

Table 2.2. The global exposure of population, land area and total income as a function of sea-level rise, calculated relative to high water. LandScan and GPW3 are different population datasets; PPP represents purchasing power parity; and MER indicates GDP measured at market rates.

Figure 2.1

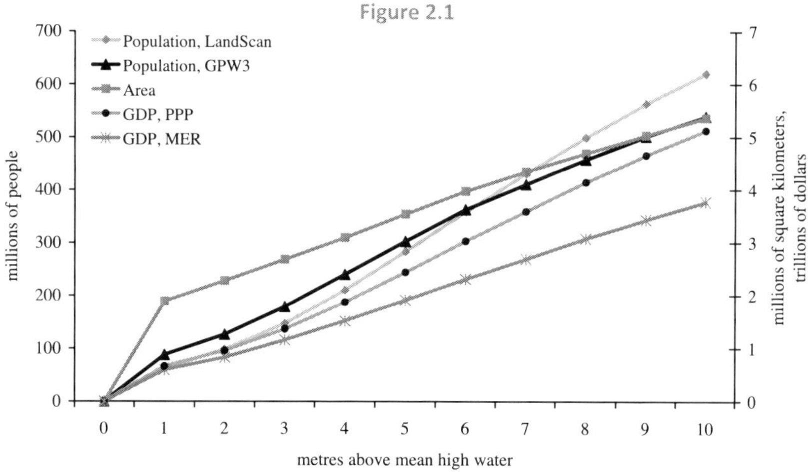

Source: Nicholls et al., 2008

Figure 2.1. Population, land area and GDP as a function of elevation above mean sea level, based on 1995 data. LandScan and GPW3 are different population datasets; PPP represents purchasing power parity; and MER indicates GDP measured at market rates.

The idea that no adaptation will occur in response to rising sea level is, of course, simplistic, since coastal defences may be erected and land excavated to create raised land for habitation. In order to evaluate the impact of sea-level rise with adaptation a number of authors have assumed a cost-benefit approach whereby the costs of the direct submergence/loss of land are compared to the costs of coastal protection (Frankhauser, 1994; Yohe et al., 1996; Nicholls et al., 2008). These analyses ignore other impacts of increased flood risk, saltwater intrusion and storm surges on land and population. Figure 2.2 shows the estimated global annual forced migration when including adaptation for various scenarios of sea-level rise. The different curves in this graph indicate the speed with which the West Antarctica Ice Sheet (WAIS) collapses, with the year indicating the end of the collapse. As can be seen from this diagram, without a WAIS collapse, 75,000 people per year are predicted to be displaced as sea level rises, but this rapidly falls to 5,000 people a year as sea defence standards progressively improve.

The risk of WAIS collapse is currently contested. There was already a large disintegration ice sheet in March 2002 that initiated the acceleration of glaciers discharging into the sea. Glacial acceleration has also been witnessed in recent years in Greenland; yet current ice sheet models incorporated into climate models do not include the processes thought to be responsible for this increased movement and discharge into the sea. Paleoclimatic studies point to past increases of several metres in less than 100 years and there remains the legitimate fear that current knowledge cannot rule out a return to such conditions (Overpeck et al., 2006). While the IPCC's latest report recognizes the possibility of a larger ice sheet contribution to sea-level rise, its main quantitative predictions of sea-level rise neglect this inclusion in its calculations. Oppenheimer et al. (2007) suggest that the omission in the latest IPCC report of a numerical estimate of a potential contribution to sea-level rise from the West Antarctica Ice Sheet, although provided in a previous report, reflects a lack of consensus arising from the inadequacy of ice sheet models compared to recent observations.

With a WAIS collapse, forced migration due to sea-level rise is predicted to increase dramatically, even with adaptation, peaking to 350,000 forced migrants a year for the decade around 2050. This figure represents a 70-fold increase of those affected, compared to a scenario whereby the WAIS never collapses. In the most extreme scenario of ice sheet collapse, 15 million people would be displaced from 2030 to 2130, but it should be noted that this is only 2 to 3 per cent of the total exposed if no adaptation occurs (Nicholls et al., 2008). These results illustrate the importance of considering adaptation in assessing impacts and, for sea-level rise, include the finding that the non-uniform distribution of population allows for relatively small lengths of defences to be effective protection for disproportionate numbers of population (e.g., coastal defences for 50 per cent of land protect 95 per cent of the coastal population). Other studies of extreme sea-level rise for specific regions indicate higher levels of migration as socio-ecological thresholds, such as an overall loss of confidence, are exceeded, triggering increased abandonment (Olsthoorn et al., 2008; Lonsdale et al., 2008; Poumadere et al., 2008). As stated above, the analyses based on the cost-benefit of sea defence adaptation also ignore low-probability extreme events above defence standards and systematic failures, as shown by Hurricane Katrina in New Orleans. In geographical terms, Asia has the highest exposure of population to sea-level rise, with its large coastal population vulnerable to coastal flooding. Africa is also likely to experience a substantially increased exposure, with East Africa (e.g., Mozambique) having particular problems due to a combination of tropical storm landfalls and large projected population growth, in addition to sea-level rise (Nicholls et al., 2008).

Figure 2.2: Global annual forced migration as a function of time and WAIS scenario

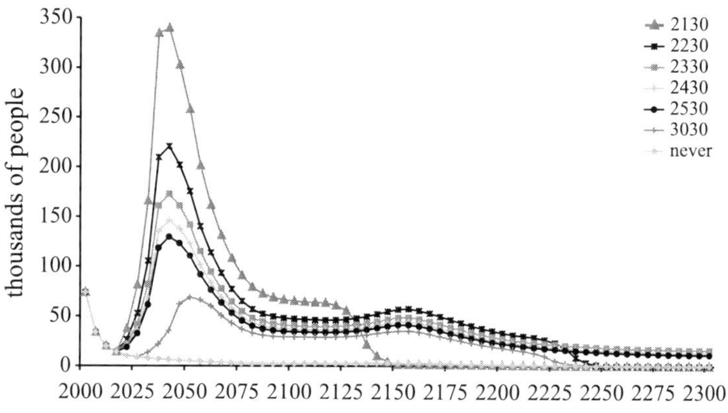

Source: Nicholls et al., 2008

In terms of the environment–migration nexus, therefore, the impact approach can provide useful estimates of forced migration from the loss of land, assuming both adaptation and no adaptation. However, these estimates do not take into account the impact of occasional flooding from storm surges where the relationship between migration and hazard are likely to be more complex. By extension, other features of climate and environmental change where the impact is less absolute, such as water stress, present difficulties to the measurement of the environment–migration nexus using current impact approaches.

Vulnerability assessments

The term 'vulnerability' was originally defined in the climate change literature as the extent to which climate change may damage or harm a system and was deemed dependent not only on system sensitivity but also on its ability to adapt to new conditions. More recently, the term has been extended to include social vulnerability and to encompass vulnerability to current climate, vulnerability to climate change in the absence of adaptation and mitigation measures, and residual vulnerability whereby adaptation and mitigative capacities have been exhausted. Various vulnerability assessment frameworks and methods have been proposed to integrate social and biophysical dimensions of vulnerability (Klein & Nicholls, 1999; Polsky et al., 2003; Turner et al., 2003; Ionescu et al., 2005; Metzger & Schröter, 2006) but these are currently considered very preliminary (Carter et al., 2007). Vulnerability assessments are often associated with assessments of adaptive capacity, in order to examine the potential responses of a system to climate variability and change (O'Brien & Vogel, 2006). Quantitatively, indicators such as national economic capacity, human resources and environmental capacities have been used for vulnerability assessments (Moss et al., 2001), with adaptive capacity being determined by measuring economic wealth, technology and infrastructure, information, knowledge and skills, institutions, equity and social capacity (Yohe & Tol, 2002).

The two key steps in both adaptation and vulnerability assessments are, first, determining the factors that are indicative of adaptive capacity and vulnerability to a particular climate hazard and, second, deciding how to combine these factors to form one index. Regional studies have included using stakeholder elicitation and surveys (Eakin et al., 2006; Pulhin et al., 2006) and multi-criteria modelling (Wehbe et al., 2006). To map vulnerability of Indian agriculture to climate change, O'Brien et al. (2004) obtained indices for adaptive capacity and sensitivity. To measure the former, they followed the method used in the United Nations Development Programme's Human Development Index (UNDP, 2005) to calculate normalized biophysical, social and technological indices. In this example, the various components of the indices included the biophysical indicators of soil conditions (quality and depth) and groundwater availability; the socio-economic indicators of adult literacy rates and gender equity, to measure human and social capitals, respectively; and the percentage of district workforce employed in agriculture, and of landless labourers in the agricultural workforce, to measure the presence of alternative economic activities. Technological indicators used included the availability of irrigation and quality of irrigation.

Climate sensitivity was expressed in this study as an index of drought and a measure of monsoon dependency, both derived from rainfall data. The factors of adaptive capacity and sensitivity were then summed up to form a map of climate change vulnerability. The same study also examined vulnerability to globalization by replacing the climate sensitivity index with one based around exposure to import competition, calculated using a basket of locally produced crops that may be subject to competition from imports and affected by the distance of the district from the nearest port. Validation of this approach was attempted through participatory rural appraisal techniques in local-level case studies. These case studies also allowed for the identification of state and local-level institutions and policies that influenced coping and adaptation strategies used by farmers.

An alternative method of determining vulnerability was illustrated by Thornton et al. (2006), who used a workshop setting to determine 14 indicators of vulnerability to climate change in Africa, which they then reduced to four orthogonal factors by 'principal components analysis' (PCA) and used these four factors to construct an overall indicator of vulnerability. PCA is used to reduce the complexity of multivariate datasets to a set of factors that are orthogonal and, thereby, avoid problems of correlation between indicators. Since the four factors are products of a statistical analysis, their interpretation is not straightforward, although Thornton et al. (2006) suggest they are related to public health expenditure and food security issues; human diseases and governance; the human poverty index (HPI), based on social capital and internal renewable water resources; and market access and soil degradation. In this study, the overall indicator was calculated from the weighted sum of the four components based on the percentage of variance explained by each component.

As an alternative to constructing the components of vulnerability indices using expert groups, Brooks et al. (2005) developed an empirical method of examining the statistical relationships between a large number of potential proxies for national vulnerability and measures of mortality from climate-related disasters. In their work, they focused on *generic* determinants of vulnerability, including developmental factors such as poverty, health status, economic inequality and elements of governance, as distinguished from *specific* determinants relevant to a particular context and hazard, such as the price of a particular crop, the number of storm shelters or the existence of regulations concerning the robustness of buildings. As in other studies, the choice of variables used was constrained by data availability but, from an original list of 46 variables, Brooks et al. used correlation analysis to identify 11 statistically significant relationships between various socio-economic and other variables and mortality risk. The resulting key indicators from the statistical analysis were: population with access to sanitation; literacy rate among 15-24-year-olds; maternal mortality; literacy rate for over 15-year-olds; calorific intake; voice and accountability; civil liberties; political rights; government effectiveness; literacy ratio (female to male); and life expectancy.

Evaluation of the appropriateness and relative importance of the indicators was carried out using a focus expert group. To combine the different indicators, Brooks et al. (2005) chose to divide the range of data for each indicator into quintiles and assign different weights to each indicator. The sensitivity of the composite to the subjectively derived weights was calculated and found to be relatively robust, although it was noted that the length of time the data for each indicator was available and the averaging of data for countries of very different socio-economic characteristics and hazard profiles restricted the interpretation of the results of the analysis. A comparison with 11 key indicators developed for vulnerability to food security in Kenya at district level from a World Food Programme study shows some identical factors of health, education and governance but also illustrates the difference between national generic indicators and more specific local-scale ones.

The abovementioned studies illustrate the use of vulnerability assessments to explore the extent to which climate change may damage or harm a particular sector of society. The studies differ in terms of which indicators were used to construct vulnerability indices according to the scale of vulnerability being assessed, the particular hazard for which the vulnerability is being exposed to, and the methods used to construct the indices. In terms of migration, some data exist on migration flows that can be explored to derive vulnerability indices, using both expert groups and statistical analyses. However, the concept of vulnerability by definition might be questioned as to its appropriateness for migration–environment studies. As noted earlier, migration is not only seen as a last resort but also as a means of wealth accumulation or a route out of poverty and can also be outside the reach of those most vulnerable to climate change impacts. However, the vulnerability lens does allow for a more complete analysis (relative to impact assessments) of the context in which environmental pressures relate to migration processes and might be applicable to situations in which populations have no choice but to migrate. Further vulnerability analyses can be used to identify locations most likely to be affected by climate change impacts and, thus, regions where the pressure to adapt is likely to be highest (with migration being one possible means of adapting).

Adaptation assessment

In recent years, adaptation assessment has shifted from being a research-driven activity to one whereby stakeholders participate in order to improve decision making. As a result, it has become more anchored in adaptation to past and present climate variability and change, which not only has the advantage of assessing climate impacts in terms of what is already known, but also helps reduce the influence of uncertainty from variable scenarios of future climate (Carter et al., 2007). Adaptation assessment incorporates a wide range of methods, including:

- the scenario-based approach, whereby most impact assessments consider future adaptation as an output;
- normative policy frameworks that explore which adaptations are socially and environmentally beneficial, and apply diverse methods, such as vulnerability analysis, scenarios, cost-benefit analysis, multi-criteria analysis and technology risk assessments (UNDP, 2005);
- indicators, employing models of specific hypothesized components of adaptive capacity (see, for example, Moss et al., 2001; Yohe & Tol, 2002; Brooks et al., 2005; Haddad, 2005; and see section 2.2 for examples of indicators of adaptive capacity);
- economic modelling, and anthropological and sociological methods for identifying learning in individuals and organizations (Patt & Gwata, 2002; Tompkins, 2005; Berkhout et al., 2006);
- scenarios and technology assessments, for exploring what kinds of adaptation are likely in the future (Dessai & Hulme, 2004; Dessai et al., 2004; Klein et al., 2005);
- risk assessments combining current risks to climate variability and extremes with projected future changes, utilizing cost-

> benefit analysis to assess adaptation (see, for example, Asian Development Bank (ADB), 2005).
>
> (Carter et al., 2007)

As discussed in the last section on vulnerability assessments, adaptation and vulnerability share some common methodologies, including the creation of indicators of adaptation capacity. Like vulnerability, adaptive capacity has generic dimensions, including such factors as education, income and health, and specific dimensions, such as institutions, knowledge and technology. However, there is a lack of consensus on the usefulness of indicators of adaptive capacity (Downing et al., 2001; Moss et al., 2001; Yohe & Tol, 2002; Brooks et al., 2005; Haddad, 2005). A comparison of five vulnerability assessments by Eriksen and Kelly (2007) revealed little consistency among the 20 countries ranked most vulnerable and showed that national indicators failed to capture many of the processes and contextual factors that influence adaptive capacity. Thus, they concluded that national indicators provided little insight at the local level, where most adaptations take place (Eriksen & Kelly, 2007).

The importance of specific contextual factors, such as the nature of the relationships between community members and the extent of access and participation in the decision making process, has been recognized in a number of recent studies of vulnerability and adaptive capacity (Leichenko & O'Brien, 2002; Allison et al., 2005; Schröter et al., 2005; Belliveau et al., 2006). Adaptive assessment studies have recognized the importance of social capital, social networks, values, perceptions, customs, traditions and levels of cognition in affecting the capability of communities to adapt to risks related to climate change. In the Cayman Islands, for example, strong local and international support networks enable communities to recover from and prepare for tropical storms (Tompkins, 2005), whereas, in Samoa, communities rely on informal non-monetary arrangements, social networks, livelihood diversification and financial remittances through extended family networks to cope with storm damage (Adger, 2001; Barnett, 2001; Sutherland et al., 2005). In these examples, the role of migration can be seen as an adaptation option. However, as studies of the 1930s USA Dust Bowl phenomenon have shown, adaptive migration options are, in turn, greatly influenced by the access households have to economic, social and cultural capital (McLeman & Smit, 2006).

The role of capital and assets in helping to define vulnerability and adaptatibility is recognized in a number of approaches to local-level analysis, the best known of which is the sustainable livelihoods approach (SLA) (UK Department for International Development (DFID), 2000). This approach relies on present-day climate and current climate variability as a proxy for near-term climate change and understands vulnerability and adaptation in terms of external stresses and shocks placed upon livelihoods and assets (DFID, 2000). The original aim of this approach was to understand people's livelihoods so that development assistance could be tailored to their individual needs. The underlying idea in this formulation was that families possess a variety of natural, physical, financial, human and social assets, which are all used to maintain a family's livelihood. If one of the assets is lost, it can be compensated for by falling back on the other available assets in the so-called asset-pentagon. External influences in the form of policies and institutions are also taken into account (DFID, 2000). Yet SLA is also concerned with the question of how vulnerable livelihoods are to shocks, trends and seasonal developments and what kinds of coping strategies are used by people in the case of one of these events (Carney, 1998). This particular aspect helps in understanding how shocks and stresses caused by climate change and variability are likely to significantly influence people's livelihoods. Thus, Ziervogel and Calder (2003) used SLA to assess the impact of climate variability on adaptive capacity in Lesotho, where this was the coping strategy of interest. The analysis of local-level adaptive capacity has revealed it to be highly heterogeneous within a society or locality and differentiated by age, class, gender, health and social status (Carter et al., 2007). The conceptualization of migration as an adaptive response to household stress has been explored in the *New Economics of Labour Migration* (NELM) (Stark & Bloom, 1985). In this approach, migration decisions are seen as being made usually within families, who expect remittances in return for investment in the initial migration of a household member (Stark & Bloom, 1985). Migration thus becomes a means of diversifying sources of household income and reducing risk (Arango, 2000).

While the pentagon of assets can help explain adaptive ability, regulations and economic policies determined at the regional or national level can also limit the freedom of individuals and communities to act and can make certain potential adaptation strategies unviable. For instance, violent conflict, the spread of infectious diseases, and social trends such as urbanization or economic consequences of trade liberalization have been shown to erode and influence (both positively and negatively) adaptive capacity (Woodward, 2002; Barnett, 2006; Pelling, 2003). There has also been recognition of the idea that the vulnerability of one region can be teleconnected (i.e., statistically connected over large distances) to another (Carter et al., 2007). For example, Adger et al. (2007) linked smallholder coffee farmers' vulnerability to, and ability to cope with, severe droughts in Central America with increased coffee production in Viet Nam.

The use of temporal or spatial analogues is widespread in assessing climate change adaptability. While analogues, unlike modelled quantitative scenarios, have the advantage of providing important information on real past climate adaptation experiences, they are fundamentally limited as they can say little about long-term climate change, since such change is likely to be beyond past experiences of variability and because analogies between cases are never perfect.

The role of psychology in determining responses to environmental stresses and shocks has previously been recognized by the literature on natural hazards, with observations of how people behave under conditions of uncertainty and the possibility that they systematically underestimate the likelihood of the hazard affecting them (Freeman & Kunreuther, 2002; Kunreuther, 1996). In terms of climate change, the importance of measurable and alterable psychological factors, such as risk perception and perceived adaptive capacity, was found by Grothmann and Patt (2005) to provide better statistical power in explaining adaptation than traditional socio-economic models. A schematic showing the process model devised by Grothmann and Patt (2005) is given in figure 2.3 below.

By acknowledging the socio-physical context of the individual, this model attempts to explain why some people show adaptive behaviour while others do not. The model begins with a climate change risk appraisal, within which are two subcomponents: perceived probability of exposure and perceived severity of harmful consequences. The second major component – adaptation appraisal – comes after the risk perception process and only comes into play if a specific threshold of threat is exceeded. Within the adaptation appraisal, three subcomponents of perceived adaptation – efficacy, perceived self-efficacy and perceived adaptation costs – govern the response. Based on the outcomes of the risk and adaptation appraisal processes, an individual responds to the threat through either adaptation or 'maladaptation' (which includes avoidant reactions and 'wrong' adaptations that inadvertently increase climate change damage). If an individual chooses to employ an adaptive response, they first form a decision or intention to take these actions. Labelled as adaptation intention, this component of the model distinguishes between intention and actual behavioural adaptation. The model also incorporates additional levels of complexity by considering the cognitive biases that affect people's perceived adaptive capacity and how their previous experience of risk affects subsequent appraisal. Permitting deeper consideration of the cognitive process of individuals, the model also includes the socio-physical context of the individual by including social discourse. Based on the framework of social amplification of risk (by Kasperson et al., 1988), the inclusion of social discourse in the model allows for the concept that people's perceptions of risk or adaptive capacity with regard to climate change may be amplified or attenuated by what they hear about the issue from the media, friends, colleagues, neighbours and public agencies. By highlighting the importance of people's perceptions of the stimuli affecting the appraisal processes, the model provides an innovative conceptual basis upon which to consider the socio-cognitive processes behind proactive adaptation to the risk posed by future climate change.

Figure 2.3: Process model of private proactive adaptation to climate change (MPPACC)

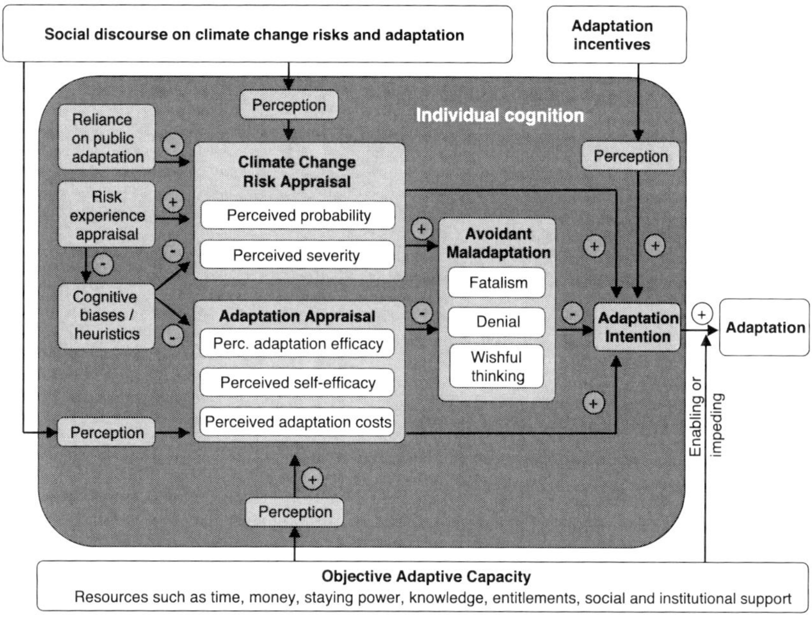

Source: Grothmann & Patt, 2005

While there has been recognition of the role of the cognitive processes in adaptation decision making in climate change research, few theories of social psychology have, so far, been integrated into these studies. In the field of social psychology, the Theory of Planned Behaviour (Ajzen, 1991) has been developed to explain how decisions are made and whether they result in action across a number of behaviours.

Consisting of the socio-cognitive variables of attitudes towards behaviour, subjective norms and perceived behavioural control, the theory has been used to understand behaviour in exercise (Nguyen et al., 1997), diet (Conner et al., 2003), entrepreneurial intentions (Krueger & Carsrud, 1993), conservation technology adoption (Lynne et al., 1995) and wastepaper recycling (Cheung et al., 1999). In the field of migration research, De Jong (2000) showed that the Theory of Planned Behaviour is useful in explaining migration decision making in Thailand, and Smith et al. (2008) have developed this framework into an agent-based model to simulate the movement of rural populations of Burkina Faso under drought conditions. Agent-based modelling (ABM) is a technique for modelling the decision-making behaviour of individuals; it captures, and is able to predict, both individual behaviour and how the interplay of individuals' actions produces large-scale outcomes.

To date, agent-based models have been used sparingly in climate change studies, but they provide an intriguing method for simulating the complex behaviour of human adaptation to climate change impacts. The behaviour of groups of individuals or households can be conceptualized and modelled in a number of ways. In conventional economic approaches, the behaviour of a group of individuals or households is often represented by a single average meta-actor (i.e., each individual or household represented is assumed to have average characteristics) and the aggregated behaviour of a group taken as the sum of individual interests. An alternative approach is multi-agent simulation, whereby a collection of autonomous and heterogeneous entities, whose behaviour is different to that of the average meta-actor, interact with each other and their environment, often to produce complex and emergent phenomena (Kirman, 1999; Boulanger & Brechet, 2005). Additionally, while traditional economic models formalize individuals as rational actors, maximizing their own well-being with unlimited cognitive resources, multi-agent-based approaches emphasize the idea that actors take decisions with limited cognitive resources, as their perception of reality is biased and incomplete. Further multi-agent approaches assume that individuals are not necessarily isolated and self-regarding, having social dimensions and interactions such as imitation, exchanges of information, mutual aid and cooperation (Jager et al., 2000; Barnaud et al., 2007).

The high level of uncertainty in scenarios of future climate and adaptive behaviour to the resultant changes in the climate system provide a wide range of possible outcomes for adaptation. Within this context, it has been argued that policy analysts need probability estimates to assess the seriousness of the implied impacts of climate change (Scheider, 2001) and so, therefore, avoid the selection of arbitrary scenarios. However, others have argued that climate change uncertainty is so considerable that a strategy of resilience and adaptive environmental management should be pursued instead (Piekle, 1998; Adger, 1999). As will be explored in section 4, uncertainty has a number of dimensions, some of which are amenable to representation with probabilities, while others are not.

In summary, there are a number of methods available that explicitly acknowledge the complex relationships existing to determine adaptive capacity. In terms of the migration–environment nexus, adaptation assessments allow for the incorporation of the role of economic, socio-cultural and even psychological aspects of determining people's behaviour, while recognizing that migration in itself can be seen as an adaptive strategy.

Integrated assessment and risk management

Integrated assessments attempt to represent the complex interactions and feedbacks between multiple drivers and impacts for national and global policy analysis, using a range of methods. In particular, integrated assessments aim to represent interactions across multiple spatial and temporal scales, processes and activities, and may involve mathematical models and/or an integrated process of assessment linking different disciplines and groups of people. The models used vary in complexity, having to balance attempts to provide realistic simulations with flexibility, while recognizing that no single model can represent all the interactions, and that no single theory describes and explains dynamic behaviour across scales in socio-economic and ecological systems (Rotmans & Rothman, 2003; Schellnhuber et al., 2004). Integrated models have been used for analysis of multiple sectors (Tol, 2002; Mendelsohn & Williams, 2004; Nordhaus, 2006) and have been found to reveal amplified global costs of climate change compared with single-sector analysis. Potentially, integrated assessments can provide a broad conceptual basis for the analysis of the interactions of multiple stressors on migration. The multiple drivers influencing migration behaviour, and the different contexts in which it occurs, lends itself to an integrated analysis; however, it is unclear how this integration should evolve, with integrated analysis currently most advanced in representing the physical aspects of the climate system and less so with regard to integrating social processes.

Risk management has been put forward as a framework for assessing adaptive responses to future climate change, assessing the limits of adaptation, linking adaptation to sustainable development, and engaging stakeholders and decision makers (Jones, 2001; Willows & Connell, 2003; UNDP, 2005; ADB, 2005). In general, risk is measured as a combination of the probability of an event and its consequences. According to Carter et al. (2007), in a CCIAV context, risk management can include:

- a scoping exercise to identify the overall approach to be used;
- risk identification to determine the location of the risk, what and who is at risk, the main climate-related and non-climate-related stresses contributing to the risk, and levels of acceptable risk, as well as the scenarios required for further assessment;
- risk analysis, with a range of methods used in mainstream risk assessment and CCIAV assessment;
- risk evaluation for the prioritization of different adaptation and/or mitigation measures;
- risk treatment, whereby the adaptation and/or mitigation measures selected from the process above are applied and monitored and progress reviewed.

As an approach orientated towards decision making, risk management importantly incorporates communication and consultation with stakeholders, ensuring that clarity and transparency are maintained around assumptions, concepts and uncertainty. Stakeholders include policy makers, scientists, communities and managers in the sectors and regions most at risk from the impacts of climate change, now and in the future (Carter et al., 2007). It is argued that the success of stakeholder involvement relies on both informing and empowering interested and affected people to act on their increased knowledge (Kasperson, 2006) and should involve an ongoing process of negotiation and modification to allow stakeholders to integrate scientific information into their own social, economic cultural and environmental context (van Asselt & Rotmans, 2002).

The concepts of thresholds, criteria for risk and coping ranges are important for assessing risk to climate change impacts. Thresholds to risk can include a non-linear change in state as a system shifts from one condition to another, called a systematic threshold, and a level of change measured on a linear scale, regarded as 'unacceptable' (Kenny et al., 2000; Jones, 2001). While the former can potentially be objectively measured, the latter tends to be value-laden or normative and best derived with stakeholder involvement (Jones, 2001; Carter et al., 2007). Thresholds and criteria for risk can be used in risk analyses to calculate the probability of them being exceeded (Jones, 2001). A coping range refers to the capacity of a system to accommodate variations in climate conditions, including using adaptation, and is best defined, once again, with stakeholders (Carter et al., 2007). Risk can then be calculated in accordance with how often the coping range is exceeded, under given conditions.

In summary, the assessment of climate change impacts, adaptation and vulnerability (CCIAV) involves a variety of approaches that can provide a conceptual and methodological basis for measuring the environment–migration nexus. Conceptually, such assessments include ideas of vulnerability, adaptation, adaptive capacity, thresholds and criteria of risk, and coping ranges that are transferable to migration–environment studies. Methods used in CCIAV vary in their aims, complexity, scale, motivation and how they incorporate the multiple contexts and stressors on society. Top-down impact studies have provided estimates of numbers of people forced to migrate from land permanently inundated by sea-level rise, assuming both on-going adaptation and no adaptation, with a large variation in numbers predicted to be affected. Interestingly, estimates of people forced to migrate, assuming adaptation, are 2–3 per cent of those predicted to be affected if no adaptation is assumed. However, for other scenarios of environmental change and other types of migration behaviour, the complexity of relationship between migration and hazard makes it difficult to measure the environment-migration nexus using current impact approaches. Thus, while impact studies integrate some of the wider economic contexts of adaptation decisions, they currently neglect more complex representations of the social, cultural and psychological environment in which decisions are made. Vulnerability assessments involve a variety of methods, including using expert groups and statistical analyses to integrate generic and specific determinants of vulnerability in order to diagnose and highlight vulnerability to climate change impacts over time and space. This form of analysis is most appropriate to studies of households forced to migrate. However, the concept of vulnerability is possibly not the most applicable to other dimensions of the migration–environment nexus, where the most vulnerable are sometimes unable to migrate due to a lack of resources and individual migratory behaviour can arise from a decision to diversify household income-generation in the context of a deteriorating environment. Vulnerability analyses do, however, provide a vehicle with which to identify locations most likely to be affected by climate change impacts and so highlight regions where the pressure to adapt is likely to be highest (with migration being one possible means of adapting). Adaptation assessments complement vulnerability studies aligned around a bottom-up approach that goes beyond a simple cause–effect relationship of climate and society and try to incorporate the myriad socio-economic, cultural and psychological connections and contexts within which decisions to adapt are made and carried out. Equally importantly, the concepts of adaptation, adaptation strategies and adaptive capacities (unlike concepts of vulnerability) allow for the investigation of the environment–migration nexus to go beyond seeing migration as a negative outcome of stress. Lastly, ideas of risk management provide a framework within which to assimilate stakeholder involvement, responsibility and activity in determining coping ranges and thresholds, marking the point where, if stress on a system is exceeded, sudden mass migration results.

3. Migration and the environment

Migration theories and the role of the environment

Studies of migration have integrated a number of different approaches from disciplines ranging from geography and economics to sociology and political science (Boswell & Mueser, 2008). In general, the key drivers of migration are considered to fall into three categories: (a) factors related to the region or country of origin, including political instability and conflict, lack of economic opportunities, and lack of access to resources ('push' factors); (b) factors related to the region or country of destination, including the availability of employment and demand for workers, higher wages, political stability or access to resources ('pull' factors); and (c) intervening factors that facilitate or restrict migration, including ease of transportation, family or social networks, government immigration or emigration policies, economic ties such as trade and investment linkages, or social and cultural exchanges (Black et al., 2008). Environmental considerations are included in a number of classic migration frameworks, including those posed by Wolpert, Speare, De Jong and Fawcett (Hunter, 2005). Table 3 below provides a summary of classic migration theories and how they incorporate the role of the environment.

One of the earliest considerations of the role of the physical environment was by Petersen (1958) who conceptualized migration in primitive times as being the movement from ecologically risky areas to safer locations. The theoretical 'stress–threshold model' developed by Wolpert in 1966 is often identified as the first migration model (of contemporary society) to incorporate non-economic aspects (Fredrickson et al., 1980). In this model, the environment is conceptualized both as a residential 'stressor' creating strain that may lead to the consideration of migration, and also as means of determining the 'place utility' of the destination location (i.e., the value of the place being migrated to) (Hunter, 2005). Speare (1974) further developed the concept of 'utility' into how an individual experiences the effects of social and contextual factors on levels of dissatisfaction and that, once a threshold of dissatisfaction is crossed, migration may be considered. In this conceptualization, the environment is considered as a 'locational characteristic' providing physical amenities or disamenities.

The 'value–expectancy' model of De Jong and Fawcett (1981) posited that individual migration is motivated by the interplay of values placed on different goals, such as wealth, status, stimulation, autonomy, affiliation and morality, and the perceived likelihood that a chosen behaviour will lead to these goals. In this conceptualization, the physical environment is included as helping to determine the goal of 'comfort' by providing a more pleasant residential location and/or a less stressful one. According to Zelinksky (1971), the role of personal preferences is central to the migration decision-making process and is primarily influenced by social and economic changes due to modernization. Within this context, the ability to fulfil the recognized preference for a less risky residential environment is suggested by Hunter (2005) to be facilitated by increases in modernization. Micro-economic perspectives on migration generally consider the environment only implicitly, as a locational characteristic. Instead, such perspectives view migration as shaped by a cost-benefit calculation between personal investment and returns. Despite a focus on human capital and the economic dimensions of migration decision making, econometric models have revealed an indication of societal value placed upon locational amenities or disamenities, which is reflected in wage differentials across locations (Knapp & Graves, 1989).

Table 3: Classic migration theories and the potential placement of environmental hazards

General typology of migration (Petersen, 1958)[2]
Potential placement of environmental hazards: as ecological "push" factor yielding migration as an "innovative" response.
Stress–threshold model (Wolpert, 1966)[3]
Potential placement of environmental hazards: as "stressors".
Mobility transition hypothesis (Zelinsky, 1971)[4]
Potential placement of environmental hazards: as related to "personal preferences'.
Residential mobility decision-making model (Speare, 1974)[5]
Potential placement of environmental hazards: as "locational characteristics".
Neo-classical migration models (various contributors)[6]
Potential placement of environmental hazards: as a 'location-specific disamenity'.
Value-expectancy model (De Jong and Fawcett, 1981)[7]
Potential placement of environmental hazards: as a personal value/goal of 'comfort'.
Macro–micro decision-making model (Gardner, 1981)[8]
Potential placement of environmental hazards: as a locational characteristic in conflict with 'what people value'.

Source: Hunter, 2005

While a number of the abovementioned approaches to migration have been focused at the micro level, Gardner's (1981) work attempted to link micro and macro characteristics by breaking the decision-making process into five stages whereby macro-level factors should be explicitly considered. The five stages were: (1) formation of values; (2) real, place-related macro-level factors; (3) factors that affect accurate perception of place-related factors and, thus, one's expectations; (4) objective constraints and facilitators to migration; and (5) factors that affect accurate perception of the constraints and facilitators (Gardner, 1981: 63–64). Within this context, Gardner argues that values shape

[2] Key citation: "A general typology of migration." *American Sociological Review*, 23: 256–266.

[3] Key citation: "Migration as an adjustment to environmental stress." *Journal of Social Issues*, 22, 4:92–102.

[4] Key citation: "The hypothesis of the mobility transition." *Geographical Review*, 61:219–249.

[5] Key citation: "Residential satisfaction as an intervening variable in residential mobility." *Demography*, 11, 2:173–188.

[6] Summary: DaVanzo, J., 1981, "Microeconomic approaches to studying migration decisions" in De Jong, G.F. & Gardner, R.W. (Eds), *Migration decision making: multidisciplinary approaches to microlevel studies in developed and developing countries*: 90–129. Pergamon Press, New York. Example: Harris, J.R. & Todaro, M.P., 1970, "Migration, unemployment and development: A two-sector analysis", *American Economic Review*, 70: 126–142. Example: Graves, P.E., 1983. "Migration with a composite amenity: The role of rents", *Journal of Regional Science*, 23(4): 541–546.

[7] Key citation: "Multidisciplinary frameworks and models of migration decision making" in De Jong, G.F. & Gardner, R.W. (Eds), *Migration decision making: multidisciplinary approaches to microlevel studies in developed and developing countries*: 13–58. Pergamon Press, New York.

[8] Key citation: "Macrolevel influences on the migration decision process", in De Jong, G.F. & Gardner, R.W. (Eds), *Migration decision making: multidisciplinary approaches to microlevel studies in developed and developing countries*: 59–89. Pergamon Press, New York.

peoples' perceived fulfilment at a location with feelings of stress and dissatisfaction, creating preferred residential locations (Hunter, 2005).

In summary, it can be seen that a number of migration frameworks include the environment as a contextual consideration. The following section explores the empirical evidence of a linkage between migration and the environment.

Evidence of a migration–environment link

Like many environmental processes, the phenomena of human migrations occur across a continuum of space and time scales. In addition, the decision to migrate or not is taken within a wide range of social, cultural, psychological and economic contexts. As a result, the derivation of a simplistic relationship between the two can be considered somewhat wishful thinking. A number of studies have explored how manifestations of environmental variability, such as droughts, earthquakes, heavy rainfall, storms and other extreme weather events, have affected people's migratory behaviour. This section seeks to provide an overview of such studies, grouping evidence for a migration–environment linkage according to whether migration is internal or international.

Given the wide media coverage of recent reports warning about the potentially massive flows of 'climate change refugees', there is a surprising lack of empirical evidence of mass overseas migration resulting directly from environmental change or variability (Black et al., 2008). Studies have shown that international migration:

- increased with loss of harvest and livestock, but decreased following a severe earthquake in El Salvador (Halliday, 2006);
- decreased in drought years in Burkina Faso (Henry et al., 2004) and Mali (Findley, 1994);
- both increased and decreased with declining rainfall in Mexico (Munshi, 2003; Kniveton et al., 2008).

These apparently contradictory findings point to a complex relationship between international migration and the environment – a relationship in which the direction of the response is determined by the social and economic conditions in which the community, households or individuals find themselves. At first glance, the limiting factor to some environmental stresses and shocks initiating international migration appears to be the various capitals required to migrate. In Burkina Faso, for instance, food scarcity during drought was found to lead to increased prices, forcing people to spend more money on their basic needs rather than on long-distance migration (Henry et al., 2004). Similarly, in El Salvador, migration was found to be more likely for wealthier households than poorer ones, which were liquidity-constrained (Halliday, 2006). However, it was also found in El Salvador that migration was reduced for both wealthier and poorer households following earthquakes, indicating that the relative magnitude of the environmental stressor is important (ibid). A possible complicating factor in the decision to undertake international migration from environmentally pressured regions comes from a micro-study in Mexico (Schmidt-Verkerk, personal communications) that revealed the importance of community conceptualizations of migration outcomes. In this study, in one community, it was found that there was a negative association with migration, due to the family breakdown of a particular household; yet, in a nearby settlement, migration abroad was seen as producing a positive outcome in terms of wealth accumulation (ibid). In summary, there is evidence of both an increase and a decrease in international migration as a result of environmental stress, with the magnitude of the stressor, and economic, social and psychological contexts of the exposed population, helping to determine whether the response is positive or negative.

Methodologically, the abovementioned studies of the international migration–environment nexus have been performed using multilevel and standard regression and correlation-based methods. As such, they fail to take into account the non-linear (and non-gradual) interactions of different factors in the migration decision, despite the widespread recognition of non-linear outcomes in social phenomena. To achieve a non-linear representation of the migration decision, the modelling techniques of neural networks and agent-based models can be used.
Within the category of international migration, the movement of people can be overseas or regional. Given that overseas migration is an expensive endeavour, with significant resources required to undertake the journey, the literature points to a likely reduction in this form of migration during periods of increased environmental stress (Black et al., 2008). By contrast, regional migration is a common phenomenon in a number of locations – for example, West Africa has a long history of migration across national borders, reflecting freedom of movement within the Economic Community of West African States. While still relatively expensive compared to internal movements, migration across international borders within a region provides a potentially more realizable adaptation route for more people than overseas migration. This is due to financial and accessibility factors and the fact that there are fewer legislative and social network barriers to movement.

While international migration remains out of reach for many of the most vulnerable to environmental stresses and shocks, the residents of some small island states are also limited in terms of their ability to undertake internal migration. For example, over 10,000 people were evacuated from Montserrat in 1995, due to the imminent eruption of the volcano on the island (Avery, 2003). While this movement was, for the majority of people, temporary, it provides an example of forced overseas displacement from a small island state. Similar concerns have been expressed for the populations of some Pacific small island developing States, such as Kiribati, the Marshall Islands, Tokelau and Tuvalu, in the context of raised sea levels and increased storm surge intensity due to climate change (Adger et al., 2007).

The number of people who undertake internal migration vastly outnumbers those moving internationally, throughout the world. Seasonal and circular migration is a crucial livelihood option for many in rural areas, providing both a means of household income diversification and a reduction in household consumption requirements in the source region. While the literature points to differences in who migrates and where they migrate to, between different regions, countries and over time, there is broad agreement that internal migration often intensifies following major droughts or famines (see Leighton Chapter 6 in this volume; Shipton, 1990; Findley, 1994; Pederson, 1995; Ezra, 2001; Perch-Nielsen, 2004). Internal circular migratory behaviour in response to environmental stress and shocks has been found especially among the poor, although not necessarily among the poorest or most vulnerable, who may not have the resources to move (Deshingkar, 2006: cited in Raleigh, 2008). Local displacement forms much of the adaptive response to environmental shocks in a number of regions. For example, 88 per cent of migrant agricultural communities in Bangladesh were found to remain within two miles of their previous residence following the erosion of land and loss of homes due to flooding (Zaman, 1989).

There was a similar lack of permanent long-distance movement after the evacuation following the fatal gas leak in Bhopal, India in 1984 (Dhara & Dhara, 2002). The disaster and hazard response communities have long recognized the common response of communities to locally relocate after disasters, with approximately 30 per cent of the affected population moving and with over 90 per cent of these people returning at some later stage (Surhke, 1994; Berry & Downing, 1993; Belcher & Bates, 1983: cited in Raleigh, 2008). However, empirical evidence gathered from the aftermath of a tornado in north-central Bangladesh illustrates that migration is not always a primary response to disaster when emergency aid compensates in monetary terms for damage caused by disasters (Paul, 2005).

Migration in response to environmental change and variability is not restricted to developing countries, as illustrated by the response to Hurricane Katrina in the USA in 2005. In this instance, the hurricane resulted in significant demographic changes, with those areas characterized as having greater social vulnerability (with greater proportions of disadvantaged populations, a higher level of housing damage, and a higher density of population) being more likely to experience out-migration (Elliott & Pais, 2006; Falk et al., 2006; Landry et al., 2007; Myers et al., 2008). One important feature of the displacement associated with Katrina was that, initially, at least, as many as 70,000 African-American residents of New Orleans (mainly the poorer residents) were unable to leave; (Landry et al., 2007: 2); yet, over time, it appears that it is the poorer, African-American residents who have been least able to return (Black et al., 2008).

In the general migration literature, internal migration is often differentiated in terms of whether it is rural–rural, rural–urban, urban–rural or urban–urban. Some studies have attempted to distinguish the relationship between the environment and migration in these terms (Barrios et al., 2006; Henry et al., 2004). Using multi-level statistical analysis, Barrios et al. (2006) found that changes in rainfall, have acted to encourage urbanization in sub-Saharan Africa, especially after decolonization, whereas Henry et al. (2004), using the same technique, failed to find any such relationship in Burkina Faso. Methodologically, the study of Henry et al. (2004) reveals an important insight into approaches to measuring the environment–migration nexus. While their study found no evidence of an effect of rainfall conditions on the likelihood of an individual migrating when considering all destinations, it did reveal that, when separating their data according to migration destination, men living in areas where rainfall is scarce and in years following poor rainfall conditions were more likely to move to another village (i.e., rural–rural migration) than those from areas of greater rainfall, particularly if the move was short term. Interestingly, it also found that, while women were also found to embark on a rural–rural migration if they lived in a drier area, they were less likely to move after poor rainfall conditions (Henry et al., 2004).

For future work on the migration–environment nexus, this shows the importance of producing separate analyses according to destination of migration and type of migrant. However, one recent study of the impact of Hurricane Mitch on Nicaragua did reveal that, in this case, the strength of the migration–environment relationship can be so strong that separation according to destination is not necessary (Caravajal & Pereira, 2008). The abovementioned study used the concept of adaptive capacity to group households and from which to explore their migration response, with the finding that households exposed to high rainfall during Hurricane Mitch had a 50 per cent increased probability of sending a migrant after the disaster compared to non-exposed households (ibid). This study was particularly fortunate in having access to detailed panel data before and after the hurricane.

While the discussion has focused on past migration flows and environmental pressures, relatively less attention has been given to how existing migration flows may be affected by environmental change in the future. One exception to this has been the work of Black et al. (2008), which sought to assess the sensitivity of the drivers of existing migration flows to future climate change, in four case study countries: Bangladesh, Ghana, Ethiopia and Sudan. In this study, the key drivers of major migration flows for each country were identified and assessed for their susceptibility to climate change and variability. A series of intervening factors that influence people's sensitivity to climate change and their ability to cope, affecting the migration drivers, were then identified and the policy implications explored. While such an approach is limited, as it is only qualitative, and does not provide detailed scenarios of future migration patterns, it does have the advantage of being able to incorporate both direct and indirect impacts of climate change.

In summary, while there is no conclusive evidence of a direct correlation between environmental stresses or shocks and migration, a number of studies have shown that 1) short-term internal migration can be initiated under varied socio-economic contexts by environmental stresses and shocks; 2) international migration flows following environmental shocks and stresses can increase or decrease, depending on the socio-economic and psychological contexts and barriers to migration of those exposed; and 3) the measurement of environment–migration linkages is sensitive to the available data and the ways in which such data are analysed. Sensitivity analyses and rigorous multi-level statistical analyses provide a mechanism for assessing the nature of the migration–environment nexus.

4. Issues of uncertainty and data requirements

Uncertainty

Accurately predicting future climate change is impossible, due to an incomplete knowledge of climate system processes affecting and affected by changes in the system, the natural variability of the climate system, and the uncertainty about future levels of greenhouse gases and aerosols. There is further uncertainty about the way in which changes in the climate will impact natural and human systems and the way in which society might perceive and respond to these changes. The issues of perceived risk and the ability to adapt were raised in section 3, with regard to individual cognition of climate change. A number of reasons are given in the literature as to why residents may opt to not migrate when exposed to an environmental hazard:

1. They are not aware of the hazard.
2. They are aware, but do not expect a disaster.
3. They expect a disaster, but do not anticipate loss.
4. They expect a loss, but not a serious one.
5. They expect a serious loss and have taken, or are planning to take, actions to mitigate such loss.
6. They expect a loss, but have accepted it as the price they pay for locational benefits.
7. They have nowhere else to go.

(Hunter, 2005)

In less developed regions, environmentally hazardous areas are often settled by poor households, sometimes after migration from a previously environmentally stressed location, because these are the only places available to them. In more developed regions, sometimes the wealthier households tend to remain in a hazardous area because they have the means to rebuild in the face of disaster or because they expect any losses to be offset by other locational benefits or because they are able to undertake mitigation strategies (Morrow-Jones & Morrow-Jones, 1991; Peacock & Girard, 1997: as cited in Hunter, 2005).

All of the abovementioned uncertainties combine to make definite estimates of climate change-induced migration unfeasible. The Intergovernmental Panel on Climate change (IPCC) recognizes two types of uncertainty in climate change studies: 'value' and 'structural'. Value-based uncertainties refer to the incomplete determination of particular values – for example, when the phenomenon of interest is not fully represented by the data or the data are inaccurate. Structural uncertainties refer to the incomplete understanding of processes, such that not all the relevant processes may be included. In particular, this second type of uncertainty requires that a judgement (a collective judgement, in the case of the IPCC), be made of the reliability of the result, which includes estimating the limits of knowledge.

Although the IPCC attempts to quantify structural and value-based uncertainty, it is less explicit in its analysis of climate change impacts in terms of the *unknowable*, such as future emission levels of greenhouse gases, apart from acknowledging different configurations of future society and emission levels. Natural and social scientists involved in climate change research often view uncertainty that relates to the *unknowable* differently (Dessai & Hulme, 2004). For example, climate scientists attempt to overcome uncertainty in predicting a chaotic atmosphere by completing ensemble simulations, whereby a climate model is run several times with varying starting conditions, and using the outputs to represent a probability distribution of outcome. In this context, the uncertainty can be better represented as time elapses and computing power and the ability to produce different model runs increases. Meanwhile, social scientists have argued that 'reflexivity', whereby humans reflect critically on the implications of their behaviour and make adjustments in the light of experience, constrains attempts to reduce the uncertainty surrounding climate predictions (ibid).

In terms of adaptive behaviour of individuals, reflexivity can be represented in complex social simulations using integrated assessment or agent-based models. Unfortunately, to date, integrated assessment has largely neglected human adaptation, and agent-based modelling is still in its relative infancy in its application to climate change. However, notable work using agent-based models in climate change research includes the work of Bithell and Brasington (2009), who combined an agent-based model of subsistence farming, an individual-based model of forest dynamics, and a spatially explicit hydrological model to investigate how demographic changes influence deforestation and assess its impact on forest ecology, stream hydrology and changes in water availability; and the work of Ziervogel et al. (2004) and Bharwani et al. (2005), who examined how seasonal climate forecasts might influence agricultural strategies and how this climate information could be improved to be more useful to farmers. As an alternative to numerical-based models, analogues of past experiences relating to adaptation to climate variability can be used to estimate future adaptation to near-term climate change. However, analogies between cases are never perfect and they can say little about long-term climate change where no past analogues exist (Dessai & Hulme, 2004).

In summary, many uncertainties surround the migration–environment nexus, due to an incomplete knowledge of how the climate system behaves, uncertainty about the future behaviour of society, and the chaotic and complex nature in which the physical system operates. Scenarios of future configurations of society and emission levels provide inputs for ensemble modelling of the physical climate system change. Simulations of future global climate change, under a variety of socio-economic scenarios, are available from the IPCC gateway.[9] However, many of these data are at a relatively low spatial and temporal resolution (with spatial resolutions of approximately 62,500km^2 and timeframes of a month). Regional climate models potentially provide more detailed spatial information of future climate change but the data from them are less readily available and are sensitive to domain positioning, size and boundary conditions, which sometimes renders their simulations of poorer quality than global models (Raghavan, 2008). Numerous studies have documented how changes in the climate will affect the various human and natural systems (see IPCC, 2007b), while integrated assessments and agent-based models provide a mechanism with which to simulate the complex social reflexive and perception-dependent adaptive behaviour of humans. However, overall, the substantial uncertainty involved in predicting climate change is likely to result in a wide range of estimated values for environmentally related migration.

Data

The analysis of evidence of how environmental stresses and shocks interact with migration processes in section 3 revealed that a number of studies showed apparently contradictory findings. This is likely to

[9] See http://www.ipcc-data.org/

stem largely from the complex, non-linear, multi-context and multi-stressor environment in which environmental and migration processes occur. Additionally, however, the apparent lack of clarity of findings for similar populations (e.g., Kniveton et al., 2008 and Munshi, 2003) could be due to the different methodologies and data sources used. The work of Carvajal and Pereira (2008) on the influence of Hurricane Mitch on migration was one study fortunate to have access to extensive field data a few months before the hurricane hit and also three years after the hurricane hit (see section 3). The data consisted of panel surveys with multi-stage stratified samples that contained data on a variety of topics – from household characteristics and demographics, to fertility, time use, migration and economic activities. Total numbers of households surveyed were 4,209 before the hurricane and 4,959 after, with 2,983 households interviewed on both occasions. This study clearly showed a strong migration–environment linkage, although it might be argued that this linkage arose, in part, due to temporary changes in the USA's migration policy in response to the disaster, rather than the direct effect of environmental change pushing people to migrate. The availability of good-quality data on migration flows and the socio-economic conditions of the populations concerned is paramount to the successful measurement and modelling of the migration–environment nexus (see also Bilsborrow Chapter 3 in this volume). Also, given the applicability of agent-based models and integrated assessments to modelling future migration scenarios, information on community-level socio-cognitive rules pertaining to how migration is viewed as behaviour is also required.

While the need for migration data is widely recognized, there are information gaps relating to both international and internal migrant stocks. Of these, data on international migration are the most readily available, with various datasets of tabulated 'macro data' on migration, including:

- the Docquier–Marfouk dataset on international migration by gender and educational attainment;
- the UN Population Division's *Trends in the Total Migration Stock* (2005 and 2007 revision);
- the Global Migrant Origin Database, assembled by the Development Research Centre on Migration, Globalization and Poverty (the 'Migration DRC'). This dataset is based on the 2000–2001 census and is the only 'complete' dataset on migrant stocks in all countries worldwide, although it is based on substantial assumptions – including, for example, the need to disaggregate migrant stocks from more than one country where

these are simply listed as 'other' (see appendix 1 for details of this dataset) (Migration DRC, 2008).

Census data can be problematic for use in migration studies, as countries often conduct censuses in different years and employ varying definitions, making cross-country comparisons imprecise and restricting time series-based work. Furthermore, census data do not distinguish between short-term and long-term migrants – a distinction that is useful for policy discussions (Migration DRC, 2008) and, as discussed earlier, might have different relationships with environmental stresses and shocks. There are several datasets of micro data on international migration, including:

- a six-country comparative survey coordinated in 2008 by the UK's Institute for Public Policy Research and the Global Development Network;
- the Mexican Migration Project (MMP), consisting of a retrospective survey built up over the last 20 years, providing insights into what influences migration, remittance flows and return/circular migration between Mexico and the USA.

Compared to data on international migration, information on internal movements is relatively poorly developed. Census data do provide some insight into internal migration, although, once again, comparisons between countries are hindered by different countries sometimes conducting censuses in different years and employing varying definitions of migrants. Even less information is available on irregular migrants and trafficked persons.

According to the Development Research Centre on Migration, Globalization and Poverty, one of the most pressing needs for migration data relates to the basic geography of migration flows (the Migration DRC, 2008). It is also suggested that better evidence-based studies of migration could be achieved by widening access to existing data sources and adding the following three questions to all new censuses: place of birth, place of residence five years previously, and country of citizenship. The addition of these questions to population censuses would allow for better tabulation of bilateral migration and would also reveal potential changes in the characteristics of migration between censuses. Other suggested ways of improving data on migration include countries providing open access to anonymized micro data on migrants; the unification of Labour Force Surveys (LFS) worldwide into a single, annually updated database; and the piloting of a core standardized migration module in Living Standards Measurement Surveys (LSMS).

Lastly, it is worth recognizing that a major Commission for International Migration Data on Development Research is also looking at migration data issues in relation to the migration–development nexus, and will launch its findings later in 2009.[10]

[10] See http://www.cgdev.org/section/topics/migration

5. Discussion and research priorities

Discussion

Migration and environment processes encompass a wide range of spatial and temporal scales, ranging from the impact of sudden, local changes in the environment, such as a tornado that causes the displacement of small populations to nearby areas while the hazard passes and livelihoods are rebuilt, to global changes, such as sea-level rise that results in large-scale inundations and permanent international migration. Simulations of future changes in the environment through human-induced climate change point to increased environmental pressure for large parts of the globe in this century. Regions most likely to be affected include the Arctic, Africa, small islands, and Asian and African mega-deltas.

A number of methodologies and approaches have been used to study society–environment relationships in the disciplines of climate science, economics, sociology, geography, disaster and hazard management, and social psychology, and these provide insights for the study of the migration–environment nexus.

Current estimates of the number of people who may become environmental migrants from climate change are, at best, guesswork because they assume that migration in areas affected by climate change are all one-way and permanent, they do not take into account the multiple and complex reasons for people migrating, and they ignore the fact that migration can be a strategy for wealth accumulation or a route out of poverty that benefits both the receiving countries and the countries or regions of origin. Approaches to understanding and measuring the impact of future climate change on society utilize concepts of vulnerability, adaptation, thresholds, coping ranges and risk management. In particular, analysis of adaptation provides a multi-level platform with which to incorporate the myriad socio-economic, cultural and psychological connections and contexts within which decisions to migrate or not are made and carried out. A framework of risk management would facilitate: stakeholder involvement; responsibility and activity in environment–migration studies; the determination of coping ranges before migration, and thresholds that mark the point at which mass migration results. Vulnerability assessments can be used to identify regions most likely to be affected by climate change impacts and, by extension, where pressure to migrate might be expected to increase, while standard impact analyses provide a relatively straightforward means of identifying populations likely to be displaced by the permanent loss of land. Methodologically, expert groups, statistical analysis and local-level studies of capitals and assets can be used to identify and integrate determinants of vulnerability and adaptive capacity, while approaches such as agent-based models and integrated assessments allow for the simulation of future flows of people, given information on the socio-cognitive rules of communities.

A number of migration frameworks view the physical environment as a contextual consideration, when exploring the causes and patterns of migration. Theoretically, environmental stresses and shocks can act as a push factor in migration decision making, while a relatively safer environment can act as a pull factor. There is evidence that short-term internal migration can be initiated under varied socio-economic contexts by environmental stresses and shocks; that international migration flows following environmental shocks and stresses can increase or decrease, depending on the socio-economic and psychological contexts and barriers to migration of those exposed; and that the measurement of environment–migration linkages is sensitive to the data available for study and ways in which the analyses are performed. Sensitivity analyses and rigorous multi-level statistical analyses provide a mechanism for assessing the nature of the migration–environment nexus.

There is a high degree of uncertainty involved in determining the impacts from future climate change, due to an incomplete knowledge of climate processes, unknowns about future emissions of greenhouse gases, uncertainty about how natural and human environments may respond to climate change, and an incomplete knowledge of how humans may adapt to these changes. Some macro and micro data on international migration flows are currently available for research into the migration–environment nexus. While such data have been collected by a number of countries, relatively fewer data are available on internal migration. This chapter has focused on the influence of the environment on migration, which has been ignored because, firstly, such a process is dependent on determining flows of migrants (which this chapter attempts to partly address) and, secondly, because a number of established environmental impact assessment techniques already exist for this purpose (e.g., UNEP, 2006).

Research priorities

The livelihoods of many people worldwide are subject to a complex interplay of economic, social, political, cultural, psychological and environmental stresses and shocks, resulting in a variety of coping strategies and decisions. One such decision made by a significant percentage of the world's population is to migrate. The motivation to migrate in the face of environmental stresses and shocks can vary from being forced to flee for survival, to a proactive strategy to diversify income and reduce vulnerability of households in affected locations. Rather, migration can be envisaged as existing on a continuum – from displacement, to an adaptation strategy in response to increased environmental stress and strain. This chapter has illustrated the complex nature of the environment–migration nexus. It has examined a number of approaches currently used to understand climate–society interactions and assessed their applicability to the measurement of the migration–environment nexus. The application of these methods is summarized in table 5.

Table 5: Characteristics of different approaches to assessing the migration–environment nexus

Type of migration	Displacement	→		Adaptation strategy
Assessment type	Impact	Vulnerability	Adaptation	Integrated
Scientific objectives	Impacts on migration under future climate	Processes affecting vulnerabilities to changes in the climate that are likely to lead to migration	Processes affecting the uptake of migration as an adaptation strategy	Interactions and feedbacks between multiple drivers of, and impacts on, migration
Practical aims	Actions to reduce risks	Actions to reduce vulnerability	Actions to improve adaptation	National, regional and global policy options and costs
Research methods	• Standard approach • Drivers-pressure-state-impact-response (DPSIR) methods • Hazard-driven risk assessment GIS	• Vulnerability indicators and profiles • Past and present climate risks • Livelihood analysis • Agent-based methods • Narrative methods • Risk perception, including critical thresholds • Development/sustainability policy performance • Relationship of adaptive capacity to sustainable development		• Integrated assessment modelling • Cross-sectoral interactions • Integration of climate with other drivers • Stakeholder discussions • Linking models across types and scales • Combining assessment approaches/methods
Spatial domains	Top-down global -› local	Bottom-up local -› regional (macro-economic approaches are top-down)		Across scales – global/regional/national
Types of direct environmental drivers	Sea-level rise	Flooding, water stress		Changes in patterns and distribution of rainfall
Types of indirect environmental drivers	Government resettlement programmes	Changes in food security		Changes in employment opportunities
Examples	Nicholls et al., 2008	McLeman and Smit, 2006; Smith et al., 2008		Black et al., 2008

(Adapted from Carter et al., 2007)

This chapter has also reviewed both the positioning of the environment in existing migration theories and the evidence of an environmental impact on migration. In the introduction (see bullet points a–d), some initial guiding objectives for the measurement of the migration–

environment nexus were provided. To meet these objectives, the following priorities are recommended.

1. *Involvement of stakeholders in research into the measurement of the environment–migration nexus.* Representatives of civil society, government and non-governmental organizations provide vital input to studies of how people cope with different environmental stresses and strains, and the thresholds that need to be crossed before a change in behaviour occurs. Their involvement throughout the research helps establish credibility to, and confer 'ownership' of, the results obtained and offers increased possibility that the research findings will be used in effective risk management (Carter et al., 2007).

2. *Documenting current levels of migration that involve environmental factors in shaping migration decisions.* A number of initiatives, such as the EACH-FOR Project (see Warner et al. Chapter 4 in this volume), have set out to produce case studies where environmental degradation has influenced migration patterns. The EACH-FOR Project used primary data from semi-structured expert and field interviews, and migrant and non-migrant questionnaires to understand the causes of forced migration in relation to environmental degradation and their association with other social, political and economic phenomena. Such studies need to be extended to other forms of migration and to produce larger and more detailed datasets. Particular emphasis should be placed on obtaining information on the socio-cognitive processes of migration and asset bases of sampled populations. Two possible approaches to data collection include the Sustainable Livelihoods Approach and the Theory of Planned Behaviour (see section 2), with the statistical analysis of data incorporating both multi-level linear-based and non-linear statistical analysis techniques.

3. *Increased accessibility to international and internal migration databases.* Currently, readily available datasets on migration are skewed towards international movements. Some countries collect information on internal migration and they should be encouraged to allow access to this information and to collate, these data into widely accessible databases. Census data can be an important source of information on migration and future censuses should be encouraged to include questions pertaining to place of birth, place of residence five years previously, and country of citizenship. Similarly, countries should be encouraged to provide open access to anonymized micro data on migrants, to create a single, annually updated database, and to pilot a core standardized migration module in Living Standards Measurement Surveys (LSMS).

4. *Assessments of the degree to which existing migration flow are sensitive to future environmental change and variability.* Research on existing migration flows often includes an understanding of the drivers of these flows. Evaluation of the sensitivity of these drivers to the impacts of climate change and comparison with predicted ranges of climate change provides a basis for estimation of future flows. Such integrated analyses should attempt to explore the interactions and feedbacks between multiple drivers of, and impacts on, migration. Currently, integrated models are most advanced in representing the physical aspects of the climate system, with some integration of economic models. Fewer advances have been made in the integrated modelling of social processes.

5. *Impact studies of possibly forced migration due to the permanent inundation of land by sea-level rise.* Current assessments of the impact of climate change scenarios on the loss of land from permanent inundation by sea water can incorporate the costs and benefits of the loss of land and adaptation through building sea defences, respectively. When incorporating adaptation, these studies predict a displacement of 15 million people from 2030 to 2130 under a worst case scenario of rapid WAIS collapse, which represents some 2–3 per cent of those people who would be inundated if no adaptation were to occur. Future research should aim to include the influence of socio-ecological thresholds, such as an overall loss of confidence, which, when exceeded, would trigger increased displacement.

6. *The mapping of vulnerable populations expected to suffer increased pressure to migrate, due to the impacts of climate change.* While not all people exposed to environmental stresses are able to move or perceive the need to do so, vulnerability assessments allow for the identification of regions where pressure to migrate might be expected to increase. There are several methods of assessing vulnerability, including, at national level, the formation of expert groups and statistical analyses to select and combine indicators of vulnerability to climate impacts where migration might be a livelihood outcome. At the more local level, the sustainable livelihoods approach provides a framework for understanding how vulnerable livelihoods are to shocks, trends and seasonal developments, and what kinds of coping strategies people use when these events occur.

7. *Assessments of migration as an adaptation strategy.* Adaptation assessments are able to incorporate the myriad socio-economic, cultural and psychological connections and contexts within which decisions to migrate or not are made and carried out. At the national

level, expert groups and statistical analyses should be carried out to select indicators of migration drivers, and these indicators should be combined to identify populations where migration as a response to increased environmental pressure might be expected to increase. Current and past conceptualizations of the migration decision making process provide a knowledge base of the drivers involved, while the existence of information on migration flows allows for the assessment of the validity of the choice of indicators. At subnational level, local studies of how assets, capitals and socio-cognitive rules relate to migration outcomes should be carried out to provide the basis for predicting future flows of migrants. Integrated assessments and agent-based models that can simulate complex emergent social phenomena should be developed and used to develop scenarios of future migration flows in the context of future climate and environmental change and variability.

6. References

ADB
2005 *Climate Proofing: A Risk-based Approach to Adaptation*. Pacific Studies Series, Pub. Stock No. 030905, Asian Development Bank (ADB), Manila. Available at: http://www.adb.org/Documents/Reports/Climate-Proofing/climate-proofing.pdf

Adger, W.N.
1999 "Social vulnerability to climate change and extremes in coastal Vietnam", *World Development* 27: 249–269.

Adger, W.N.
2001 "Scales of governance and environmental justice for adaptation and mitigation of climate change", *Journal of International Development*, 13: 921–931.

Adger, W.N. et al.
2007 *Nested and networked vulnerabilities in South East Asia. Global Environmental Change and the South-east Asian Region: An Assessment of the State of the Science*, Lebel, L. et al. (Eds), Island Press, Washington, DC, in press.

Ajzen, I.
1991 "The Theory of Planned Behaviour", *Organizational Behavior and Human Decision Processes*, 50: 179–211.

Allison, E.H. et al.
2005 *Effects of climate change on the sustainability of capture and enhancement fisheries important to the poor: analysis of the vulnerability and adaptability of fisherfolk living in poverty.* Fisheries Management Science Programme, London, Department for International Development (DFID) Final Technical Report. Project No. R4778J. Available at: http://p15166578.pureserver.info/fmsp/Documents/r4778j/r4778j_1.pdf

Arango, J.
2000 *Explaining migration: a critical view*, Blackwell Publishers, Oxford.

Avery, J.G.
2003 "The aftermath of a disaster. Recovery following the volcanic eruptions in Montserrat, West Indies", *West Indian Medical Journal*, 52 (2):131–135.

Barnaud, C. et al.
2007 "Multi-agent simulations to explore rules for rural credit in a highland farming community of Northern Thailand", *Ecological Economics*, 66 (4): 615–627.

Barnett, J.
2001 "Adapting to climate change in Pacific island countries: the problem of uncertainty", *World Development*, 29: 977–993.

Barnett, J.
2006 "Climate change, insecurity and injustice", *Fairness in Adaptation to Climate Change*, Adger, W.N. et al. (Eds), MIT Press, Cambridge, Massachusetts: 115–129.

Barrios, S. et al.
2006 "Climatic change and rural–urban migration: The case of sub-Saharan Africa", *Journal of Urban Economics*, 60(3): 357–371.

Belcher, J.C. & Bates, F.L.
1983 "Aftermath of natural disasters: coping through residential mobility", *Demography*, 7(2): 118–128.

Belliveau, S. et al.
2006 "Multiple exposures and dynamic vulnerability: Evidence from the grape and wine industry in the Okanagan Valley, Canada", *Global Environmental Change*, 16: 364–378.

Berkhout, F. et al.
2006 "Learning to adapt: organizational adaptation to climate change impacts", *Climatic Change*, 78: 135–156.

Berry, L. & Downing, T.E.
1993 "Drought and famine in Africa, 1981–86: A comparison of impacts and responses in six countries", in *The challenge of famine: Recent experience, lessons learned*, John Osgood Field (Ed.): 35–58. Kumarian Press, West Hartford, CT.

Bharwani, S. et al.
2005 Multi-agent modelling of climate outlooks and food security on a community garden scheme in Limpopo, South Africa. *Philosophical Transactions of the Royal Society B – Biological Sciences*, 360 (1463): 2183–2194. Available at: http://rstb.royalsocietypublishing.org/content/360/1463/2183.full.pdf+html?sid=9df2dad7-985b-4af7-afe0-451e31d1ac24

Bithell, M. & Brasington, J.
2009 "Coupling agent-based models of subsistence farming with individual-based forest models and dynamic models of water distribution", *Environmental Modelling & Software*: 24(2): 173–190.

Black, R.
2001 *Environmental Refugees: Myth or reality?* UNHCR working paper 34, UNHCR, Geneva.

Black, R. et al.
2008 *Demographics and climate change: future trends and their policy implications for migration.* Report prepared for the Department for International Development (DFID) by the Development Research Centre on Migration, Globalization and Poverty, University of Sussex. Available at: http://www.migrationdrc.org/publications/working_papers/WP-T27.pdf

Boswell, C. &Mueser, P.R.
2008 "Introduction: Economics and interdisciplinary approaches in migration research", *Journal of Ethnic and Migration Studies*, 34(4): 519–529.

Boulanger, P.M. & Brechet, T.
2005 "Models for policy-making in sustainable development: The state of the art and perspectives for research", *Ecological Economics*, 55(3): 337–350.

Brooks, N. et al.
2005 "The determinants of vulnerability and adaptive capacity at the national level and the implications for adaptation. *Global Environmental Change*, 15: 151–163. Available at: http://www.uea.ac.uk/~f030/papers/gec2005.pdf

Carney, D. (Ed.)
1998 "Sustainable rural livelihoods: What contribution can we make?" Papers presented at the Department for International Development's Natural Resources Advisers' Conference, July 1998, Department for International Development (DFID), London.

Carter, T.R. et al.
2007 "New assessment methods and the characterization of future conditions", *Climate Change 2007: Impacts, Adaptation and Vulnerability.* Contribution of Working Group II to the Fourth Assessment Report of the Intergovernmental Panel on Climate Change, Parry, M.L. et al. (Eds), Cambridge University Press, Cambridge, UK, 133–171. Available at: http://www.ipcc.ch/ipccreports/ar4-wg2.htm

Caravajal, I. & Pereira, L.
2008 "Evidence on the link between migration, climate disasters and human development", Environment, Forced Migration and Social Vulnerability International Conference, 9–11 October 2008, Bonn. Available at: http://www.efmsv2008.org/file/D2+HDRO

Cheung, S.F. et al.
1999 "Reexamining the theory of planned behavior in understanding wastepaper eecycling", *Environment and Behavior*, 31(5): 587–612. Available at: http://www.citeulike.org/user/sfcheung/article/4498209

Conner, M. et al.
2003 "Environmental Influences: Factors Influencing a Woman's Decision to Use Dietary Supplements", *Journal of Nutrition*, 133: 1978S–1982S. Available at: http://jn.nutrition.org/cgi/content/full/133/6/1978S

De Jong, G.F. & Fawcett, J.
1981 "Multidisciplinary frameworks and models of migration decision making", in *Migration Decision Making: Multidisciplinary Approaches to Microlevel Studies in Developed and Developing Countries*, De Jong, G.F. and Gardner, R. (Eds), Pergammon Press, New York: 13–58.

De Jong, G.F.
2000 "Expectations, gender, and norms in migration decision-making", *Population Studies*, 54: 307–319.

Dennell, R.W.
2008 "Human migration and occupation of Eurasia", *Episodes*, 31(2): 207–210.

Deshingkar, P.
2006 *Internal Migration, Poverty and Development in Asia*, Overseas Development Institute (ODI), Briefing Paper. Available at: http://www.odi.org.uk/resources/download/29.pdf

Dessai, S. & Hulme, M.
2004 "Does climate adaptation policy need probabilities?" *Climate Policy*, 4: 107–128.

Dessai, S. et al.
2004 "Defining and experiencing dangerous climate change", *Climatic Change*, 64: 11–25.

DFID
2000 *Sustainable Livelihoods Guidance Sheets*, Department for International Development (DFID), London.

Dhara, V.R. & Dhara, R.
2002 "The Union Carbide disaster in Bhopal: A review of health effects", *Archives of Environmental Health*, September/October 57(5): 391–394.

Downing, T.E. et al.
2001 *Vulnerability Indices: Climate Change Impacts and Adaptation*. UNEP, Nairobi.

Downing, T.E. & Patwardhan, A.
2005 "Assessing vulnerability for climate adaptation", in *Adaptation Policy Frameworks for Climate Change: Developing Strategies, Policies and Measures*, Lim, B. et al. (Eds), Cambridge University Press, Cambridge and New York: 67–90.

Eakin, H.
2006 "Institutional change, climate risk, and rural vulnerability: cases from Central Mexico", *World Development,* 33: 1923–1938.

Eakin, H. et al.
2006 *A comparison of the social vulnerability of grain farmers in Mexico and Argentina*, AIACC Working Paper No. 29, Assessment of Impacts and Adaptation to Climate Change in Multiple Regions and Sectors Program, Washington, DC. Available at: http://www.aiaccproject.org/working_papers/Working%20Papers/AIACC_WP29_Eakin.pdf

Elliott, J.R. & Pais, J.
2006 "Race, class and Hurricane Katrina: social differences in human responses to Disaster", *Social Science Research*, 35: 295–321.

Eriksen, S.H. & Kelly, P.M.
2007 "Developing credible vulnerability indicators for climate adaptation policy assessment", *Mitigation and Adaptation Strategies for Global Change*, 12: 495–524.

Ezra, M.
2001 *Ecological Degradation, Rural Poverty, and Migration in Ethiopia. A contextual analysis*, Working Paper No.149, Policy Research Division, Population Council, New York.

Falk, W. et al.
2006 "Hurricane Katrina and New Orleanians' Sense of Place – Return and Reconstitution or "Gone with the Wind"? *Du Bois Review*: *Social Science Research on Race,* 3(1): 115–28.

Fankhauser, S.
1995 "Protection versus retreat – the economic costs of sea-level rise", *Environment and Planning A*, 27(2): 299–319.

Findley, S.E.
1994 "Does drought increase migration? A study of migration from rural Mali during the 1983–1985 Drought", *International Migration Review*, 28(3): 539–53.

Fordham, M.
1992 Choice and constraint in flood hazard mitigation: the environmental attitudes of floodplain residents and engineers. Unpublished PhD thesis, Middlesex Polytechnic, Enfield.

Fredrickson, C. et al.
1980 "Residential preferences in a model of migration intentions," *Population and Environment,* 3(3-4): 280–297.

Freeman, P.K. & Kunreuther, H.
2002 "Environmental risk management for developing countries", *Geneva Papers on Risk and Insurance – Issues and Practice*, 27(2): 196–214.

Gardner, R.
1981 "Macrolevel influences on the migration decision process", in *Migration Decision Making: Multidisciplinary Approaches to Microlevel Studies in Developed and Developing Countries*, De Jong, G.F.& Gardner, R. (Eds), Pergammon Press, New York: 59–89.

Grothmann, T. & Patt, A.
2005 "Adaptive capacity and human cognition: The process of individual adaptation to climate change", *Global Environmental Change*, 15: 199–213.

Haddad, B.M.
2005 "Ranking the adaptive capacity of nations to climate change when socio-political goals are explicit", *Global Environmental Change*, 15: 165–176.

Halliday, T.
2006 "Migration, risk, and liquidity constraints in El Salvador", *Economic Development and Cultural Change*, 54(4): 893–925.

Henry, S. et al.
2004 "The impact of rainfall on the first out-migration: A multi-level event-history analysis in Burkina Faso", *Population and Environment*, 25(5):423–60.

Hoozemans, F.M.J. et al.
1993 *A global vulnerability analysis: vulnerability assessment for population, coastal wetland and rice production on a global scale*. 2nd Edition ed., Delft Hydraulics and Rijkswaterstaat, Delft and the Hague.

Hunter, L.M.
2005 "Migration and environmental hazards", *Population and Environment*. 26(4): 273–302.

Ionescu, C. et al.
2005 "Towards a formal framework of vulnerability to climate change", NeWaterWorking Paper 2. Available at: http://www.newater.info

IPCC
2007a *Climate Change 2007: The Physical Science Basis*. Contribution of Working Group I to the Fourth Assessment Report of the Intergovernmental Panel on Climate Change, Solomon, S. et al. (Eds), Cambridge University Press, Cambridge and New York.

IPCC
2007b *Climate Change 2007: Impacts, Adaptation and Vulnerability*. Contribution of Working Group II to the Fourth Assessment Report of the Intergovernmental Panel on Climate Change, M.L. et al. (Eds), Cambridge University Press, Cambridge.

Jacobson, J.
1988 *Environmental Refugees: A Yardstick of Habitability*, Worldwatch Paper 86, Worldwatch Institute, Washington, DC.

Jager, W. et al.
2000 "Behaviour in commons dilemmas: *Homo economicus* and *Homo psychologicus* in an ecological-economic model", *Ecological Economics*, 35(3): 357–379.

Jones, R.N.
2001 "An environmental risk assessment/management framework for climate change impact assessments", *Natural Hazards*, 23: 197–230.

Kasperson, R.E.
2006 "Rerouting the stakeholder express", *Global Environmental Change*, 16: 320–322.

Kaperson, R.E. et al.
1988 "The social amplification of risk: a conceptual framework", *Risk Analysis*, 8: 177–187.

Kates, R.W.
1962 *Hazard and Choice Perception in Flood Plain Management*. Research Paper No. 78, Department of Geography, University of Chicago.

Kenny, G.J. et al.
2000 "Investigating climate change impacts and thresholds: an application of the CLIMPACTS integrated assessment model for New Zealand agriculture", *Climatic Change*, 46: 91–113.

Kirman A.
1999 "Aggregate activity and economic organization", *Revue Européenne des Sciences Sociales,* 37(113): 189–230.

Klein, R. & Nicholls, R.J.
1999 "Assessment of coastal vulnerability to climate change", *Ambio*, 28: 182–187.

Klein, R.J.T. et al.
2005 "Integrating mitigation and adaptation into climate and development policy: three research questions", *Environmental Science & Policy*, 8: 579–588.

Knapp, T. & Graves. P.
1989 "On the role of amenities in models of migration and regional development", *Journal of Regional Science*, 29(1): 71–87.

Kniveton, D.R. et al.
2008 *Climate Change and Migration: improving methodologies to estimate flows*. IOM Migration Research Series, No. 33, International Organization for Migration (IOM), Geneva.

Krueger, N.F.J. & Carsrud, A.L.
1993 "Entrepreneurial intentions: Applying the theory of planned behaviour", *Entrepreneurship and Regional Development*, 5: 315–330.

Kunreuther, H.
1996 "Mitigating disaster losses through insurance", *Journal of Risk and Uncertainty*, 12 (2–3): 171–187.

Landry, C.E. et al.
2007 *Going Home: Evacuation–Migration Decisions of Hurricane Katrina Survivors*, Center for Natural Hazards Research Working Paper. Available at: http://econ.appstate.edu/RePEc/pdf/wp0703.pdf

Leichenko, R.M. & O'Brien, K.L.
2002 "The dynamics of rural vulnerability to global change: the case of Southern Africa", *Mitigation and Adaptation Strategies for Global Change*, 7: 1–18.

Lonsdale, K.G. et al.
2008 "Plausible responses to the threat of rapid sea-level rise in the Thames Estuary", *Climatic Change*, 91(1–2): 145–169.

Lynne, G.D. et al.
1995 "Conservation technology adoption decisions and the theory of planned behavior", *Journal of Economic Psychology*, 16: 581–598.

McLeman, R. & Smit, B.
2006 "Migration as an adaptation to climate change", *Climatic Change*, 76: 31–53.

Mendelsohn, R. & Williams, L.
2004 "Comparing forecasts of the global impacts of climate change", *Mitigation and Adaptation Strategies for Global Change*, 9:315–333.

Metzger, M. & Schröter, D.
2006 "Towards a spatially explicit and quantitative vulnerability assessment of environmental change in Europe", *Regional Environmental Change*, 6: 201–206.

Migration DRC
2008 *Measuring the Migration-Development Nexus: An Overview of Available Data*, Development Research Centre on Migration, Globalization and Poverty Policy Briefing 13. Available at: http://www.migrationdrc.org/publications/briefing_papers/BP13.pdf

Morrow-Jones, H.A. & Morrow-Jones, C.R.
1991 "Mobility due to natural disaster: Theoretical considerations and preliminary analyses", *Disasters*, 15(2): 126–132.

Moss, R.H. et al.
2001 *Vulnerability to Climate Change: A Quantitative Approach*, Pacific Northwest National Laboratory, Richland, Washington. Available at: http://www.globalchange.umd.edu/data/publications/Vulnerability_to_Climate_Change.PDF

Munshi, K.
2003 "Networks in the modern economy: Mexican migrants in the US labour market", *Quarterly Journal of Economics*, 118 (2): 549–599.

Myers, N.
1997 "Environmental refugees", *Population and Environment*, 19(2): 167–82.

Myers, N.
2002 "Environmental refugees: A growing phenomenon of the 21st century", *Philosophical Transactions of the Royal Society B, Biological Sciences*, 357(1420): 609–13.

Myers, C.A. et al.
2008 "Social vulnerability and migration in the wake of disaster: the case of Hurricanes Katrina and Rita", *Population and Environment*, 29(6): 271–291.

Nicholls, R.J. & Tol, R.S.J.
2006 "Impacts and responses to sea-level rise: A global analysis of the SRES scenarios over the 21st century", *Philosophical Transactions of the Royal Society B, Biological Sciences*, 364:1073–1095.

Nicholls, R.J. et al.
2008 "Global estimates of the impact of a collapse of the West Antarctic ice sheet: an application of FUND", *Climatic Change*, 91(1-2): 171–191.

Nguyen, M.N. et al.
1997 "Regular exercise in 30- to 60-year-old men: combining the stages-of-change model and the theory of planned behaviour to identify determinants for targeting heart health interventions", *Journal of Community Health*, 22: 233–246.

Nordhaus, W.D.
2006 "Geography and macroeconomics: new data and new findings", *Proceedings of the National Academy of Sciences*, 103: 3510–3517.

O'Brien, K.L. & Vogel, H.C. (Eds)
2003 *Coping with Climate Variability: The Use of Seasonal Climate Forecasts in Southern Africa*, Ashgate Publishing, Aldershot.

O'Brien, K. et al.
2004 "Mapping vulnerability to multiple stressors: climate change and globalization in India", *Global Environmental Change*, 14: 303–313.

O'Brien, K. & Vogel, C.H.
2006 "Who can eat information? Examining the effectiveness of seasonal climate forecasts and regional climate-risk management strategies", *Climate Research*, 33: 111–122.

Olsthoorn, X. et al.
2008 "Neo-Atlantis: The Netherlands under a 5-m sea level rise", *Climatic Change*, 91(1–2): 103–122.

Oppenheimer, M. et al.
2007 "The limits of consensus", *Science*, 317: 1505.

Overpeck, J.T. et al.
2006 "Paleoclimatic evidence for future ice-sheet instability and rapid sea-level rise", *Science*, 311(5768): 1747–1750.

Patt, A. & Gwata, C.
2002 "Effective seasonal climate forecast applications: examining constraints for subsistence farmers in Zimbabwe", *Global Environmental Change*, 12: 185–195.

Paul, B. K.
2005 "Evidence against disaster-induced migration: the 2004 tornado in north-central Bangladesh", *Disasters*, 29(4): 370–385.

Peacock, W.G. & Girard, C.
1997 "Ethnic and racial inequalities in hurricane damage and insurance settlements", Chapter 9, in *Ethnicity, Gender and the Sociology of Disasters*, Peacock, W.G. et al. (Eds) Routledge, New York, 171–205.

Pederson, J.
1995 "Drought, migration and population growth in the Sahel – the case of the Malian Gourma, 1900–1991", *Population Studies*, 49: 111–126.

Pelling, M. (Ed)
2003 *Natural Disasters and Development in a Globalising World*, Routledge, London.

Petersen, W.
1958 "A General Typology of Migration", *American Sociological Review*, 23(3): 256–266.

Perch-Nielsen, S.
2004 *Understanding the effect of climate change on human migration: The contribution of mathematical and conceptual models*, MSc Environmental Physics, Department of Environmental Studies, Swiss Federal Institute of Technology, Zurich.

Piekle, R.A., Jr.
1998 "Rethinking the role of adaptation in climate policy", *Global Environmental Change* 8(2): 159–170.

Polsky, C. et al.
2003 *Assessing Vulnerabilities to the Effects of Global Change: an Eight-Step Approach.* Belford Centre for Science and International Affairs Working Paper, Environment and Natural Resources Program, John F. Kennedy School of Government, Harvard University, Cambridge, Massachusetts.

Poumadere, M. et al.
2008 "Worst case scenario as stakeholder decision support: a 5- to 6-m sea level rise in the Rhone delta, France", *Climatic Change*, 91(1–2): 123–143.

Pulhin, J. et al.
2006 *Vulnerability of communities to climate variability and extremes: the Pantabangan- Carranglan watershed in the Philippines*, AIACC Working Paper No. 44, Assessment of Impacts and Adaptation to Climate Change in Multiple Regions and Sectors Program, Washington, DC.

Raghavan, S.
2008 *Quantifying uncertainties in regional climate predictions over southern Africa using nested regional climate models*, PhD thesis, University College London.

Raleigh, C. et al.
2008 *Assessing the Impact of Climate Change on Migration and Conflict*. Working paper for World Bank programme on Exploring the Social Dimensions of Climate Change. Available at: http://siteresources.worldbank.org/EXTSOCIALDEVELOPMENT/Resources/SDCCWorkingPaper_MigrationandConflict.pdf

Renaud, F. et al.
2007 "Control, adapt or flee – How to face environmental migration?" *InterSecTions*, No. 5, United Nations University Institute for Environment and Human Security (UNU-EHS), Bonn.

Rotmans, J. & Rothman, D. (Eds)
2003 *Scaling Issues in Integrated Assessment*, Swets and Zeitlinger, Lisse.

Scheider, S.H.
2001 "What is 'dangerous' climate change?" *Nature* 411: 17–19.

Schellnhuber, H.J. et al.
2004 "Integrated assessments of benefits of climate policy", in *The Benefits of Climate Change Policies: Analytic and Framework Issues*, Morlot, J.C., Agrawala, S. (Eds), Organization for Economic Cooperation and Development (OECD), Paris: 83–110.

Schröter, D. et al.
2005 "Ecosystem service supply and vulnerability to global change in Europe." *Science,* 310: 1333–1337.

Shipton, P.
1990 "African famines and food security", *Annual Review of Anthropology*, 13: 353–39.

Smith, C.D. et al.
2008 "Climate change, migration and agent-based modelling: modelling the impact of climate change on forced migration in Burkina Faso", *Forced Migration Review*, 31: 58–59.

Speare, A. Jr.
1974 "Residential satisfaction as an intervening variable in residential mobility", *Demography*, 11(2): 173–188.

Stark, O. & Bloom, D.E.
1985 "The New Economics of Labour Migration", *The American Economic Review*, 75(2): 173–178, Papers and Proceedings of the Ninety-Seventh Annual Meeting of the American Economic Association (May 1985).

Stern, N.
2006 *The Economics of Climate Change – The Stern Review*, Cambridge University Press, Cambridge.

Suhrke, A.
1994 "Environmental degradation and population flows", *Journal of International Affairs*, 47: 473–496.

Sutherland, K. et al.
2005 "Vulnerability to climate change and adaptive capacity in Samoa: the case of Saoluafata village", *Tiempo*, 54: 11–15. Available at: http://www.adaptation2005.ca/abstracts/sutherland_e.html

TERI
1996 *The economic impact of a one-metre sea level rise on the Indian coastline: method and case studies*, Tata Energy Research Institute (TERI), New Delhi.

Tompkins, E.
2005 "Planning for climate change in small islands: insights from national hurricane preparedness in the Cayman Islands", *Global Environmental Change*, 15: 139–149.

Thornton, P.K. et al.
2006 *Mapping climate vulnerability and poverty in Africa*, report commissioned by the Department for International Development (DFID); produced by International Livestock Research Institute (ILRI), Nairobi/Tata Energy Research Institute (TERI), New Delhi: 171.

Tol, R.S.J.
2002 "New estimates of the damage costs of climate change. Part I: Benchmark estimates", *Environmental Resource Economics*, 21: 45–73.

Turner, B.L. et al.
2003 "A framework for vulnerability analysis in sustainability science", *Proceedings of the National Academy of Sciences*, 100: 8074–8079.

UN Population Division
2006 *Trends in the Total Migration Stock*, 2005 Revision, Department of Economic and Social Affairs, UN Population Division, New York. Available at: http://www.un.org/esa/population/publications/migration/UN_Migrant_Stock_Documentation_2005.pdf

UNDP
2005 *Adaptation Policy Frameworks for Climate Change: Developing Strategies, Policies and Measures*, Lim, B. et al. (Eds), Cambridge University Press, Cambridge and New York.

UNEP
2006 *Environmental considerations of human displacement in Liberia*, United Nations Environment Programme (UNEP), Nairobi.

Unruh, J. et al.
2004 "Environmental change and its implications for population and migration", *Advances in Global Change Research*, Vol. 20, Springer, New York.

van Asselt, M.B.A. & Rotmans, J.
2002 "Uncertainty in integrated assessment modelling: from positivism to pluralism", *Climatic Change*, 54: 75–105.

Vitousek, P.M. et al.
1997 "Human domination of Earth's ecosystems", *Science*, 277 (5325): 494–499.

Warren, R. et al.
2006 *Understanding the regional impacts of climate change: research report prepared for the Stern Review on the Economics of Climate Change*, Tyndall Centre for Climate Change Research, Working Paper 90, University of East Anglia, Norwich.

Wehbe, M. et al.
2006 *Local perspectives on adaptation to climate change: lessons from Mexico and Argentina*, AIACC Working Paper No. 39, Assessment of Impacts and Adaptation to Climate Change in Multiple Regions and Sectors Program, Washington, DC. Available at: http://www.tyndall.ac.uk/publications/working_papers/twp90.pdf

Willows R.I. & Connell, R. (Eds)
2003 Climate Adaptation: Risk, Uncertainty and Decision-Making. UK Climate Impacts Programme (UKCIP), Oxford. Available at: http://www.ukcip.org.uk/images/stories/Pub_pdfs/Risk.pdf

Wilbanks, T.J. et al.
2007 "Industry, settlement and society", *Climate Change 2007: Impacts, Adaptation and Vulnerability*. Contribution of Working Group II to the Fourth Assessment Report of the Intergovernmental Panel on Climate Change, Parry M.L. et al. (Eds), Cambridge University Press, Cambridge: 357–390.

Wolpert, J.
1966 "Migration as an adjustment to environmental stress", *Journal of Social Issues*, 22(4): 92–102.

Woodward, A.
2002 "Epidemiology, environmental health and global change", in *Environmental Change, Climate and Health: Issues and Research Methods*, Martens, P., McMichael, A.J. (Eds), Cambridge University Press, Cambridge: 290–310.

Yohe, G.W. et al.
1996 "The economic costs of sea level rise on US coastal properties", *Climatic Change*, 32: 387–410.

Yohe, G.W. & Tol, R.S.J.
2002 "Indicators for social and economic coping capacity: moving toward a working definition of adaptive capacity", *Global Environmental Change*, 12: 25–40.

Zaman, M.
1989 "The social and political context of adjustment to riverbank erosion hazard and population resettlement in Bangladesh", *Human Organization*, 48: 196–205.

Zelinksky, W.
1971 "The hypothesis of the mobility transition", *Geographical Review*, 61: 219–249.

Ziervogel, G. & Calder, R.
2003 "Climate variability and rural livelihoods: assessing the impact of seasonal climate forecasts in Lesotho", *Area*, 35(4): 403–417.

Ziervogel, G. et al.
2004 "Agent-based social simulation: a method for assessing the impact of seasonal climate forecast applications among smallholder farmers", *Agricultural Systems*, 83(1): 1–26.

7. Appendix 1 (from Migration DRC, 2008)

The Migration DRC and data accessibility

The Migration DRC has pursued two projects aimed at making existing macro data related to migration more acceptable: the Global Migrant Origin Database and the Migration in National Surveys (MiNS) catalogue. The Global Migrant Origin Database is the first 'complete' dataset on international migration, listing bilateral migration stocks, based on the 2000–2001 round of population censuses. There are four different versions of the dataset available, ranging from raw data derived from censuses, to complete matrices of international flows, which required large assumptions in order to disaggregate some countries' census data on 'foreign-born' stocks. The Global Migration Origin Database has been used by the World Bank to underpin a substantial analysis of the global impact of migration on well-being: *Global Economic Prospects 2006: Implications of Remittances and Migration*.

MiNS is being compiled to improve the availability and use of data on migration that can be extracted from national surveys. In addition to providing direct links to each country's most recent survey data, it provides summary information on each survey, including how it defines terms related to migration. The catalogue includes Living Standards Measurement Surveys (LSMS), Demographic and Health Surveys (DHS), Integrated Surveys (IS), Population and Household Censuses (PHC) and Child Labour Surveys (CLS). The construction of MiNS began in 2006, with an additional emphasis on children added in 2007.

MiNS will soon be expanded to include Labour Force Surveys (LFS). Both databases are available online: MiNS at: www.migrationdrc.org/publications/resource_guides/Migration_Nationalsurveys/index.html; and the Global Migrant Origin Database at: www.migrationdrc.org/research/typesofmigration/global_migrant_origin_database.html.

8. Appendix 2 (from the Mexican Migration Project (MMP) website: mmp.opr.princeton.edu)

The Mexican Migration Project (MMP) is a collaborative research project, based at the Princeton University and the University of Guadalajara.

Data overview

The MMP118 Database is the result of an ongoing multidisciplinary study of Mexican migration to the USA. It contains data gathered since 1982 in surveys administered every year in Mexico and the USA.

The MMP118 Database contains general demographic and migratory information for each member of a surveyed household, as well as general characteristics of the household, its members, and other holdings and labour histories for each head of household and each spouse. Starting in 2005, the MMP has gathering detailed information on migratory experience of all household heads that have migrated to Canada.

In addition, supplementary data have been created to provide researchers with additional information that may be useful in analyses of migration. For instance, for all the communities surveyed by the Mexican Migration Project, data at the community and municipal level have been collected and compiled. Detailed environmental data are also available at state level, including information on such variables as type of weather, land use and degradation, and historic monthly rain (from 1941 to 2004).

Collecting data on the migration–environment nexus[1]

Richard E. Bilsborrow[2]

"The increasingly complex interplay of social and economic factors in the environment is exacerbating the vulnerability of both people and environments and intensifying the impacts of such changes when they occur" (Oliver-Smith & Shen, 2009: 9).

[1] The author is grateful to Christine Aghazarm and Clark Gray for their valuable comments and editorial suggestions, and to the National Institutes of Health (Grant no. R21HD052092-01A2) for supporting the Ecuador project described in Ecuador case study.

[2] Professor of Biostatistics, Adjunct Professor of Ecology, Economics, Geography and City and Regional Planning Fellow, Carolina Population Center University of North Carolina at Chapel Hill; Email: richard_bilsborrow@unc.edu

1. Introduction

Given the increasing concerns about migration and growing environmental problems worldwide, it is not surprising that the scientific community and policy makers have become quite focused on the linkages between the environment and migration, and on how these linkages tie in with concerns about poverty and prospects for sustainable development. The quotation above reflects this increased focus on migration–environment linkages, and the importance of understanding them better. Indeed, it has become fashionable to view migration as a major proximate cause of environmental degradation in the world, and as a major anthropogenic factor involved in the complex human population–environment nexus. At the same time, environmental factors are often considered, especially in the popular media, as major factors in inducing or forcing population movements. However, there is almost no solid empirical evidence on the latter, and even the former, which has been the subject of recent research, has rarely adequately separated out the impacts of human population movements from those of many other coterminous factors hypothesized to affect the environment, including other sources of human population growth. This uncertainty is due to the lack of adequate data on migration and the environment, as well as the lack of a clear theory for causally linking them.

This chapter examines some existing methods of collecting data on migration (both internal and international) and on the environment, and ways of measuring both, for the purpose of investigating their interrelationships. The lack of reliable evidence on the relationships, and other challenges, are highlighted through a review of types of studies in the scientific literature, resulting in recommendations for better methods of data collection and analysis. The latter will be illustrated by an experimental survey designed and carried out in rural Ecuador in 2008. To put that discussion in context, and to promote improvements in relevant data collection, some pertinent theoretical perspectives are first reviewed, as are issues in the definition and measurement of migration and environment, and data collection experiences in the few published studies that attempt to quantitatively investigate migration–environment interrelationships.

This chapter focuses on developing countries – particularly rural areas – since that is where the main issues are, with respect to human relationships with the environment, along with the related key issues of poverty, human welfare and threats to biodiversity. There will, therefore, be little discussion here of the direct impacts of migration on urban environments, or of the impact of urban environments (e.g., lack of space, greenery, but plenty of pollution, in many forms) on migrants. The reason is that the focus in this publication is on the relationships between human migration movements and the natural environment, rather than the artificial or man-made environments that dominate cities and smaller urban areas. However, the chapter will deal with the relationships in *both* directions – from the natural environment to migration (attracting or detracting migrants), to the impacts of population movement on the natural environment. It will also look at migrants' areas of origin and destination.

The chapter does not consider relationships between sudden natural disasters and migration or how human habitation or manipulations of the physical environment can make ongoing climactic events or natural disasters much worse than they would otherwise be[3], as they are covered elsewhere in this volume (see Naik Chapter 5; Warner et al. Chapter 4 in this volume). Instead, the focus is on the *gradual and cumulative* environmental impacts of increasing human populations resulting specifically from migration, rather than from other causes of population growth (such as natural population growth, defined as fertility minus mortality) – affects the productivity of land, resulting from deforestation, soil erosion and other forms of land degradation. Similarly, with respect to the impacts in the other direction, the discussion concentrates on the impact of *gradual* declines in land productivity on out-migration.

The topic of global climate change is also only briefly addressed here (see Kniveton et al. Chapter 2; Leighton Chapter 6 in this volume), partly so as to focus on specific, measurable linkages in particular environments where sufficiently detailed data can be collected and measured to isolate migration–environment linkages from other factors affecting the environment for particular populations. That the Earth has been experiencing significant anthropogenic transformation since the industrial revolution and the related population explosion of

[3] For example, a greater human population (which may be due partly to in-migration) living in or near to areas susceptible to flooding, volcanic eruptions, hurricanes/typhoons, and earthquakes guarantees a greater disaster measured as human populations displaced or killed by such disasters.

the past two centuries, and that this is causing major climate changes and reductions in the habitability of parts of the planet (Oliver-Smith & Shen, ibid:, p. 12) is now widely accepted by scientists. The effects of climate change on human migration may eventually come to constitute "the greatest single impacts of climate change" (Intergovernmental Panel on Climate Change, 1990, cited in Brown, 2007).[4] Even now, significant future impacts on human mobility and displacements of populations are beginning to be seen and anticipated in many places, such as the Himalayas, the Ganges and Mekong deltas, the Nile delta, Mexico and Central America, the Sahel, and Island states (Warner et al., 2009).

Finally, this chapter does not engage in the controversial and emotionally charged debate about 'environmental refugees' – how many there are, how important they are, or even if they exist (see Zetter Chapter 8 in this volume). By adopting the term *refugee*, that debate has given those "*compelled mainly by sudden or gradual environmental change*" implied rights of refugees, even though the United Nations High Commissioner for Refugees does not recognize the term. Indeed, the lack of common definitions has contributed to the confusion about, and exaggerated estimates of, present and possible future 'environmental refugees' (El Hinnawi, 1985; Hugo, 1996; Myers, 1993, 2002; Black, 2001; Bates, 2002; Lonergan, 1998). A much better and more general term is *'environmental migrants'*, defined by IOM as:

"persons or groups of persons who, for compelling reasons of sudden or progressive changes in the environment that adversely affect their lives or living conditions, are obliged to leave their habitual homes, or choose to do so, either temporarily or permanently, and who move either within their country or abroad".

(IOM Discussion Note: Migration and the Environment, MC/INF/288, 1 November 2007).[5]

[4] The impact of greenhouse gases on global warming and thereby on melting of glaciers and the polar ice packs, which causes sea-level rise, have attracted great attention. Models are estimating the impacts on coastal flooding and population displacement, and even the possible disappearance of island states below the sea. The Prime Minister of the Maldives is considering using state resources to purchase land on a continent (Schmidle, 2009). Among the many recent references on this issue in the professional literature is Dasgupta et al. (2009).

[5] A glossary of terms related to migration is available at: http://www.iom.int/jahia/webdav/site/myjahiasite/shared/shared/mainsite/published_docs/serial_publications/Glossary_eng.pdf.

In this chapter, the focus is on relationships between gradual changes in the environment and migration which involves a change in residence across administrative boundaries, whether internally or internationally. Both the effects of environmental factors on the move, and the consequences of the move for the environment are considered.

Thus this chapter addresses the major form of human population redistribution which has been occurring across the globe over the past century – the exodus of people from rural areas –and its linkage to the environment. It therefore focuses on the effects of environmental factors on inducing out-migration from rural areas, and the impacts of that migration on areas of destination, also rural locations. The focus on rural areas reflects the focus on the natural environment. In terms of evaluating the environmental effects on out-migration, therefore, only rural areas of origin are considered. Similarly, regarding the effects of migration on the environment, the focus again is on rural areas. Thus, although the relevance of environmental factors on any type of out-migration from rural areas of origin to any type of destination is of interest, the chapter is concerned with impacts only on rural areas of destination, whether domestic or international, including migration to the agricultural frontier.[6] Indeed, migration to the frontier, to establish new farms, has been a dominant aspect of population distribution in the past four decades in much of Latin America, Africa and Asia, and is of particular interest in terms of its environmental impacts, due to its association with deforestation of tropical forests and the resulting serious loss of biodiversity (Cincotta & Engleman, 2000; Oglethorpe et al., 2007).

In sum, the main purpose of this chapter is to review recent research and to suggest better methods of data collection that will facilitate improved understanding of the migration–environment nexus. Key areas of research for understanding and investigating the linkages between migration and the environment will be reviewed first, focusing on rural areas. The neglect of environmental factors in research on human migration will later be seen to be part of a larger shortcoming in migration research: the very limited research on the impacts of contextual factors on migration.

6 The agricultural frontier is where the land area used for agriculture (crops or animal husbandry) is being increased at the extensive margin – that is, at the edges of the existing lands used for agriculture. This usually means the expansion of lands used for agriculture, or agricultural extensification, into tropical forest, swamplands or dryland areas not previously used for agriculture.

2. The current state of knowledge

Three bodies of knowledge or areas of research are particularly relevant and are reviewed below. They provide guidance on the kinds of information to be collected when investigating migration–environment linkages: (1) the impact of migration on land use, including the environment, in areas of both destination and origin; (2) the determinants of migration; and (3) the role of environmental factors in inducing out-migration from rural areas.

Changes in land use/land cover and the impacts of migration

Going beyond the classic studies of population impacts on land by Malthus (1798, 1960 ed.) and Boserup (1965), significant research has been carried out in recent years on land use and land cover change (LULCC), using methods from both the social and geographic or spatial sciences (see, for example, reviews by Geist & Lambin, 2001, 2002; Walsh & Crews-Meyer, 2002; Entwisle & Stern, 2005). Research on LULCC in tropical forest areas has identified a number of key drivers of land use change, including government policies, such as tax subsidies to ranchers or requirements that a certain minimum percentage of land be cleared in order to receive a land title (Hecht, 1985; Schmink and Wood, 1992; Kaimowitz et al., 1999), national and international markets (Walker at el., 2000), logging (Sierra & Stallings, 1998; Mertens et al., 2001), large-scale agriculture (Walker et al., 2000), fires (Cochrane et al., 1999; Nepstad et al., 1999), mineral and oil extraction (Pichon & Bilsborrow, 1997; Walsh et al., 2002), and road construction (Heckandon & McKay, 1984; Joly, 1989; Rudel & Horowitz, 1993; Chomitz & Gray, 1996; Nelson & Hellerstein, 1997; Pfaff et al. 2007).

A commonly cited key driver is the migration of colonists to the forest frontier, especially small-holder agriculturalists (World Bank, 1991; literature reviews in Lambin et al., 2001; Geist & Lambin, 2002). In Latin America, this migration has occurred throughout Central America and Mexico (Shelhas, 1996; Bilsborrow & Stupp, 1997; Sader 1997; Carr, 2008) as well as in the Amazonian countries, including Bolivia

(Kaimowitz et al., 1999), Brazil (McCracken et al., 2002) and Ecuador (Rudel & Horowitz, 1993; Bilsborrow et al., 2004). Studies on Africa and Asia also find large impacts of migrants on forest cover – in Kenya (Little, 1987); Sudan (Ibrahim, 1987; Bilsborrow & DeLargy, 1991); Senegal (Sessey & Mohamed, 1997); Nepal (Shrestha, 1990); Thailand (Cropper & Griffiths, 1999; Panayotou & Sungsuwan, 1994), the Philippines (Cruz, 1997) and Indonesia (Whitten, 1987). However, it should be noted that very few of these studies directly relate migration or migrants to the loss of forest cover: most just correlate the latter with total population size or change. One exception is a study in the canton of Sarapiquí, Costa Rica, where migrants moved into the area to plant crops and raise cattle, causing the population to quadruple between 1963 and 1983, and forest cover to decline from 70 per cent to 30 per cent. Another study, also at an aggregate level, relates to the Brazilian Amazon, where Wood and Skole (1998) found deforestation at the municipal or district level linked to in-migration to rural areas. In the northern Ecuadorian Amazon, following the discovery of oil in 1967, forest cover fell from virtually 100 per cent in 1970 to 56 per cent in 1990, on a representative sample of farm plots, falling further to 45 per cent by 1999 (Bilsborrow et al., 2004). Similar dramatic declines in forest cover have occurred in various states of the Brazilian Amazon, characterized by heavy in-migration to frontier areas, including Acre, Mato Grosso, Pará, Rondonia and Santarem.[7] Despite this evidence, the origins of spontaneous migration to frontier areas and the factors pushing people to migrate from other rural areas to the frontier have received almost no attention in the literature, even though such migration is a key contributor to the deforestation.

The above-mentioned literature is the most developed with respect to providing evidence on the impacts of migration on the environment – principally via its effects on the expansion of the agricultural frontier through clearing for farms. While this is the key area of recent research on migration impacts on the environment, the hypothesis that migrants tend to have greater impacts on the natural environment than non-migrants in areas of destination is not new, and arises out of the belief that migrants do not care as much about the destination area since

[7] But there is by no means a one-to-one correspondence between human migration to a frontier area and loss of forests or other vegetation cover. The latter depends on what land use practices are adopted by the migrants. In southern Honduras and in the Brazilian Amazon, cattle farming, which need not involve many people, is the main cause of deforestation (DeWalt, 1985; Hecht, 1985; Humphries, 1998; Walker et al., 2000). For a region-wide study on Latin America, see Bilsborrow and Carr (2001).

it is not their home, or because they bring concepts and practices of resource use from their own areas of origin which are inappropriate for the area of destination. In the case of the Amazon, Browder (1995), Pichon (1997) and Perz (2003) suggested this, while Bilsborrow (1992) found state-sponsored, organized transmigrants had much smaller deleterious impacts on their destination environments than spontaneous migrant settlers. Similarly, based on data collected for 599 households in 17 coastal villages in northern Sulawesi, Indonesia, Cassels et al. (2005) found that migrants are more likely to live in villages with low environmental quality due to their poverty but have fishing practices that are no more ecologically destructive than those of native residents. The authors conclude that the topic would be much better addressed with longitudinal data, to determine whether migrants degraded the resources after arriving, or whether they were simply forced by poverty to settle in the more degraded environments.

A much broader and more compelling research topic relates to the effects of international migration, particularly from developing to developed countries, on households, communities and countries of origin: does such migration contribute to economic development and, if so, how – via remittances, transfers of technology brought by emigrants during visits or brought by returning migrants? Another distinct body of literature has linked international migration to resource degradation in destination countries. Some of this literature is fundamentally anti-immigrant (and some is Neo-Malthusian – against *any* population growth, whatever the cause), and does not specifically relate degradation in any particular location to an increased population of immigrants but rather only to a general increase in population. Another body of literature, with limited research evidence, links large refugee settlements to environmental deterioration around the settlements, due to depletion of forests and vegetation for fuel and mining of surface and ground water.

Since migration to the frontier is predominantly from other rural areas of the country, as in the case of the Ecuadorian Amazon, where 85 per cent of the migrants up to 1999 were from rural origins (Bilsborrow et al., 2004), then the declining rural population density often observed in migrants' areas of origin may, depending on adaptations in land use by the remaining population, result in less pressure on the land and, hence, spontaneous reforestation. Similar population movements out of rural areas have occurred throughout the developing world, led by Latin America where the absolute size of the rural population has been declining for several decades in most countries (Bilsborrow, 2002; United

Nations, 2008). Several interesting studies exist on the impacts of out-migration on origin area environments. Thus, in the Camacho valley of Bolivia, out-migration led to less grazing of animals and less pressure on the environment (Preston et al., 1997). In China, the massive rural–urban migration of the so-called 'floating population' (those who did not have a *hukou* or residence permit) has led to population declines in many rural origin areas of Szechuan and other central provinces, although at the same time, farmland and green vegetation are lost as urban areas expand onto nearby land (Heilig, 1997; Seto et al., 2002).

There are a few studies on the consequences of out-migration from rural areas for the re-growth of vegetation in areas of origin, based on international migration. This is a fruitful area for research, and could be carried out at the micro level using data from household surveys, satellite imagery or both (see sections 3, 4 and 5 below). One study that does examine precisely such impacts of migration on a rural area of origin, albeit for a very limited dataset, is that by Radel & Schmook (2008). The authors collected data in 2003 on 203 households in 14 *ejido* communities[8] in the southern Yucatan peninsula of Mexico (states of Campeche and Quintana Roo, around the Calakmul Biosphere Reserve), on rural households with and without international out-migrants to the United States of America (USA). They found that migrant households cultivated significantly less farmland than non-migrant households, leading to some forest recovery on their lands. The authors attribute this to the effects of receiving large remittances, which makes the hard labour of cultivation less necessary for supporting the origin household. If the total per capita incomes of the two types of households are similar, it would imply that households without remittances must work harder to earn the income equivalent to the value of remittances received by the other households. Still, the economic welfare of migrant households improved, as they had more assets and were not working so hard. The authors concluded elsewhere that the case "illustrates the need to incorporate the role of globalizing household economies into forest transition theory" (Schmook & Radel, 2008: 891, 905).[9] Indeed, international market factors affect the

[8] *Ejidos* are rural communities with communal landholdings that account for about 70 per cent of the Mexican farm population and half of Mexican farmland.

[9] Forest transition theory has been developed largely to describe the historical experience of developed countries of first clearing forests for fuel, house construction and especially agricultural expansion; however, as these countries developed economically and fertility and population growth declined, they have come to intensify their agricultural production (increase output from land through increased inputs of capital, irrigation and chemical inputs) and also import more of their natural-resource-based products from developing countries, allowing their own forests to regenerate (Perz & Skole, 2003).

prices of locally produced commodities, which affects land and labour use decisions of populations in even the most isolated parts of the globalizing world. But while the study is interesting and important, focusing on a neglected topic of growing importance worldwide, it has several methodological limitations. First, the sample of communities is stratified on the basis of distance from the main north–south road through the main local town, Xpujil, though that is not representative of the local *ejido* population (unless the data were weighted, which is not stated). Probably more important is the fact that the sample itself is very small, with only 48 of the 203 households having an international out-migrant – equivalent to 3.4 per community. Finally, the methods of analysis are purely descriptive: in the absence of multivariate analysis, it is not possible to firmly attribute the causes of forest re-growth to the effects of remittances on reducing farm effort of recipient families, rather than to other factors that differ between households with and without out-migrants, even though it seems plausible.

A similar, earlier study by Jokisch (2002) investigates the effects of high levels of out-migration to the USA from two rural communities in Canar province of the Ecuadorian Sierra (a province of high emigration to Spain as well as the USA, especially since 1996). Jokisch hypothesized that emigration could have led to either agricultural abandonment or agricultural improvements, but found no evidence of either in the two communities. Instead, international migration had led to large remittances that were invested in luxury houses and purchases of land around urban areas, which were often converted to peri-urban housing. Despite the author's conclusions, the impact of the emigration on the natural environment would be considered negative, given that the area of farmland was reduced as a result of the emigration. Jokisch attributed the lack of investment in agricultural crops to the low returns to cultivation prevailing in the area, due to the lack of irrigation. He notes that, for decades before, temporary migration and off-farm employment had long been important ingredients in household livelihood maintenance, and that, once international migration emerged as significant, remittances came to be depended upon as a source of household income in origin households, which used them to build nice houses as symbols of success of the emigrant, leading to "rural gentrification" of the countryside (ibid: 547). Jokisch viewed this process as fomenting further emigration, and possibly eventually leading to increased replacement of crops by pasture as a less intensive form of land use, creating a new twenty-first century generation of '*hacendados*' (owners or managers of a hacienda).

Out-migration from rural areas may have not only environmental impacts but also impacts relating to gender. Collins (1986) found that male out-migration resulted in the deterioration of stone terraces used for agriculture (and, hence, agricultural production) in the Peruvian Andes, since the maintenance of terraces is a highly labour intensive activity usually carried out by men. The same phenomenon was observed on an island in Lake Victoria, Kenya (Conelly, 1994).

A recent paper on China (Song et al., 2008) investigates the relationships between migration and the environment at the macro (provincial) level. It illustrates both the advantages and limitations of using existing data for large areas from satellite imagery and population censuses or surveys to study the relationships between migration and the environment. In this study, the environment is characterized by vegetation cover, so the change was the change in vegetation cover, measured by the NDVI.[10] Vegetation cover in 8 km x 8 km pixels or square areas on the ground were measured from satellites, and used to quantify the change in the mean (across all pixels) provincial level vegetation in 1995–2000 compared to 1985–1990. Changes in mean annual temperature and mean annual precipitation were controlled for in the statistical model in which the dependent variable is the change in the mean coverage of vegetation for the province. The key independent variable of interest was migration, which was measured as the sum (based on the 1990 and 2000 census data, each of which included a question on place of residence five years earlier) of in-migration, out-migration, and intra-province migration. Given that most internal migration in China is rural–urban, then, of these three types of migration, out-migration from rural areas will tend to reduce pressures on land and thereby potentially lead to re-growth of vegetation. In-migration is usually to urban areas, which involves a physical expansion of the urban land area onto the surrounding countryside and, hence, urban consumption of green areas. *A priori*, the effect of the urban sprawl in areas of destination of migrants might be expected to be less than that of out-migration from rural areas, in which case a province with high levels of both in- and out-migration would be expected to experience a higher level of increase in vegetation over time than one

[10] The abundance of vegetation was measured by the Normalized Difference Vegetation Index (NDVI), which is a widely used measure of greenness of vegetation on the ground, as detected by satellites – specifically the Advanced Very High Resolution Radiometer, on National Oceanographic Atmosphere Administration (NOAA) satellites. The spatial resolution is crude (8 km x 8 km) but this facilitates covering broad geographic areas (Tucker et al., 2005).

with low levels of both. But the authors did not examine this, instead analysing the effects of the three types of migration on vegetation separately. Surprisingly, it was the effect of *in-migration* that had the strongest and statistically most significant linkage to the change in vegetation – in other words, the greater the in-migration to a province, the greater the loss of vegetation. Even more surprising is the fact that the greater the out-migration in a province, the greater the *negative* impact on vegetation cover – which is the opposite of what is expected (ibid: 5075), although the effect was not statistically significant. But this unexpected result may have a logical explanation. Migrants to urban areas often sent back considerable remittances that could have been used to build larger houses and change land use (ibid: 5076). For example, remittances could have been used to buy farm machinery and even increase the area used for agriculture (at the expense of forests). But there may be another explanation for the lack of an observed effect of out-migration (which decreased rural population density) on the regeneration of vegetation: it takes time for vegetation to re-grow, so the effect may not show up in satellite imagery in such a short ten-year reference period, but may appear in the future. In contrast, the effect of urban sprawl is immediate. Thus, further studies are needed, at not only an aggregate level and over a longer time period, but also at the smaller/micro area and household level, in China as well as in other countries. Indeed, the authors conclude that "migration should be an important factor in making environmental policies" in China and that it is "important for policy-makers in China to take the impacts of migration on vegetation growth into account while making policies aiming at sustainable human–environment relations" (ibid: 5078–9).

The determinants of migration

A large body of work dating back to Ravenstein (1885, 1889) has considered the many economic, social, demographic and other factors that can influence migration. Migration theories are summarized in Ritchey (1976), De Jong & Gardner (1981), Bilsborrow et al. (1984), Massey et al. (1993), and White & Lindstrom (2005). Migrants are generally considered to be motivated by differences in economic opportunities and amenities between origin and destination areas, or by differences in 'place utility' (Wolpert, 1965), which traditionally meant mainly opportunities for employment and land ownership. Early micro-level research highlighted the importance of personal characteristics in determining which persons in a household migrated, such as age, sex, education and work experience or human capital. Prospective migrants have been hypothesized by economists to compare the economic costs and returns of migration (Sjaastad, 1962; DaVanzo, 1981), i.e., estimating the expected returns (or income) of alternative places of residence compared to the current place of residence, and then moving if a destination offered a higher income than the present place of residence, taking into account the costs of migration. Todaro (1969) extended the human capital model of Sjaastad by noting that migration is more responsive to *expected* than actual wage differentials – in other words, that prospective migrants take into account the probability of being employed as well as the expected wages. In contrast to the focus of economists on income, sociologists have considered broad sets of push–pull factors (Lee, 1966) and the roles of social networks or social capital (see references below). Finally, geographers focus on locational factors. Indeed, since Ravenstein (1885, 1889), distance to potential destinations has been thought to have a significant negative effect on migration (e.g., Henry & Bilsborrow, 2007).

In recent decades, migration researchers have begun to investigate the implications of the fact that individuals do not make migration decisions in isolation, but in a broader context. To begin with, the *household context* is seen as particularly important for migration in the developing world, where the migration of household members is seen as reflecting household-level survival and income diversification strategies (Mincer, 1978; Arguello, 1981; Hugo, 1981; De Jong & Gardner, 1981; Ibrahim & Ruppert, 1991). Thus, in the 'household survival strategy' approach (Arguello, 1981), households are seen as allocating household members (labour) to alternative uses/ employments and locations, such as farming their own land or that of others, off-farm employment, and even sending a household member to another location to work. The latter reduces consumption needs/ pressures on the household, but it may also raise its income if the migrant sends remittances that are greater in value than the migrant's prior contribution to household income. Similarly, the 'new economics of labour migration' (Stark & Bloom, 1985; Stark & Lucas, 1988; Stark & Taylor, 1991) views migration as a way in which households diversify risk, by allocating labour to different kinds of work.

Household characteristics that are consistent with these theories and are thought to influence migration include household size, household income and assets, land ownership, perceived relative deprivation, and previous migration experience and migration networks.

However, migration decisions are increasingly recognized as being made in a broader socio-economic-institutional context, as seen in political ecology (e.g., Blaikie & Brookfield, 1987) and focuses on structural factors (Wood, 1982). Social relationships and networks, of the community or ethnic group as well as the household – both in the community of origin and in potential destination areas – may also have effects (Massey & Espinoza, 1997; Davis et al., 2002; Cassels et al., 2005). Many other aspects of the local origin community and the larger regional context in which it is imbedded may also influence out-migration. In the case of international migration, the relevant context is the nation state, including policies and regulations of the origin country pertaining to emigration and of the destination countries considered with respect to immigration. Such national contextual factors include wage levels and employment opportunities (consistent with Sjaastad and Todaro, above); geographical accessibility to roads and cities; land size and ownership (the focus of Shaw, 1974, and others, including Bilsborrow et al., 1987); community infrastructure endowments, including schools, health clinics and government services; population size and density; and markets. Yet, despite the strong theoretical case for the incorporation of contextual factors in studying individual and household migration decisions (Wood, 1982; Bilsborrow et al., 1984; Findley, 1987; Massey, 1990), there have been few studies in which this has been done using appropriate statistical methods (Boyle & Shen, 1997; Zhu, 1998; Henry et al., 2004, Kulu & Billari, 2004; Barbieri et al., 2005, 2009). Thus, contextual factors have been much discussed but little studied in relation to the causes of migration, with environmental conditions in origin areas hardly studied at all (see next section).

Effects of the environment on migration

Literature from cultural ecology (e.g., Sandor & Furbee, 1996) and agricultural and development economics (e.g., Reardon & Taylor, 1996) documents the direct dependence of rural households on the natural environment (land, soil, forest and water resources) for their livelihoods, since the time of hunters and gatherers to the beginning of agriculture some 10,000 years ago. Even though, since 2008, over half of the world's population now lives in areas defined as urban (UNFPA, 2008), humankind is still dependent on the natural environment for its food, water and recreation, even while seriously degrading the environment across the planet. Degradation of land resources, including soil degradation (Zweifler et al., 1994, on the Dominican Republic; Kalipeni, 1999, on Malawi), deforestation (Sessay & Mohamed, 1997; Godoy et al., 2002), floods (Kayastha & Yadava, 1985), and drought (Findley, 1994; Henry et al., 2004; Owens et al.,

2003), undermine the sustainability of rural production systems and, hence, livelihoods, which is likely to stimulate rural out-migration, once a threshold or tipping point is reached.

Thus, these kinds of gradual environmental deterioration are likely to induce out-migration from rural areas, whether to other rural areas (rural–rural migration, which has received little attention in the literature despite its still being greater than rural–urban migration, in many developing countries[11]) or to other destinations. The generally negative environmental impacts of this type of out-migration, when it becomes *in-migration* at agricultural frontiers, have received the bulk of researchers' attention (Bilsborrow & Hogan, 1999; Geist & Lambin, 2001; Bilsborrow, 2002). Thus rural–rural migrants have often escaped environmental degradation in areas of origin, only to induce it elsewhere, creating an 'environmental cascade' (Charnley, 1997).

An extension of this discussion is found in the literature on 'environmental refugees', mentioned in the introduction in section 1 above, which has built upon good studies on the short-term effects of major natural disasters (Hugo, 1996; Hunter, 2005) and anecdotal evidence from case studies (Hamilton et al., 2004; Jiang, 2005; Carr, 2008) to claim that millions of persons become 'environmental refugees' each year. While there are instances of major sudden natural disasters (earthquakes, floods, etc.) forcing thousands of persons to leave their homes and livelihoods suddenly, these are usually temporary population displacements. Up to now, there have been few *quantitative* studies that have explicitly tested the effects of *gradual environmental degradation* on out-migration, such as deforestation, climate change, or soil degradation, leading some authors to question whether such an environment–migration link exists. In a review of the literature from the past 50 years, of 321 publications, including 153 articles in peer-reviewed journals and 29 books, Moriniere (2009: 26) found only two articles in which the authors investigated the effects of environmental factors on out-migration, based on quantitative multivariate methods.

[11] In a review of the 14 developing countries with available data, mostly from the 1970 and 1980 rounds of censuses of population, rural–rural migration was greater than rural–urban in ten of the 14 countries. In fact, urban–urban migration was the largest of all of the four mathematical types of migration (the fourth being urban–rural) and, hence, larger than either rural–urban or rural–urban, in eight countries, with rural–urban and rural–rural each largest in three. As countries become more urban, this is an inevitable consequence, with urban–urban migration being the largest in all four Latin American countries in the study (Bilsborrow, 2002; United Nations, 2001a).

In fact, there are more, but most of the few quantitative studies on the environmental effects on out-migration have examined only the effects of rainfall or lack thereof (Faulkingham & Thorbahn,1967; Findley, 1994; Henry et al., 2004; Gutmann et al., 2005; Schmidt-Verkerk, 2009), and find that, while drought usually stimulates out-migration because of its effects on depressing household production and incomes (Findley, 1994; Roncoli et al., 2001; Henry et al., 2004; Meze-Hausken, 2000), in a few cases it has suppressed out-migration (Henry & Bilsborrow, 2007). While most of these studies on climate effects use data on rainfall, from either country sources or global precipitation data archives, qualitative methods were used in a study of two villages in the Zacatecas state of Mexico, where Schmidt-Verkerk (2009) found that drought provoked out-migration to other internal destinations in Mexico but had no effect on international migration.

Most of the few studies that have examined the effects of environmental factors other than rainfall/drought have significant methodological limitations, including both unclear conceptual approaches and clear measurement problems. For example, Ezra and Kiros (2001) studied the factors involved in out-migration from 2000 households[12] in 40 villages in a drought-prone area comprising three northern regions of Ethiopia, based on a 1994–95 survey of households and communities. The 40 villages were classified as 'vulnerable' (to drought) and 'less vulnerable', based upon opinions of Ministry of Agriculture and local government officials. Data were collected using a migration history format (as in the Ecuador study described in section 5 below). Using a multi-level statistical model, Ezra and Kiros found that there is more out-migration from the vulnerable communities. But community vulnerability is defined in a totally subjective manner, there being no objective measures of either household or community environment or vulnerability. Nevertheless, out-migration was highest by far precisely in the year of the drought (1984), indicating a strong immediate impact of the environment on out-migration.

[12] Fifty households were selected from each village. Any time the number of households is such an exact round number, and especially when coverage appears to be 100 per cent, one must be suspicious: it would appear that non-found or non-responding households were simple replaced, though this is not specified. It also raises questions about whether the sample is a probability sample, in which the probabilities of each household being selected is fixed and known, a priori. Any replacement house had more than one chance of being selected. The author also uses the term, "new economics of migration", omitting labour.

In another study, focusing on sample attrition, Rindfuss et al. (2007) examined the effects of local forest cover on out-migration from Nang Rong, Thailand, with forest cover measured from satellite data as the percentage of the land within a 2 km buffer around the village centre. Villages with more forest cover were found to have more rather than less out-migration. But there are other possible explanations for this unexpected finding: areas with more forest are on steeper slopes, are less adequate for agriculture, and are less accessible to roads (there were no controls for topography, duration of settlement or accessibility). In another study based on Asia, on data from 1974 households in the Chitwan Valley of Nepal, Shrestha & Bhandari (2007) found that the greater the time it took to collect firewood, the more likely was out-migration from the household. They interpreted this as indicating the importance of "environmental insecurity" on out-migration, although it is difficult to believe that firewood shortages alone could be a significant cause of out-migration. In fact, they had no objective measures of the environment or even land use in their study, the sole environmental variable being the reported time it took to collect firewood. In another study, Chopra and Gulati (2001) investigated whether environmental degradation drove out-migration in semi-arid central India, but used only crude indirect, aggregate measures of both out-migration (whether the sex-ratio changed over time, with a decline providing an indirect indicator of male out-migration) and environmental change (a change in livestock composition). Another study on Nepal yielded different results: Massey et al. (2007) investigated the relevance of origin environmental factors to out-migration (referring to the migrants as "environmental refugees", although they were only 'environmental migrants', since they remained within Nepal), with the environment measured by *perceived* declines in land productivity over time and the *perceived* increased time that it took to gather firewood. Both were linked to increased mobility within the immediate vicinity, but were not associated with long-distance mobility – which is a much more significant type of migration.

In fact, there have been very few studies that focus specifically on the effects of environmental factors in rural areas on out-migration. In most of the studies on environment–migration linkages mentioned above, it was not a central issue from the beginning of the data collection, but rather an afterthought. An important study by de Janvry et al. (1997), supported by the Natural Heritage Institute (NHI) (1997), illustrates this point, in a study of the relevance of "environmental degradation" to international migration from rural areas of Mexico to the USA. The NHI (1997) report concluded there is a "strong correlation between environmental stress, poverty and population pressure, which can lead to migration" (p. i), later clarifying (p. iv) that "high environmental stress is associated with poverty ... and poverty with out-migration". While the latter two clauses are widely believed, that does not prove that environmental stress causes out-migration. Indeed, it is worth examining this unique study in more detail. The NHI report is based on the empirical work of de Janvry et al. (1997), which used a household survey, designed mainly for other purposes (the norm in this field of research up to now), to investigate the relationships. The data are from a 1994 survey of 1,543 households in 276 *ejidos* and indigenous communities covering most of Mexico. From these households, 950 out-migrants were reported (de Janvry et al., 1997: 8). The section entitled 'Some hypotheses about the role of environmental factors' does not, in fact, propose any hypotheses, nor does it mention any environmental variables, or even use the word environment, instead discussing poverty and employment. Later, four environmental variables, all measured at the municipal or district level, rather than at the level of the household, are hypothesized to be linked to out-migration from households: (a) the percentage of households in the municipality living in highly degraded (according to independent soil studies) environments; (b) the percentage loss of forest cover from 1980 to 1990; (c) mean corn yield, as an 'indicator of soil quality'; and (d) a 'population pressure' variable defined as the inverse of the average income from corn of households in the municipality (1 divided by the product of mean corn yield and mean household size). The latter variable cannot be construed as either a population pressure or an environmental variable. The statistical results showed that the effects of deforestation on out-migration from households were only marginally significant (at the 10% level) and that there were no effects from the other variables mentioned above.

While the model studies the effects of municipality or contextual variables on household migration, it would be better to investigate the effects of environmental factors at the *household* level, since that is the level at which migration decision are made (contextual variables reflecting the environment and other factors could also be included in the statistical model). The results for environmental factors are quite weak and thus do not support the conclusion of the parallel NHI (1997) report quoted above. This example illustrates the pitfalls of studying the environmental impacts on migration from data collected for other purposes – leading to the use of very poor indicators of environmental factors, which makes it difficult to draw clear conclusions about whether there are environmental effects or not.

Apart from forest cover, other aspects of household farm or community land use may affect out-migration tendencies from rural areas Barbieri et al. (2005, 2009) studied the factors associated with out-migration in 1990–1999 from about 760 household farms in the northern Ecuadorian Amazon, based on survey data originally collected in 1990 and 1999 to study land use. Out of 1,458 persons aged 12 to 59 present in the study households at some time during the decade, 398 left the migrant settler household to live elsewhere, mainly within the Amazon region; about two-thirds moved to rural destinations, which would usually result in increased deforestation, but the point here is the association between land use and whether someone out-migrated or not. Barbieri et al. (2009) observed that households with large areas under crops had less out-migration, which was interpreted as reflecting the greater demand for labour needed to raise crops, compared to having land in pasture for raising cattle (many other variables, such as household composition, education, farm size and location, were controlled for, statistically). But, in this situation, it is not possible to establish clear causality, since households that experienced out-migration may have switched from crops to pasture *after* migration, due to the loss of household labour. In any case, land use per se is not an environmental variable; forest cover would have to have been used instead if the purpose of the research was to study the environmental effects on migration.

One study that was designed from the outset to collect data at the micro level to assess the role of environmental factors on out-migration from rural areas is that by Gray (2008). Gray collected data on 397 households in five contiguous drought-prone cantons (similar to US counties) in the southern part of Loja province in Ecuador. The data collected were from 18 census sectors comprising 36 communities. This study served as a kind of pre-test for the larger study described in section 5 below. Data were collected from the rural household respondent on all persons in the household (migrants from the household and non-migrants), who were aged 14 to 49 at any time in the ten-year reference period, 1996–2006. Data were collected on the person's age, marital status, education, places of residence for over six months, etc. Gray examined the effects of a number of variables in the household survey on out-migration and the choice of destination (internal or international), including five 'environmental variables' – household total land area; land under corn; flatness of topography; slope (linked to soil erosion); and precipitation, obtained from global data on rainfall in one-kilometre pixels . Clearly, the first two are not environmental variables, but rather a measure of household resource endowments. Of the other three, only precipitation was found to have a statistically significant relationship to out-migration, with higher rainfall linked to less out-migration to both internal and international destinations (ibid: 464). While this does suggest some linkage between the environment of origin and out-migration, the lack of compelling results led to a study based on a much larger and more geographically diverse dataset, together with a full set of satellite data imagery to help measure certain aspects of the natural environment not measured in household surveys (see section 5).

In most of the studies mentioned in sections here, environmental factors were either measured unreliably or represented by unrealistic proxy variables. In addition, in almost all cases, the data collection was not specifically designed in order to study environment–migration linkages, leading to many 'proxy' variables. In addition, a number of social, economic and political factors possibly affecting out-migration and its consequences were usually omitted, which has made it difficult to accurately separate out environmental factors. Since efforts to disentangle the many factors have been scant and rarely successful, more research is required.

Returning to theory…

A flowchart can tie together the factors affecting migration and the environmental consequences. Figure 1 below illustrates the multiple ways in which a rural household in a developing country can respond to pressures on its living standards, whether from health problems, additional children resulting from high fertility, price declines in major products, loss of a job, or environmental degradation. The multiple possible responses are described by Bilsborrow (1987), who expands on the work of Davis (1963). The left box refers to the farm household, which has a given endowment of land, a house and other assets, and is inhabited by household members of a given number, age, gender, education, skill level, etc. In the event of any significant threat to its living standard, it can react in multiple ways – a reaction referred to as the 'multiphasic response'. The more one response occurs, the less likely it is that other responses will occur, since there is less pressure for other responses. Thus, the farm household, if it has unused (forest or fallow) land under its control, may simply clear more land (land extensification) if its living standard is threatened. But this may have unfortunate environmental consequences, increasing soil erosion or future flooding as the cleared vegetation may have previously retained water runoff from rainfall. Another possible response would be to intensify agricultural production, by decreasing fallow time or using more irrigation or fertilizer, etc. This can have deleterious environmental consequences as well, as indicated in the lower right box. A third option would be for someone in the household to seek, if available nearby, off-farm wage employment, whether urban or rural. A possible fourth response – one that is more transcendental and disruptive to their family life – would be for the whole household to migrate to an urban area to seek employment, which has lesser and only indirect environmental consequences, depending on urban and rural consumption levels and patterns, and where resources are used to satisfy them.[13] Finally, a fifth option would be rural–rural migration, involving the extensification of agriculture to lands elsewhere in the country, provided such lands are available. The latter shows the fundamental importance of roads and road building: if roads are being built to open up or expand previously unused lands, such as frontier areas, or if concessions have been granted to loggers or other business firms to build access roads to extract timber, petroleum or minerals (upper right box), this opens up new areas for land use. Which of the five possible responses[14] is adopted by a particular farm household depends on its own resource endowments and human capital, as well as the contextual factors surrounding the household, including resource availability, infrastructure and government policies regarding land tenure, access to open or public lands, credit services, taxes and subsidies, road construction, etc. While the environmental consequences of rural–rural migration and agricultural extensification usually appear to be greater because they involve opening up new areas for agricultural use (especially frontier areas), land intensification and extensification in situ, and rural-urban migration can also have environmental impacts, as noted above.

[13] If migrants to urban areas are successful in increasing their incomes and then adopting more meat-intensive consumption patterns, as has been widely observed in recent decades, and if meat is produced using land-extensive methods such as grazing in the open, and if this, in turn, requires clearing forests to increase the area in pasture, then rural–urban migration can have overall *negative* implications for the environment. This is a cycle that warrants further study.

[14] Depending on the time frame under consideration and local conditions, other responses exist, including postponing marriage and reducing fertility (longer-term responses, and the focus of Davis, drawing on Malthus), and international migration.

Figure 1: Rural household decision making, migration and the rural environment

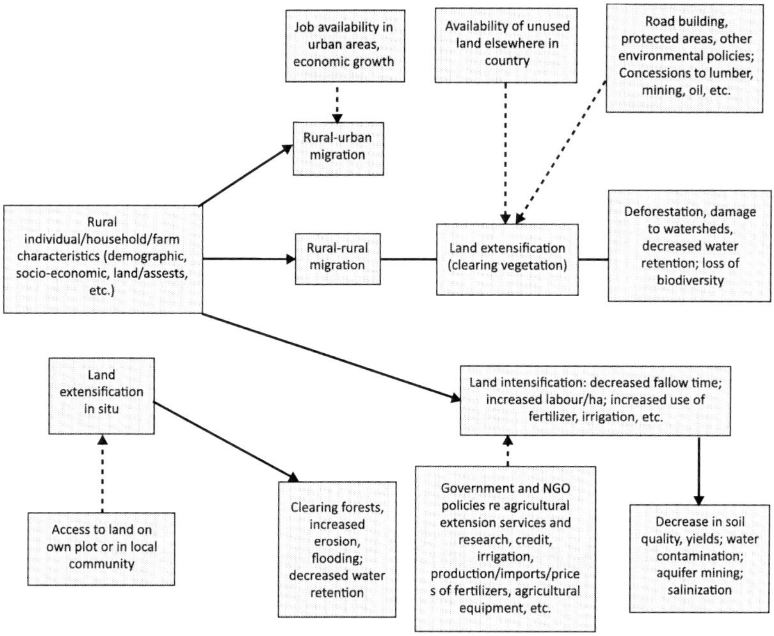

Environmental degradation is one of the possible household-level or contextual factors (not specifically shown in figure 1) in the community that could cause the rural household to respond in some fashion. In the short term, for example, as land productivity declines due to soil degradation (soil erosion, desiccation, nutrient loss, surface soil loss from wind or water), the household could engage in any of the three in situ responses indicated – extend the land area, intensify land use (e.g., restore nutrients with fertilizers), or seek off-farm employment. Indeed, these responses are preferable, since no one in the household has to leave the home or community (Bilsborrow, 1987). But, as soil degradation becomes more severe, only the latter option is viable, and if that is not sufficient for the household to achieve its livelihood aspirations, then one of the two forms of out-migration is likely.

Thus, migration may be viewed as part of a household survival strategy but, within this strategy, who migrates depends on 1) the characteristics of the individuals available in the household and their perceived employment and wage opportunities as migrants in destinations being considered; and 2) the characteristics of the household, including size and composition and perceived needs or sense of deprivation. The size and composition determine the presence of other household members who could be alternative candidates for out-migration, or could replace the out-migrant in carrying out important household functions, such as child care or farmwork. Seeing households as engaging in a household 'survival' strategy, whereby migration is an integral component, is consistent with human capital, household survival, and income-risk diversification theories, such as the new economics of labour migration, discussed above. However, households are embedded in a larger socio-environmental context, including the local community, which is important for not only providing economic and social opportunities for them but also shaping values and tastes. Other factors, such as the maximum reasonable distance children would travel to get to school, or the distance that members of the labour force could reasonably go daily or weekly for work, depend on the transportation resources of households (whether they have a vehicle or not) and the local transportation infrastructure. These factors define and limit the opportunity sets for community residents in the absence of out-migration. Environmental aspects of the local community or region may also be important, since the loss of forests, degradation of soils, depletion of watersheds, or reduction of vegetation due to drought, for example, in the larger community, will affect local livelihood prospects. Soil degradation, for example, reduces the agricultural productivity of land, reducing the demand for labour in the community as a whole, which may lead to out-migration. Similarly, deforestation increases the time/cost of fuelwood collection, with implications for the time and energy remaining for household members to engage in agricultural production, take care of their health, and maintain a good quality of life.

It is the overall effect of these factors that, at any given time, causes individuals and households to choose a particular set of livelihood strategies and, if necessary, out-migration. The discussion here therefore links 'determinants of migration' theory with the literature on rural livelihood diversification strategies in the developing world (Ellis, 2000; Reardon et al., 2001), which emphasizes the importance of diversifying sources of income for risk avoidance and food security. It is also consistent with the household survival strategy of Arguello (1981) and the new economics of labour migration approach, described above. Thus, the potential impacts of environmental factors in origin areas on out-migration, and the potential impacts of that migration on land use and the environment in destination areas, may both be seen in the context of rural livelihood and survival strategies in low-income countries.

To summarize, the factors that *may* influence rural populations to migrate from their places of origin include environmental factors, while, at the same time, migrants may have impacts on their rural destination areas. The former draws on the theory of migration, the latter on the theory of land use.

Finally, those who migrate must choose from various types of destination – other settled rural areas, urban destinations, frontier areas, or international destinations. Migration to each type of destination has different environmental implications. The risks and opportunities presented by the various types of destinations are quite distinct, and each type comprises many possible destinations, though each household usually considers only a small number. Given the differences in the types of destinations, the individual, household and contextual characteristics that affect migration decisions are likely to differ according to the types of destinations considered. This is why a full model of the determinants of migration should take into account, and model, the choice of destination type. This has implications for questionnaire design in migration surveys, which is addressed in the next section.

3. Data collection

What to measure

Migration

In investigating the relationships between migration and the natural environment, there are various kinds of migrants that may be of interest: internal, international and return migrants, long-term/permanent versus short-term/seasonal/temporary migrants, and even circular migrants and daily 'commuting migrants'. These can be narrowed down if a migrant is defined as someone who moves across a political or administrative boundary to change his/her place of residence.[15] Obviously, international migration involves a movement across a country border, whereas moves *within* a minor political unit (as defined by each country), such as a district, municipality or county, are *not* considered to be migration. This definition also excludes temporary and circular migration, both of which may have effects on, and are influenced by, the environment and may be substitutes for so-called 'permanent' migration – migration involving a change of residence. Assessing the effects of these forms of temporary population mobility, which may differ from those leading to a change of residence, is beyond the scope of this chapter. However, it is possible that temporary migrants are less concerned about their impacts on destination areas, and therefore more destructive, than those permanently changing their place of residence. Seasonal agricultural workers, for example, may exploit the forests, water and other resources of the places where they are working temporarily, with little concern for the environmental consequences, since they have no sense of belonging or residing there. This could also be true for refugees, even though they may be forced

[15] Social scientists have debated and invented definitions of 'migrant', as can be found in Guy Standing's *A Typology of Migrants,* which provides a rather complete and imaginative 'typology' (Chapter 3 in Bilsborrow et al., 1984). Definitions may also be found in documents of the UN statistical office, in the IOM Glossary cited earlier, and various textbooks, including Siegel and Swanson (2004), which defines migration as "a change of residence between clearly defined geographic units ... or between specifically designated political or statistical areas or ... type-of-residence areas (e.g., rural to urban movement)" (p. 453).

by circumstances to live for years in a country other than that of their citizenship or previous residence, and have de facto changes of residence during those times.

For most purposes, especially for policy formulation, we are primarily interested in *recent migrants* – those who migrated within the past five to ten years. Nevertheless, for studying the relations between migration and the environment, a longer time perspective is also useful, since migrants' impacts on the environment in the place of destination are cumulative over time. Similarly, people may be forced to move due to gradual degradation of the environment over many years, such as the loss of soil fertility, which renders the land less productive.

For studies of the determinants or consequences of migration, it is crucial to obtain data on the key characteristics of migrants and non-migrants at both the individual and household levels in the same locations. At the individual level, this includes age, sex, educational attainment, marital status, economic activity/employment, and motives for migration (and for not migrating, for non-migrants). At the household level, it includes household size and composition, education of the household head and other members (beyond the person in question), household assets/wealth and income, employment or other income-generating activities of all members, previous out-migration of household members (which create household migration networks), and the location of the house relative to transportation networks and sources of employment, such as urban areas. Finally, it includes information on remittances received, and their use by receiving households.

This is the minimum information required for a study of the relationships between migration and the environment at the household level, although additional questions should, ideally, be included in order to probe those relationships directly. These are referred to in the following discussion.

Environment

The focus here is on aspects of the natural environment that may be linked causally (in either direction) to human population movements, rather than other demographic and socio-economic factors, such as population size, population growth or production/consumption patterns in general. This excludes measures of global climate change, air pollution, issues of food security and generalized famine, loss of biodiversity (although it is a direct consequence of loss of, or damage to, habitats), health and infectious diseases, urban pollution (air, water or noise), water pollution or chemical contamination of land and waterways from excessive use of fertilizers, pesticides or herbicides, damage from mining or petroleum activities, mining or contamination of underground water aquifers, etc., as well as all sudden natural disasters.

Particularly relevant in a study of linkages between internal or international migration and the natural environment are:

- land use, including forest cover;
- energy use by rural populations;
- use of common property resources;
- intrusions into protected areas;
- soil erosion and soil degradation;
- flooding/drought (and precipitation excesses and deficits).

For each of these variables, the focus is on changes that occur gradually, over time, in an area or at the farm household level. The types of change of interest will vary with the type of area: for example, semi-arid areas, mountain environments, and rainforests are quite different in their ecologies and anthropogenic risks.

Of particular interest is whether *changes* in any of the above environmental variables influence out-migration from rural areas, or are attributable to in-migration. One common scenario worldwide may be soil degradation stimulating out-migration or the movement of persons and households living in well established areas to other areas, such as frontier areas, where they clear the forest to establish new farms. Migrants who move to areas they are not familiar with may have particularly injurious effects, such as in protected areas, either directly by settling in or exploiting those areas (see Grandia el al., 2001; Oglethorpe et al., 2007) or indirectly, by providing a market for the sale of products by others who are exploitative (Liu et al., 2003). Also, the influx of people may increase demands there for fuelwood for cooking/heating or for trees for house construction, either of which can significantly impact forest or vegetation cover (see citations in Marquette & Bilsborrow, 1999). In-migration to rural areas will likely also result in changes in land use – from clearing forests/vegetation to growing food or cash crops, or raising cattle or other animals. If migrants cannot find land in the destination area for their use, they (as well as others already living in the area) may have to resort increasingly to using common property resources, or exploiting protected areas. In settled areas with agricultural households, in-migration increases pressures on the land, which, in the absence of significant compensating changes in agricultural practices (e.g., improvements in crop rotation, use of new hybrid seeds) or nutrient replenishment (from fertilizers), may lead to soil erosion or degradation, from excessive land intensification (see lower right of figure 1 above).

Similarly, the deterioration of the natural environment, as captured by changes in any of the abovementioned variables, may stimulate the out-migration of people who see the productive capacity of their land, forests and water resources declining (see Textbox 1: Carrying capacity). However, it is important to note that hypotheses about the impacts on out-migration of some of the environmental variables listed above are ambiguous. Thus, the loss of forest cover is likely linked to increasing land in agricultural use, and hence to in-migration rather than out-migration, regardless of its implications for ecological loss. Changes in land use may also have complex effects – for example, a switch from land- and labour-intensive short-cycle crops to pasture results in a decrease in demand for labour and, hence, out-migration. While increased population pressures may stimulate positive technological change (Boserup, 1965),[16] in the short term, the accumulation of these pressures over time and the environmental degradation they cause may eventually lead to a threshold or breaking point, inducing households to migrate away.

Textbox 1: Carrying capacity

An intriguing but complicated concept is that of carrying capacity, which may be defined as the productive capacity of the land, with a given technology. This capacity is evidently impaired by any of the processes outlined above, linked to increasing human population pressures (in the absence of changes in technology). This capacity may be declining or not, in the absence of migration, but is more likely to suffer with migration if the total population rises. In addition, the increased population matched up against the constant or decreasing carrying capacity implies an increasingly unsustainable situation, with further land clearing or other land degradation occurring, or a migration reversal (out-migration). The classic empirical study of carrying capacity, carried out with the United Nations Food and Agricultural Organization, in Higgins et al. (1983), showing large areas of the developing world, especially in Africa, South Asia and Central America, as either already having populations beyond their carrying capacity or likely to soon exceed it (based on their prevailing low or intermediate levels of agricultural technology). The concept has been

[16] Also known as a Boserupian response, as described by Boserup (1965). See also Bilsborrow (1987) and Bilsborrow and Geores (1994).

> criticized as being static, since technology can change, increasing agricultural production, and for assuming a closed economy – in other words, absence of international trade, as a means whereby food deficit countries could trade for food from surplus countries. The counter arguments are (a) is more productive technology being adopted quickly enough in developing countries to keep them ahead of the demands for food resulting from ongoing population growth combined with urbanization and changing tastes towards more land-using meat; and (b) have food deficit or marginal countries in the early 1980's really come to be successful in finding other export commodities to use to earn foreign exchange to import the necessary food? A dramatic change in the past quarter century (since the work of Higgins et al.) is the decline in fertility rates and hence in population growth in most of the low-income countries of the world, which reduces the extent to which food production needed to rise to keep ahead of demand. But it may be precisely those countries where the answer to the two questions above are negative that are lagging behind the demographic transition to lower fertility, and becoming increasingly dependent on food aid. In fact, it may be time to revisit this issue, to again reassess the food security situations of developing countries around the world, in view of ongoing trends in population growth and agricultural technology adoption.

How to measure migration and the environment

From household surveys

Migration

As noted above, questions are needed to clearly identify migrants of interest – and their main characteristics, as well – for inclusion in multivariate models to investigate either the environmental determinants of migration or the environmental impacts of migration. This leads to the development of several modules comprised of a minimal number of questions, described below. It should be noted that this discussion refers to the design of a project that seeks to determine the relationships between migration and the environment based on data from a single cross-sectional survey in which retrospective data are collected about (1) the last move of a person from a household surveyed, and (2) circumstances pertaining to that one move, including environmental conditions. There is an alternative methodology of data collection, described in the case study for Ecuador in section 5, below, which essentially collects migration histories for current and former members of study households.

Most questions proposed here are not discussed below, since their relevance is self-evident, in the context of migration surveys. More detailed modules and discussions of questions in the modules for surveys of internal and international migrants are found in Bilsborrow et al. (1984) and Bilsborrow et al. (1997), respectively. Only a few key points that may not be clear are discussed below, following the prototype questions. The assumption here is that the household survey questionnaire will already include the usual information on all current members of the household through a household roster, plus data on ownership of the house, housing quality and conditions, ownership of household assets, employment of household members above age 15, ownership of land, agricultural production, etc. All of these topics are potentially relevant in determining migration decisions.

Module to identify out-migrants from survey household (h/h)

- Has anyone who used to live in the household left to live elsewhere since Y (e.g., 5, 10) years ago, whether in another part of this country or in another country?

  If no, there are no out-migrants, so skip subsequent questions.

  If yes, name of X, sex, current age, relationship to household head, education level at time of departure. Did X leave with anyone else, or was X joined by anyone later? Month and year left.

- For those aged 15+ at time of leaving: marital status at time of departure.
- *First* place of residence of migrant (for at least 6 months) after leaving this community?

  If internal, name of province, district, city, and whether urban or rural.

  If abroad, name of country, urban or rural destination, state/province, and city, if urban.

- Place or country of current residence. state/province and city of destination.
- How long (years, months) has X been living in current country of residence?

  If international migrant, is X a citizen of current country of residence? Of any other country?

- How well does X know the language of the country?
- If married at time of departure, did X leave with spouse, or was X joined later by spouse?

This module above is key to the entire data-collection process, since it is based on a clear definition of migrants of interest. The focus in studies of the determinants or consequences of migration, in general, and with respect to the environment, in particular, should be on migration involving a change of residence, with a time cut-off to focus on recent migration movements, for which there is not only more policy and research interest but also greater ability of respondents to accurately recall important details.

Module on work activity of migrant prior to departure

- Was X mainly working, studying, looking for work, doing housework, or other [specify], during the 6 months before leaving? (Skip rest of module, if not working or looking for work at that time.)
- What kind of work did X have? Occupation; economic sector; work status (employee, manager, day labourer, own account worker, unpaid family worker, housemaid, other).
- If mostly not working, had X been looking for work, or was X entering the labour force to seek work for the first time? If looking for work, for how long?

Module on reasons for migration

- Why did X leave here to move to another place of residence?
- Why did X choose to move to that particular (initial) destination?
- Had X ever visited there before? Did X migrate alone or with someone else? If with someone else, who? Relationship to X.
- Did X have any relatives or close friends living in [destination country] before moving there?

  If yes, did X expect to receive assistance when he/she first arrived?
- Was any assistance actually provided, as far as you know? What kind(s)?

Module on work activity in current place of destination
• Does X *currently* work or have a farm or business in his/her place of residence? If not, is he/she looking for work, studying, retired, disabled, otherwise not working, other [specify]? • If working, branch or sector of economic activity, occupation, income or wage rate, fringe benefits, etc. • If has business or farm, type of business, whether owns land or building, rents either or not (sells in street, or has no fixed site), whether has any paid employees, permanent or temporary, and number of each. • Approximate monthly or weekly gross sales and net income or profits.

Module on remittances received
• Did X send any money in the past 12 months to anyone in this household? When was the last time? How much was received the last time? • How many times did the h/h receive money from X in the past 12 months? How much was received in total? • What was it mainly used for? List by category. If any was invested in a business, in what economic sector [branch]? Where?

The above-mentioned mini-modules are concise and seek data pertaining to the situation of the migrant *and* the household of origin on marital status, education and employment *at the time of migration*. This is appropriate for the study of the *determinants of migration*, since it is those circumstances that may affect migration, rather than circumstances at the time of the survey. In fact, the latter could have been affected by the prior migration and, hence, be *consequences* of migration rather than causes. Note that the current economic activity of the migrant and remittances relate only to the previous 12 months and pertain only to the *current* country of residence. Questions pertaining to the work of the migrant, posed to the proxy respondent, are best asked about work during *the three-to-six months before* migration rather than for the whole year (12 months) prior to out-migration, which is too imprecise, or during the month before, which could well be atypical (and not reflect the migrant's underlying employment situation) as the migrant could have given up work by then to prepare for the move. To keep the questions as simple as possible, few questions are included on wages of persons or individuals or household incomes, although such data are crucial when investigating the determinants or consequences of migration (and have a strong justification in theory), and *should* be asked, if possible. Such questions relate to wage rates of the migrant before and after migration, wages of other household members in migrant and non-migrant households, and overall household income.

The questions about the *use of remittances* are important but are considered optional since (a) remittances may be reported to be used in a far more 'socially correct' way than was really the case (e.g., for school fees instead of a party); and (b) remittances are 'fungible', in the sense that, while their ostensible use is known, the real issue is what the remittances made possible that would not otherwise have happened. The remittances can free up *other* funds in the household for something else, which would then be considered the main impact. An alternative recommended by some economists is to just ask the respondent directly whether the receipt of remittances made something else possible in the household, but this question will make no sense to most respondents. In a repeated or panel survey, household consumption and investment could be measured before and after the receipt of remittances, but this would require far more resources than a single-round retrospective survey and is therefore rarely done. In the absence of a panel survey, the simple question, "*What did you do with the funds?*" is usually the most practical one, despite its limitations.

Environment

Data on relevant aspects of the natural environment in which the household lives can be measured from household surveys using the following questions:

a. How much land do you have (under your management/use, whether owned or not)? (Responses should be in hectares or in local units, which may sometimes differ from one part of the country to another.)

b. How much land do you currently have in annual or short-cycle crops, such as corn, rice, wheat?
c. How much in perennial crops, such as coffee, cacao, fruit trees?

d. How much in pasture for animals?

e. How much is in fallow, for future use?

f. How much is in swampland, or land too steep or too arid for agricultural use?

g. How much is in secondary forest, regenerating itself?

h. How much is in essentially intact forest?

i. How much have you produced of each crop named in b and c in the past 12 months?
(If the household has separate parcels, this should be asked separately for each parcel. Note that some land may be used for more than one crop in the past 12 months, which is known as multiple cropping.)

j. How much of each crop have you sold (or given away) in the past 12 months?

k. Have you used any agricultural inputs, such as fertilizer, herbicides or pesticides, in raising the crops?
(Optional) What did you use, and what brand? What quantity? How much did it cost you, altogether (in the 12 months)?

l. (Optional) Has the amount of land in any of the other categories e–h above changed recently (e.g., since 5 years ago)? That is, has land in any of the categories been switched into production in categories b–d, or vice versa? If so, why?

m. Within b–d, has land use in particular crops or animals changed (e.g., from corn to beans, or cattle to goats)? If so, why?

n. Have yields changed for any crops in b or c that you grew *on the same parcel* in the past 12 months as, say, 5 years ago? Why do you think that is?

o. (Optional) Apart from your own land, did you use/rent any land from any neighbours in the past 12 months? Did any of them use any of your land? If so, how was this use compensated? Did you use or rent any land from non-neighbours? If so, how much land? How much did you pay?

p. Did you use any common or public lands, national forests or protected areas, in the past 12 months, such as for grazing animals, obtaining firewood or timber, gathering foods or medicine, hunting, etc.? Has this changed over time? Explain.

q. For the land that you use in crops, do you give it a rest after some years of use? After how many years of use, usually? Has this changed over time, compared to, say, 5 or 10 years ago? In what way and why? How long do you let it rest (fallow) before using the land again? Has this changed, compared to, say, 5 or 10 years ago?

r. Have there been changes in the weather (e.g., rainfall) that have affected your crops in recent years? Explain.

s. Have soil conditions on the plots you use for crops or raising animals changed? Explain.

(Optional) Repeat question for each plot in use.

The questions above on land use indicate both the sophistication of land use practices (changes in crops over time, crop rotation, etc.) as well as the effects of market conditions, such as changes in prices, which farmers are quite aware of and generally respond to quickly, although certain crops may be grown out of tradition, sometimes for years, even after prices decline (e.g., corn in Mexico and Guatemala, rice in South-east Asia). Increased areas in use (crops or pasture) and declines in land in the non-current-use categories indicate pressures on the farm household to extend the area in use or increased consumption desires, and amount to expanding or *extensifying* agricultural production (see figure 1 above). Changes in inputs indicate land intensification.

The questions above on changes in the number of consecutive years a plot of land is used before being left fallow, and on the number of years it is left fallow before use is resumed (whether for the same purpose or a different one) are crucial indicators of the intensity of land use, and have implications for soil degradation to the degree that nutrient replenishment is not practised or is insufficient. Low-income farm families, in particular, often cannot afford fertilizers to restore nutrients, so the land becomes progressively more degraded over time, resulting in declining crop yields per hectare.

From community surveys

Data on migration and the environment can usually also be collected at the level of the village, town or even urban neighbourhood – essentially, at the level of the local community relevant to household behaviour. It is possible to prepare estimates of some community characteristics (e.g., age distribution, land ownership, main economic activities, education level) by aggregating data from households in the community included in the household survey (called *contextual* community variables), but this requires a large number of cases to provide variables independent of the component household variables for statistical analysis (Blalock, 1985). Even if there is a large and representative sample of observations for every study community, contextual variables cannot provide many variables of interest (e.g., population size, whether various kinds of infrastructure exist and since when, transportation linkages to cities, age of community). These types of information must instead be collected independently, usually from community leaders or others knowledgeable about the community. The best way to do this is to collect such data directly from the leaders/respondents together in a small group, which reduces the possibility of biased responses from a single respondent and also draws on the knowledge of more people.

The rationale and procedures for conducting community-level surveys to study migration are described in detail in Bilsborrow el al. (1984, Ch. 13). Although it is possible to undertake community-level surveys in urban neighbourhoods (ibid.), the lack of clear boundaries or a sense of neighbourhood and the ease of transport and communications across urban neighborhoods (e.g., in a city) often reduces the value and meaning of community-level surveys in urban neighbourhoods. In any case, the focus of this chapter is squarely on rural areas, where all but the most isolated households identify themselves with a local community by name. That community we call the main *reference* community of the household. Rural communities differ greatly in their physical characteristics (e.g., they may be in mountainous, coastal or rainforest areas, or on semi-arid plains, etc.), their form of land tenure (common property, private property, public property, open access), their main form of economic life (mining, small farm or large farm *hacienda* rural community, small town, ecotourism community, etc.), and their political decision-making structure (elected officials, appointed officials, officials elected by consensus or a community meeting, etc.). These differences require different adaptations of the prototype questions below. In any case, implementing a community questionnaire makes most sense if rural dwellings are clustered in the centre, with lands used for agriculture radiating outwards; or when the population lives in more spatially scattered dwellings but with some minimal core area, with perhaps no more than a primary school, store, athletic field, and/or religious or community centre. Communities with populations varying from a few dozen people in a handful of households, to no more than a few thousand people lend themselves best to community surveys, and to the study of the effects of community factors and attributes on the behaviours of community members.

The information directly related to migration that one can reasonably seek from a community-level survey includes the following:

a. What is the population size of the community now? What was it in the recent past (perhaps at the time of the last population census, or a community census)? (Compute population growth.)

b. Have any *households* left to live outside the community (and outside the subdistrict) in the last 12 months, or during the reference period? If so, how many? Where did they mostly migrate to? Have any new households moved into the community? If so, how many? Where did they mostly come from?

c. Have any *individuals* left their households in the community to live outside the community (and outside the subdistrict) in the last 12 months, or during the reference period? If so, how many? Where did they mostly migrate to? Have any individuals moved into the community? If so, how many? Where did they mostly come from?

d. (If relevant) Why did whole households leave?

e. (If relevant) What attracted whole households to come?

f. Why did certain individuals leave?

g. What attracted individuals here?

h. For each of the four types of individual/household migration that occurred, what do you think are the main consequences for the community (e.g., reduces/increases population pressure on resources, land clearing, limits access to common property, presents problems or advantages of having people come who are different; reduces/increases unemployment and labour surplus/scarcity; increases/depresses local wage rates; reduces/increases pressures on social services such as schools and health clinics; leads to more/less pollution of local water sources/rivers/lakes; other: specify)?

i. Are there consequences of in- or out-migration (as the case may be, in the community) for loss of skilled labour, or for changes in agricultural technology? Explain.

A community survey should seek data on the *environment* using questions such as the following:

1. Total community area (which is by no means easy to collect, as the boundaries may not be precise or well known), in hectares or square kilometres. This should include any community forests or common lands, urban areas, bodies of water such as streams and lakes, land areas in schools and government facilities, etc.

2. Community land use, in private farms, public lands, forests, urban, etc. What are the main crops or agricultural activities? Any significant change in recent years (or since a cut-off time, such as 5 or 10 years ago)?

3. Community land distribution. Numbers of families/households with no land, with less than 1 hectare, 1–2.9 ha, 3–4.9 ha, 5–9.9 ha, etc. (country-specific categories) ? Any significant change recently?

4. How many households have agricultural land, have no land but primarily work on lands of others, depend mainly on non-agricultural activities (shopkeeping, teaching, government work, business enterprises, etc.)?

5. Have any changes in land quality been observed in recent years, such as soil erosion, soil degradation, deforestation, etc? Caused by what factor(s)? What are the effects on the local population?

6. Have there been any noticeable changes in precipitation over time and, if so, due to what, and with what consequences?

7. Have there been any natural disasters, such as earthquakes, hurricanes, volcanic eruptions, floods or drought in the last Y (= 5, 10) years? If so, with what consequences for the local population?

From remote sensing

Most of the environmental variables cited above can also be measured approximately using remote sensing data.[17] If remote sensing data are available from a series of satellite images, it is possible to examine *changes* in land use and land cover, over time and at a variety of scales – from the parcel to the farm, the community, the landscape, and the entire region. For small farms, or small agricultural plots, satellite imagery may not provide accurate data, depending on the resolution of the imagery, and is then most useful at scales larger than those of the individual farm household (Liverman et al., 1998; McCracken et al., 1999; Walsh & Crews-Meyer, 2002; Geoghegan et al., 2001; Fox et al., 2003). Thus the repetitive and broad area coverage of satellite systems, with their multiple spectral signatures, has been found useful when measuring and studying forest and land use dynamics in many settings, including tropical forests, mountains and coastal areas. At the level of the household farm or parcel, changes in forest cover are in small patches, which are not easy to capture in medium/coarse-resolution landscape-level remote sensing systems, such as Landsat Thematic Mapper (e.g., Landsat 5), which provide data for pixels of 900 m^2 (30 m x 30 m). There is other multi-spectral imagery with a higher resolution but it is much more expensive (e.g., QuickBird, Ikonos, Hyperion). In any case, spectral resolutions of multi-spectral sensor systems constrain the number of classifications that can accurately be distinguished: for example, in the Amazon, it has usually been difficult to reliably distinguish land use in the form of forest from tree crops such as cacao or coffee, or to distinguish particular crops, or bare ground (e.g., following a harvest or a wind storm) from pasture land or urban or other built-up areas. Distinguishing secondary forest from primary forest also becomes more difficult as the secondary forest grows over time.[18] Thus, the ability to classify land use into, for example, 17 classes instead of five, requires higher quality, more expensive imagery, with newer hyper-spectral data (such as Hyperion), to discriminate between subtle differences in land cover (Arroyo-Mora et al., 2005). In all cases, fieldwork is desirable for the collection of ground control points that facilitate the accurate processing of images.

[17] Other units of measurement and land size categories may need to be used in some countries.

[18] One way to deal with this is to obtain data in the household survey on the age of a secondary forest parcel, take a GPS reading or readings for the parcel, and then compare them with the spectral signature of the same location in the satellite image. Once this is done for a sufficient number of parcels, the rest of the areas in the images can be assigned the same secondary forest age (see text also).

This is done by sending people to collect GPS (global positioning system) readings in the middle of patches of distinct types of land use important for the project (e.g., corn, rice, beans, pasture or coffee). This allows for calibration of the satellite imagery so that the spectral imprint of those forms of land use at those locations can be identified, and that same spectral imprint can then be used to determine land use in other locations.

The previous paragraph refers to a pixel-based approach, which has been used for decades to classify land use based on pixels of identical size and shape. More recently, Object-Based Image Analysis (OBIA) has been developed to characterize the landscape into land use/land cover types using any type of geometric shape rather than identical square pixels to distinguish contiguous forms of land use. This logically leads to more accurate classifications of land use (Burnett & Blaschke, 2003) (see box on measuring vegetation).

Textbox 2: Measuring vegetation

The abundance of vegetation can be measured from satellite imagery using the Normalized Difference Vegetation Index, or NDVI (e.g., see study on Thailand in Walsh et al., 2001). NDVI varies from -1 to +1, with higher values indicating more green vegetation, based on reflectance measured in the red and near-infrared spectra from the Advanced Very High Resolution Radiometer on NOAA[19] - series satellites (Tucker et al., 2005). In the case of China, as discussed above, Song et al. (2008) computed NDVI values at a spatial resolution of 8 km x 8 km pixels, illustrating the use to roughly measure vegetation levels and changes over time for large geographic areas, such as provinces or even countries. However, NDVI values can also be calculated from other multi-spectral satellite images (including Landsat) at a higher resolution of 30 m x 30 m, to study levels of, and changes in, vegetation at the community and even farm household level. Indeed, this has been done in a number of projects studying land use in the Amazon and elsewhere (McCracken et al., 1999; Geoghegan et al., 2001; Walsh and Crews-Meyer, 2002; Fox et al., 2003; Pan et al., 2004) and will be done for the Ecuador project, described in section 5.[20]

From other sources

Other useful sources of data on migration are population censuses, continuous population registers or other registration systems, and

[19] The *National Oceanic and Atmospheric Administration* (*NOAA*) is a federal agency that focuses on the condition of the oceans and the atmosphere.

[20] Another measure of vegetation cover used in forest areas is the leaf area index, which measures the extent to which a pixel is covered by tree cover canopy. As with land use and land cover (LULC) data, this approach requires ground-truthing to calibrate satellite imagery for the types of trees prevailing in the study area.

(in the case of international migration) border or admission statistics. However, the latter are rarely reliable in developing countries, and censuses often exclude migration, in listing the current members of households and their characteristics at a particular time, hence providing data on the population stock in general. They are also customarily implemented only every ten years, in most countries, so the timeliness of their data declines with each year after the census. Whether they provide useful data for particular areas of study, such as local communities, must be checked out in each case. Population registers have the advantage of collecting data continuously, and can therefore provide accurate data on changes in population – specifically on in- and out-migration. But there are very few countries outside Europe with high-quality, virtually complete population registers, so they can only rarely be counted on to provide useful data on migration. In any case, they collect no data on the environment and would therefore have to be used in conjunction with other data sources to provide any kind of useful environment–migration information.
In terms of the environment, data on rainfall/precipitation are often available from the National Meteorological Institute of many countries. The utility of these data depends on the number and location of sites where rainfall data are collected and the quality of the equipment and frequency of monitoring data collection. In the absence of country data, global data are available, though usually at a lower resolution.

Many countries conduct occasional soil surveys for large areas of the country, sometimes at various points over time, allowing for changes in soil attributes to be measured for some areas. But since soil quality and properties can vary significantly, even between two contiguous one-hectare plots, studies of migration–environmental linkages focusing on particular micro areas or communities would benefit from collection and analysis of soil samples for the specific study sites, if resources are available. This would be particularly useful for monitoring changes over periods of time greater than a decade. Unfortunately, collecting and assaying soil samples is not cheap, so there should be good reason to focus on soil changes to justify the cost.

Aerial photography can be used to observe changes in vegetation cover – in some ways similar to what is observable from satellite data, but based only on visual inspection. This approach can detect specific trees and distinguish natural forest species from perennial tree crops (such as coffee, cacao, fruit trees), depending on the elevation of the photography and skill of the viewer, but it does not detect some of the spectral signature differences available from satellite imagery, again depending on the type and resolution of the satellite imagery.

The types of aerial photography available are called geo-referenced videography and geo-referenced aerial photography, with the latter involving the use of a high-resolution still camera. The former is being used by Rodrigo Sierra (University of Texas and Ecociencia, Quito) in the Ecuadorian Amazon to observe land use by indigenous populations, in order to assist them in developing long-term land use plans.

A series of aerial photographs were used to study the 50-year cumulative process of human habitation and population growth from internal migration in the Dominican Republic over a 50-year period (Zweitler et al., 1994). A community called Las Ayumas experienced dramatic changes in land use over time. First, the original subsistence crops were replaced by less nutrient-demanding crops following the depletion of soil nutrients. As world coffee prices rose, up to 63 per cent of total land use was committed to coffee. The switch to coffee as well as cattle (both of which are less labour-intensive forms of land use) coincided with increasing opportunities in the local city, which led to young males leaving rural areas for urban employment.

4. Case study on Petén, Guatemala: adding questions to an existing survey

In an unusual and innovative experiment, questions were added to a Demographic and Health Survey (DHS) on migration and the environment in 1999, illustrating the potential value, as well as the limitations, of adding questions on these topics to an existing survey (Grandia et al., 2001). This example was in Guatemala, for the department of Petén in the north, which has experienced substantial net migration from elsewhere in Guatemala, combined with massive deforestation, since the 1970s. Indeed, half of the department's forests that existed in 1950 had already been cleared by 1985. A previous study linked this deforestation to population pressures on the land, associated increasing fragmentation of plots, and poverty in other parts of Guatemala (especially the *altiplano* or highlands, west of Guatemala City), which were hypothesized to have driven people to out-migrate to seek land elsewhere, primarily in Petén (Bilsborrow & Stupp, 1997).[21] This situation made it an ideal site for collecting data to study aspects of the relationships between migration and the environment. This was done by adding questions to the 1999 DHS, which would be used only for Petén, along with expanding the sample size in the department by 20 per cent (beyond that corresponding to the national sample).

The questionnaire (see Grandia et al., 2001, appendix 4) contains modules added on migration to Petén (20 questions, including background questions on date of birth, education, etc.; 18 questions on migration within the department, and nine on future migration intentions; nine on sources of family sustenance; 17 on each plot of land; 46 on land use; two on attitudes, and one on the environment. Apparently, only women were interviewed (consistent with the focus of DHS surveys on fertility and health), which is a significant limitation, since men are commonly more involved in making household economic

[21] Subsequent research revealed that the situation was more complex, as the majority of migrant settlers in the Petén came from regions of Guatemala other than the *altiplano* (cf. Carr, 2008).

decisions, especially concerning farming, land use and migration. The module on migration to Petén includes questions about place of birth of person and parents, place of previous residence, when the person came to Petén, why they came, whether they came with any family members, whether any person or institution helped her settle, etc. The questions on migration within the region are important, since most people had moved at least once within the region after arriving. Household sustenance questions relate to occupation, identification of major type/source of household income, who in the household is responsible for that income, and whether the household has its own plot devoted to corn. The questions on land area are about how many parcels the household has and then, for each parcel (up to three), what the land area, location and ownership status is, how long the household has had the parcel, whether it is located in a national park or buffer area, and whether the production from the parcel(s) is sufficient to sustain the family. The questions on land management are about corn and other crops raised on the parcel (whether for the household's own consumption or for the market), how many consecutive years the plot is used for that crop, how many years it is then left fallow, whether crops are interplanted (thought to be better for the soil and to mimic, to some extent, the biological complexity of the forest), the use of natural/chemical fertilizers and other inputs, whether there were any problems raising crops due to nature (poor soils, insect infestations, weeds, animals destroying the corn plants, lack of rain, etc.), whether they collect any products from the forest (wood, medicinal plants, food, chicle, etc.), the number of head of cattle and other domestic animals, and whether fuelwood is used for cooking and, if so, who collects it. The single question about environmental attitudes is, unfortunately, phrased as a leading question, asking what the woman thinks is "the best use of the Petén forests" – to preserve them, to use them for tourism and natural forest products, to clear them for more agriculture, or to clear them for more pasture for cattle.

The questions on migration, household sustenance and land area are straightforward, although there are no questions at all about the household situation in the *previous* place of residence, making it impossible to know either the determinants of consequences of migration (see section 3) or whether environmental factors played any role in the household decision to migrate to Petén. The only relevant datum was in response to the question of why they left their previous place of residence; one of the response codes was that their land in the previous residence was not fertile, but only 6 per cent gave this reason. On the other hand, the extensive questions on current access to land

and land management in Petén include several useful ones relating to the environment, beginning with whether the household has any land in forest and whether it is using any land in the national park (which is not legal) or the buffer zone. In fact, the survey found that, in recent years, a higher proportion of new settlers have been using land in the national parks – usually via large plots – which has therefore become a growing environmental problem (Grandia et al., 2001, op. cit.: 66, 68). How the land is prepared for planting (burning vegetation, ploughing with oxen or with tractor, or fumigating) is relevant for the environment since hand clearing and burning, or using natural fertilizer, is less harmful to the soil than ploughing with a tractor, cutting with a chainsaw and burning, or using fumigation. The questions on how many consecutive years a plot is planted with the same crop, and then how many years it is left idle, are both relevant for the environment – the latter for restoration of soil nutrients. The questions on the use of chemicals would be better if they were for each parcel, so that the name(s) and approximate quantities of the chemical(s) used could be obtained, allowing for a more accurate assessment of the environmental impact. Finally, the survey found that farmers report the climate is becoming drier over time, as deforestation spreads (op. cit.: 59).

Unfortunately, there are no questions at all on the quantity of land the household has in each form of land use, including forest, or on whether this has changed, over time. But, to be fair, such questions would have detracted the data collection effort from its overall focus on the demography and health of the household. However, without these additional key data, land use cannot be linked to household size or to in- or out-migration of persons from the household (in contrast to the Ecuador case study, which follows).

This example illustrates not only the advantages (the most obvious of which is the low marginal cost) but also the significant limitations of adding questions to an existing survey (the main focus of which is something else) to collect data to study migration–environment linkages. The limitations are evident because of the complexity of those linkages and, therefore, the need to collect more and better quantitative data to identify and measure them.

5. Ecuador case study[22]

Overall project purpose and design

This project was designed from the beginning to collect data from multiple sources and fields to investigate whether environmental conditions have had any effects on out-migration from rural areas of Ecuador. For this purpose, data were collected in 2008 from a scientific probability sample of households and their communities in four provinces, via surveys, complemented by satellite imagery and ground-truthing of land use and geodetic control points.[23] The timeframe of the study was 2000 to mid-2008, so the data collection goal was to collect data referring to this 8.5-year time period. The steps taken to collect the data are described briefly below.

It was determined that, in order to study the possible linkages between the environment and out-migration, the following steps were required:

a. Clearly define migrants as persons who have left the household (or households that migrated altogether[24]) to live outside the *parroquia,* or parish (the smallest administrative unit in Ecuador, akin to a US township, or a subdistrict, in other countries), in the previous 8.5 years (those leaving the household to live elsewhere in the local community are also identified). The data were collected from June to October 2008, and enquired

[22] Based on the project, Frontier Migration and the Environment in Ecuador, led by R. Bilsborrow, Principal Investigator, and funded by the U.S. National Institutes of Health, 2007–2009.

[23] Geodetic control points refer to the use of GPS receivers to document, on the ground, key locations observable from satellites, which can be used to precisely fit satellite images to ground observations, such as road intersections, rivers, lakes, bridges, large parcels in particular crop use, etc.

[24] Reliable data on whole households migrating away cannot be obtained from surveys conducted in origin areas only. We therefore asked community leaders about whole households departing since 2000 – name of head, number of persons leaving, when last person in household left, and destination. In small rural communities, community leaders are often cognizant of most activities and the whereabouts of community residents, including when whole households leave.

about persons out-migrating since 1 January 2000 – an easily remembered date.

b. Consider the *full range* of types of destinations of migrants, including both internal (urban and rural) and international destinations. This helps in distinguishing the differences in the determinants and consequences of all types of migration by destination type.

c. Use specialized sampling and survey design methods to ensure finding sufficient numbers of *recent migrants* (along with non-migrants), addressing the "rare elements" problem (Bilsborrow et al., 1997: 267–288). This required the use of the two methods discussed above: (i) stratification of primary sampling units (PSUs) and oversampling PSUs from strata with relatively high expected proportions of migrants; and (ii) two-phase sampling at the last stage, in sample census sectors.

d. Develop and test measures of natural resource endowments of farm households and communities and of environmental change, as well as of other socio-economic and infrastructure factors, to investigate the *relevance of context* (including environmental factors) to out-migration from areas of origin.

e. Recognize the inherent limitation of studies based on surveys conducted only in areas/countries of origin: the lack of data on the out-migration of whole households. If the characteristics of persons moving as households are different from those of persons moving as individuals, if the factors leading to migration are different, and if the consequences for households are different, there can be a potentially serious selection bias in identifying the characteristics of migrants (compared to non-migrants) from origin survey data and, therefore, in statistically assessing the determinants and/or consequences of migration.

The sample design, the content of household and community questionnaires, and the spatial data collection procedures are summarized below. Since the data were collected only in 2008, no results are available yet.

Sample design

The first step in designing a survey is to create a sampling frame. Data were tabulated in early 2008 by the National Statistics and Census Office (INEC) from the most recent (2001) population census on the *proportion of individuals who out-migrated* in the five-year reference period prior to the census. This, and budgetary limitations, led to the selection of three study areas for the project (rather than a national sample), comprising a primarily coastal province and three highland provinces. Within these four provinces, the next lower level political units – *cantons* (like districts or US counties) – were identified with relatively high proportions of out-migrants, and 17 were selected, based on the proportion of persons who out-migrated in 1996–2001. In these 17 cantons, the proportion of the population that out-migrated was then tabulated from the census for all rural parishes (two to about 12 parishes per canton studied) and listed. Systematic sampling was then used to select 30 parishes, with the probability of selection of each parish determined by the proportion of the population out-migrating.[25] These parishes constituted the PSUs, and were the lowest-level political units with data on migrants from the census. Census sectors were then selected randomly from the list of parishes, with up to two *rural* census sectors from each parish (51 census sectors) ultimately selected, constituting the Ultimate Area Units (UAU).[26]

In each UAU, interviewers and a supervisor used two-phase sampling. In the first phase, all occupied residential structures were listed according to whether the household contained any persons aged 15–39 (the population whose migration was of interest) and whether it had someone in that age range who had left to live elsewhere (and not return) since 2000, by type of destination. A page of sampling procedures was provided to the supervisors (they were trained on the procedures prior to fieldwork) so they could select households in the field, on the spot, rather than returning all the way to the central project office in Quito, thereby saving considerable travel funds. These procedures had to take into account the total number of eligible households (with a member aged 15–39 at the time of the

[25] There was no attempt, at this stage, to select areas with more out-migrants of one type than another, though this could have easily been done.

[26] These are the last-stage or smallest area units from which households are sampled.

survey or at any time since 2000)[27], and the number of households with international migrants (emigrants), with out-migrants to urban destinations in Ecuador, and with out-migrants to rural destinations in Ecuador. Households with out-migrants to international and rural destinations were oversampled compared to those with migrants to urban destinations, since they were less numerous, but all three types were oversampled compared to households with no recent out-migrant. The procedures ensure that (a) the results are generalizable to the population of the 17-canton study area of four provinces and (b) sufficient numbers of households with recent out-migrants (including migrants to each destination type of interest) are captured.

Given the goal of studying the effects of origin area environmental conditions (see below) on out-migration, besides choosing four provinces with very different environmental conditions, stratifying parishes, based on elevation, was also considered. That could have been done from an overlay of parish boundaries with a national digital elevation map. Elevation is a potentially key environmental variable, linked to variations in forest and land cover, slope, and soil type, so selecting the sample of parishes stratified by elevation would further guarantee that a wide variety of environmental conditions are represented. Nevertheless, an inspection of the sample of 17 parishes resulting from the stratification based only on the more usual demographic criterion (percentage of population out-migrating from the parish in 1996–2001 – the best available indicator of what might be the propensity to out-migrate from the sample parishes in 2000–2008) indicated a wide range of elevations in the sample.

Household survey questionnaire design

The household interview contained a household questionnaire (HQ), administered to the head of the household or his/her proxy, and an individual questionnaire (IQ) for each household member aged 15–39 at the time of the survey and each out-migrant, regardless of age at migration. Information on out-migrants since 2000 was obtained from the proxy respondent, who was expected to be the person in the

[27] This means that if the household did not have anyone aged 1539 in it at the time of the listing in 2008, it would still be an eligible household for sample selection if it had had such a person living in it during the previous 8.5 years who migrated away and did not return.

household most knowledgeable about the out-migrant. To capture the circumstances pertaining to the decision to migrate at the time the decision was made, key data were collected for both the household and individuals, not only on their current situations but also on their situation *each year* during the reference period since 2000. This differs from the questionnaire format described in section 3 above (and in Bilsborrow et al., 1997), which focuses on obtaining data pertaining to the *last* migration of the out-migrant (rather than for *all* migrations away, or back to the household, during the 8.5-year reference period). Data for non-migrant households and individuals were also obtained. Retrospective data collection was limited to the 8.5 years before the interview, to minimize recall error (Som, 1973). To statistically analyse the effects of household and individual factors on migration decisions, it is necessary to have data for *each* year of the study period on all individuals likely to migrate – in this study, taken to be those aged 15–39.[28] The explanatory or independent variables include both time-varying (e.g., household size, age composition, land area and land use) and time-invariant (e.g., education of head, location of house) variables.

The HQ includes a minimal household roster listing current members of the household (by age, gender and relationship, dwelling conditions, geographic location (using GPS) and road access, household assets as an indicator of wealth, and the major source (but not amount) of household income (agriculture, family business, agricultural or non-agricultural labour, remittances from migrants, etc.) each year from 2000 to 2008. A screening question inquired about the number of land plots or parcels the household owned or managed each year, since many own or use more than one. For *all parcels together,* the household respondent is asked about the total area planted in the past 12 months, production and sales, whether they used modern agricultural inputs, when each input was first used since 2000, and whether the area planted by the household in each crop has changed up or down since 2000 and, if so, why. Among the reasons considered were several environmental ones – favourable/unfavourable climate, good/bad soils, use of modern inputs (fertilizer, etc.), irrigation, etc. Further questions are then asked about whether there have been particularly good or bad years for agriculture since 2000 and, if so,

[28] Young individuals (under age 15 in 2000) age into the relevant age pool during the reference period, while others (in their 30s in 2000) age out.

why. Similar questions are asked about animal raising and the income it generated.

Then, for each parcel, the household is asked whether it is owned or rented, what size it is, and how the parcel was used for each year since 2000 (forest, fallow, pasture, crops, other). Follow-up questions are asked about the intensity of land use on the parcel, including crops grown and technology used (irrigation, fertilizer, or hybrid seeds), perceived quality of the soil, and whether there were any changes in yields and soil quality since 2000 and, if so, why. Questions are also asked about access to and use of communal or public lands and resources (including national parks), including for grazing animals and collecting fuelwood. Enquiries were also made about the household's experience with natural disasters, drought and floods in the past ten years, which can vary across households within the same community. Assessing the economic status of the household requires obtaining data also on off-farm employment and any non-farm business or own-account income, which was done for each member of the household above age 14 in the individual questionnaire.

The IQ obtains data for each year since 2000 for all persons aged 14+ that year – both current household members and out-migrants (provided by the household respondent) – constituting a rectangular data file. It is necessary to collect these data on both migrants and non-migrants, so that their characteristics can be compared for each year since 2000 and, hence, *at the time of migration*. These characteristics include age, education, school attendance, marital status, employment, etc. When these data are pooled for migrants and non-migrants aged 15–39, statistical models can be used to estimate the factors that determine why some people migrated and others did not – factors at the individual, household and community levels. Data are also sought on any remittances the out-migrant may have sent. To control for the effects of migration networks, data on where all close relatives were living at the time of migration were also obtained.

It is important to note that the format of the IQ here (and also in the Mexican Migration Project of Massey et al. and the Burkina Faso studies of Henry et al., 2004) differs from the perhaps more customary approach described in section 3. Data are collected over time for all eligible persons (15–39, here), which requires a totally different questionnaire layout: for each person, out-migrant or not, the questions are listed in the usual way, to the left of the page going down in rows, with the years, from 2000 to 2008 listed across the top, creating a rectangular data collection page as well as a rectangular data file.

The following data are thus recorded for each relevant individual, *for each year,* in row cells going across the page, under the 'year' column: age of person; place of residence (including elsewhere in the local community or not, elsewhere in country, foreign country, with a drop-down panel to record a few details of each place outside the community: location and whether urban or rural); level of education and whether attending school; marital status; whether the migrant sent money to anyone in this household or community (leading to another set of follow-up questions for details); whether the person worked on household lands; and whether he/she (also) worked for wages, agricultural or non-agricultural. A similar format is used in a land use questionnaire, which is part of the HQ, for each agricultural plot of land the household owns or manages for each year from 2000 to 2008, including whether it is owned, rented or otherwise accessed; size of parcel; whether it had irrigation; and its principal land use. More detailed data on specific crops raised, output, market sales and income, use of inputs, labour, etc. are asked only for the current period and the previous 12 months.

The format of the IQ (and part of the land use questionnaire) creates a rectangular array of data useful for statistically investigating the factors associated with migration movements through event history analysis, using proportional hazard models. It has the advantage of facilitating the analysis of all changes in residence over a given time period and linking those changes to an array of other changes in the person's situation, such as marital status and education. It has the disadvantage of not lending itself to as much detailed data on the situation or events just prior to (or following) any migration, such as income, migration networks, and the household context (size, composition, assets, etc.). It represents a philosophically distinct and useful approach for data collection and analysis.

Overall, the IQ format used in Ecuador obtained all of the data on migration as well as data on a number of explanatory variables relating to an individual's characteristics, which may influence his/her migration, while the HQ obtained data for estimating explanatory variables at the household level, including environmental variables.

Community survey data collection

The population in the selected sample census sectors comprised small local rural communities – that could usually be identified on census sector maps of the National Institute of Censuses and Statistics (INEC). In each community, geo-referenced with GPS, a questionnaire was implemented to obtain data from community leaders and informants on the number of households and estimated population size in 2008 and 2000; out-migration of individuals and of entire households since 2000; destinations; and perceived effects on the community, if any. Data were also collected on the three main sources of income for households in the community, as well as on local agricultural wage rates, seasonal labour migration of community residents outside the community, and whether the community had many types of infrastructure in the survey year, and if so, when it became available. This provided data on whether the infrastructure was available for each of the eight years of interest, from 2000 to 2008. Related to the environment wee questions on the total land area, including areas in forests and public lands; typical farm size and largest farm, number or proportion of households with no land; prevalence of irrigation in the community; principal crops grown and whether this had changed since 2000 and reasons for any change; use of modern agricultural inputs (fertilizers, etc.), cattle raising and changes over time; quality of soils and changes over time; prevalence of use of fuelwood for cooking and any change since 2000; occurrence of major natural disasters since 2000 (drought, flooding, etc.) and number of households affected; transportation and communications linkages with other places; etc.

These data, as well as spatial data (see below), will be drawn upon to explore the determinants of out-migration, including the effects of a panoply of environmental variables. The data sought on whole households that left included the name of the head, the number of persons in the household in 2000, the year in which the last member left, the place of destination of the last one who left, the education of the household head, and whether the household owned a house, land or cattle before leaving. The smallness and intimacy of the rural communities ensures that community leaders and informants usually know about recent out-migrating households, which are missed in traditional surveys on migration conducted only in origin areas and, therefore, in this Ecuador household survey.

Spatial data collection and measures of environmental conditions

Landsat TM and ASTER satellite imagery were acquired for the study areas in the four provinces for the baseline year 2000 and for subsequent years, as available, up to the time of the survey data collection in 2008. Standard atmospheric and geometric pre-processing was performed, and radiometric correction used to compare land use and land cover (LULC), as shown by images over time (Song & Woodcock, 2003). The image time-series is being spectrally normalized, and LULC changes are being classified through a hybrid approach (Walsh et al., 2003) and enhanced using several vegetation indices, including the NDVI, the Soil-Adjusted Vegetation Index (SAVI), and the Tasseled Cap wetness-greenness-brightness index (Jensen, 2000). GPS points were collected in the field by the interviewing teams on the locations of communities and sample households and agricultural parcels, as well as separately by geographers on key roads, rivers, road intersections, and forms of LULC in the four provinces, to assist in both geodetic control and for collecting ground-truth points for land-cover classification of satellite imagery in 2008. Thus, forms of land use on the ground (bare ground, water, urban, forest, pasture and various types of crops, from annual crops, such as corn, rice, and beans, to perennials, such as coffee, cacao and sugar cane, are identified with GPS and matched to the spectral imprint of the satellite image at that same precise location to 'teach' the satellite what kind of land use that imprint is. Other locations (pixels) in the image with the same spectral imprint can then be assigned that land use – a procedure that 'classifies' land use in satellite images.

A GIS (geographic information system) was used to encode digital spatial observations from Landsat, at a 30-meter pixel resolution, to characterize the resource endowments of study communities (soils, terrain slope, hydrography, potential soil moisture), land use/land cover (e.g., composition, spatial pattern metrics[29], greenness-wetness-brightness indices, fractional cover), and geographic access and connectivity (via roads) for survey households and communities. LULC change patterns are being generated using the classified image time-series and the derived vegetation indices. Pattern metrics computed at

[29] Pattern metrics measures were used as dependent variables to capture aspects of land use change and fragmentation at the farm level in Pan et al. (2004) for the Ecuadorian Amazon. Metrics used measured the extent of fragmentation of land use into the different forms being considered/classified, patch density (numbers of distinct [to the satellite] patches), and total edge length per unit area, for each farm.

the landscape and lower scales (largely at the community level) will be calculated and used to describe the spatial structure of LULC, including land fragmentation patterns and trends, which are important indicators of landscape structure and function (Read & Lam, 2002; Walsh et al., 2003), and also capture forest fragmentation and patches of secondary growth after land abandonment. Measures of patch form and change over time reflect aspects of the environment at the household and community levels that may be linked to out-migration. The goal is to experiment with various combinations of bio-physical measures, with various time lags, as alternative measures of environmental quality in statistical models of out-migration, to explore the impacts, if any, of origin-area environmental conditions on out-migration. The focus will be on measures or landscape features that seem most likely to be linked to rural livelihoods – e.g., land cover, climate, and soil/moisture quality on agricultural lands. Both objective and subjective (as perceived by the farmer, as reported in the household interview) measures of environmental conditions will be used to examine possible links to out-migration, using a multivariate statistical model that also includes individual, household and other non-environmental contextual (i.e., community-level) factors.

6. Conclusions and recommendations

Up to now, research on the linkages between migration and the natural environment in developing countries has been advancing well – or poorly – depending on which direction of the linkage one is looking at. There are many good studies, and more underway, on the impacts of migrants on the environments – on land use, soil degradation, deforestation, etc., at both macro and micro or household levels. On the other hand, despite widespread claims in the media and mostly grey literature, there is almost no reliable evidence o the effects of environmental factors (statistically controlling for other influences) on out-migration, particularly from rural areas. In fact, data limitations, as well as failures to take advantage of, or seek out, existing sources of data, have impeded progress on both sides of the environment–migration nexus. But there is another reason for limited progress: the disconnect between the disciplines that focus on only one of the two. Thus, few social scientists specializing in migration, and relying on data from censuses and household surveys, have ever been engaged in data collection or research relating to the environment. Similarly, few of those who focus on the environment, whether in the social or natural sciences, are interested in migration or know about migration data.

Fortunately, this is beginning to change, for several reasons. One is the mounting public and international interest in environmental issues and their relationship to development and poverty alleviation, at local, regional and country levels, as well as at the global level, where climate change dominates the discussion. The need for the collection and use of better data on the environment, in order to integrate environmental issues into national development plans and sustainable development agendas was recognized at the UN Conference on Environment and Development in Rio de Janeiro in 1992, and reiterated at the World Summit in Johannesburg in 2004. Meanwhile, for the first time in the series of decennial UN international population conference meetings since 1954, the international population policy community explicitly recognized the importance of integrating environment and development into population plans and policies, at the International Conference on Population and Development in Cairo in 1994.

Another reason is the independent increase in the quantity, quality and availability of satellites and the imagery they provide of the earth and its resources and their ongoing deterioration. This technology was not available until around four decades ago, and has steadily improved. It has provided a wealth of data on the environment, increasingly of higher measurement value and at low cost (although the highest quality imagery is still quite expensive). A new generation of geographers, other social scientists, and ecologists and environmental science students is being trained in the use and manipulation of these data to better measure changes in the environment. Landsat images have been problematic since 2004; although better-quality, new imagery is expensive, global concerns about the environment demand that evolving satellite imagery data be analysed.

At the same time, data on migration are improving, albeit at a slower, uneven pace. More countries asked about internal and international migration in the last (2000) round of national population censuses than in previous decades, and more still will be collecting at least some basic data on migration in the so-called 2010 round[30] currently underway. Most countries now ask the UN-recommended questions for measuring migration in their censuses, based on the place of previous residence one, five or ten year(s) ago (the most common census interval being ten years). This provides data on in-migration, including immigration. At the same time, more developing countries are asking in their census whether anyone has left the household since x years ago and, if so, for what destination. More importantly, household survey data that include migration are being increasingly collected, through new programmes in Europe, Latin America (e.g., the Latin American Migration Project, housed at Princeton University), the Middle East and North Africa (MEDSTAT, funded largely by the European Union), sub-Saharan Africa (funded by the World Bank), etc. Indeed, it is specialized migration surveys that can best obtain data on migration, as discussed in major research forums, meetings of international organizations, and several books (e.g., Bilsborrow et al., 1984, 1997); the latter would surely include discussions of the environment, if they were updated now. At the same time, specialized surveys can collect data not only on migration but also on the environment – on land ownership and use,

[30] The 2010 World Programme on Population and Housing Censuses is primarily aimed at ensuring that each Member State conducts a population and housing census at least once in the period from 2005 to 2014 and disseminates the results. Details available at: http://unstats.un.org/unsd/demographic/sources/census/2010_PHC/more.htm.

use of public lands, experience with soil degradation, behaviour and attitudes related to environmental policies and protected areas, etc.

Since new surveys are expensive, the possibilities for exploiting data from existing surveys to analyse relationships between migration and the environment should be further explored as well. But while there are doubtless some excellent sources of data that have not been exploited for this purpose, it is crucial that the survey questions be examined carefully first to ensure that they have actually used appropriate questions to collect the data necessary to develop good measures of both migration and the environment, so that 'proxies' that are not really acceptable are not used (as was the case in some of the studies described in section 2).

In this chapter, two alternative ways of collecting data on migration in surveys are discussed: one that obtains data for each year on every eligible person who has recently been a member of the household since, for example, ten years ago – age, education and school attendance, marital status, place of residence (so as to capture those leaving the household and those returning), work status and occupation, and ownership of land. These data constitute what is called an 'event history' database, and facilitate the use of discrete time hazard models to analyse the linkages between events (e.g., marriage/divorce and migration, unemployment and migration) and, indeed, the determinants of migration (and non-migration) in the time interval. This kind of dataset makes it possible to investigate a larger number of moves over a period of time. However, it is limited by the difficulty of obtaining detailed data on the moves, which can be numerous, particularly for a longer time interval and for large households, which can lead also to respondent fatigue and non-cooperation. It is therefore not practical to obtain *detailed* data on the circumstances preceding or following every move, especially for moves that occurred more than a few years ago. The alternative survey approach is to ask only about the *last* move of each out-migrant (and former member) in a household, restricting the migrants of interest to, say, the past five years. For such a recent move, much more detailed and better-quality data can be obtained, allowing for a more *intensive* examination of a wider range of potentially relevant causative factors determining migration, albeit for a smaller number of moves per migrant. Which approach is better for studying the determinants of migration, or the linkages between the environment and migration, is not known. Only future research will tell. In any case, each has clear advantages and disadvantages compared to the other. If the sample size is large, data on the last move will allow for a richer investigation of the determinants or consequences of migration, based on the considerably larger number of explanatory variables that can be formulated. But if the sample size is small, the event history approach increases the number of migration moves that can be studied, albeit in clusters of events per migrant.

A low-cost alternative to a new survey is adding questions to an existing survey before it is implemented, or to a planned survey. This has been debated widely across the international community, which has been seeking more cost-effective ways of collecting meaningful data on international migrants, particularly in developing countries. This community includes the World Bank, the International Labour Organization (ILO), the European Union and Eurostat, the Organization for Economic Co-operation and Development (OECD), the InterAmerican Development Bank (IDB), the Asian Development Bank, the US Census Bureau, and government statistical offices and research institutes in many countries. However, the limitations of adding questions on migration to a questionnaire designed for another purpose have been noted elsewhere (Bilsborrow, 2008). Adding questions on migration *and* the environment to an existing survey with a different focus would be even more problematic. The one study that has recently attempted to add questions on migration and the environment to an existing DHS survey is reviewed in section 4 on the Petén region of Guatemala, where both the advantages and limitations of the effort are described.

The issue of gender has not been specifically addressed in this chapter, although it is relevant to much of the discussion and is addressed in a few of the studies reviewed. There is considerable research on gender differences in the determinants and consequences of migration, both internal and international, and there is also some literature on gender and the environment, on whether women are more protective of the environment, or are more or less affected by environmental disasters or the out-migration of men for work elsewhere. Unfortunately, these bodies of literature do not address the specific interrelationships between the environment and migration. One topic on which the literature is relevant is the collection of fuelwood – an activity usually carried out by women, so in areas where fuelwood is scarce, and/or is made scarcer by deforestation, women are the most affected. This is true of much of Africa and Asia, although in Latin America fuelwood scarcity is rarely severe, and its collection is mainly a shared activity. Section 2 discuss some results along these lines, but this topic is quite distinct from the more general one on whether environmental degradation in rural areas is more likely to induce male or female out-migration or whether male or female in-migration is more likely to cause deforestation or other forms of degradation in destination areas. If males (or females) were more dependent on natural resources for their livelihood activities, then that sex would more directly affected by environmental degradation, although all members of the household are ultimately affected.

Should the focus of research be at the macro or micro levels? While much can be learned from studies of migration and environment relationships at the *macro* level of countries, provinces or districts, those relationships are usually difficult to interpret unambiguously. If, for example, a measure of migration and a measure of the environment are correlated, what is the direction of causation? Since many other factors may affect both migration and the environment, with some being common to both, it is usually very difficult to reliably separate out the specific causal relationship between the environment and migration, based on macro data – partly because important variables are inevitably not available or have not even been identified. While this is always a risk at the *micro or household* level as well, a considerably larger number of relevant variables can more easily be identified, to guide data collection, and measured more or less reliably. Extensive experience with household surveys all over the world in recent decades indicates that questionnaires can be designed and good interviewers recruited and trained to collect almost anything – including good data on migration and the environment, at the individual, household and community levels. Collecting data on households in rural areas is fundamentally important, since households are the major decision makers about resource use as well as household labour allocation and migration.

Finally, studies are needed on the impact of migration (separated from the effects of natural population growth) on the environment in different rural areas of developing countries, as well as of the effects of environmental factors on out-migration. In both cases, it is desirable to conduct multivariate quantitative analyses of individual and household behaviours, taking into account the context of those behaviours – by using multilevel models, estimated by drawing on rich datasets at both the household and higher levels, such as the local community. A key issue is how to design samples of households and communities for studying the migration–environment nexus. The samples should cover a range of environmental conditions, from degraded to less degraded (according to whichever environmental variable(s) may be relevant in the particular study site – e.g., deforestation, soil erosion/degradation, drought or soil desiccation, etc.); this is necessary in order to tease out the effects of the differences in environmental conditions on out-migration. This means that the study area should either be highly diverse, in terms of those conditions, or that several different, spatially dispersed study areas should be included in the study. In addition, the sample sizes of both communities and households should be large enough to allow for estimating statistically significant relationships when they exist. Most of the studies mentioned in this chapter that seriously address the nexus are based on samples of, at most, a few hundred households, which may impair the search for statistically significant relationships; in addition, many are based on geographic areas that are too small or are not sufficiently diverse in terms of physical and environmental conditions to detect the effects of differences on those conditions. For measuring environmental conditions, it is also highly desirable to draw on data from remote sensing as well as from surveys, as discussed in this chapter. Finally, it is crucial that research projects seeking to entangle and better estimate the relationships between the environment and migration, in either direction, be specifically designed for that purpose from the outset. The data to be collected and the measures to be calculated can then be specifically identified.

As noted in this chapter, some studies have recently been undertaken on population and the environment at the micro level, based on survey and/or remote sensing data (e.g., Moran, 1994; Rindfuss et al., 1998, 2003; Geoghegan et al., 2001; Pan et al., 2004; Mena et al., 2006; other studies include Wood & Porro, 2002; Walsh & Crews-Meyer, 2002; and Fox et al., 2003). However, none of these studies has specifically examined migration. The Ecuador study described in section 5 represents a novel, comprehensive approach to this, focusing on the determinants of migration from rural areas, including the effects of the environment on out-migration.

7. References

Adhana, A.H.
1991 "Peasant response to famine in Ethiopia", *Ambio*, 20: 186–188.

Arguello, O.
1981 "Estrategias de supervivencia: un concepto en busca de su contenido" [Survival strategies: a concept in search of its content), *Demografia y Economia,* 15(2).

Arroyo-Mora, J.P. et al.
2005 "Secondary forest detection in a neotropical dry forest landscape using Landsat 7 ETM+ and Ikonos imagery", *Biotropica*, 37(4): 497–507.

Barbieri, A. & Carr, D.L.
2005 "Gender-specific out-migration, deforestation, and urbanization in the Ecuadorian Amazon", *Global and Planetary Change* 47(2–4): 99–110.

Barbieri, A. et al.
2009 "Migration within the frontier: The second generation colonization in the Ecuadorian Amazon", *Population Research and Policy Review,* 28: 291–320.

Bates, D.
2002 "Environmental refugees? Classifying human migrations caused by environmental change", *Population and Environment,* 23(5): 465–477.

Bilsborrow, R.E.
1987 "Population pressures and agricultural development in developing countries: a conceptual framework and recent evidence", *World Development*, 15(2): 183–203.

Bilsborrow, R.E.
1992 *Rural Poverty, Migration and the Environment in Developing Countries: Three Case Studies*, World Bank Policy Research Working Paper (WPS 1017), The World Bank, Washington, DC.

Bilsborrow, R.E. (Ed.)
1998 *Migration, Urbanization and Development: New Directions and Issues*, Kluwer Academic Publishers, Norfolk, and United Nations Fund for Population Activities, New York.

Bilsborrow, R.E.
2002 "Migration, population change and the rural environment", *Environmental Change and Security Project Report,* 8: 69–94.

Bilsborrow, R.E. et al.
1984 *Migration Surveys in Low-income Countries: Guidelines for Survey and Questionnaire Design*, Croom-Helm, London.

Bilsborrow, R.E. et al.
1987 "The impact of origin community characteristics on rural-urban out-migration in a developing country (Ecuador)", *Demography,* 24(2): 191–210.

Bilsborrow, R.E. & Ruiz, L
1990 "Los efectos demográficos de proyectos de desarrollo rural integral en el Ecuador" [Demographic effects of integrated rural development projects in Ecuador], *Forum Valutazione* (Rome), 0(1): 33–61.

Bilsborrow, R. E. & DeLargy, P.
1991 "Land use, migration and natural resource deterioration: the experience of Guatemala and the Sudan", *Population Development Review,* 16(Supp.): 125–147.

Bilsborrow, R.E. & Geores, M.
1994 "Population change and agricultural intensification in developing countries", in *Population and the Environment: Rethinking the Debate*, Arizpe, L. et al. (Eds), Westview Press, for Social Science Research Council, Washington, DC: 171–207.

Bilsborrow, R.E. & Stupp, P.
1997 "Demographic Processes, Land and the Environment in Guatemala", in *Demographic Diversity and Change in the Central American Isthmus*, Pebley, A, Rosero-Bixby, L. (Eds), The Rand Corporation, Santa Monica: 581–623.

Bilsborrow, R.E. et al.
1997 *International Migration Statistics: Guidelines for Improving Data Collection Systems*, International Labour Organization, (ILO), Geneva.

Bilsborrow, R.E. & Hogan, D. (Eds)
1999 *Population and Deforestation in the Humid Tropics*, International Union for the Scientific Study of Population (IUSSP), Liège.

Bilsborrow, R.E. & Carr, D.
2001 Population, Agricultural Land Use, and the Environment in Developing Countries, in *Tradeoffs or Synergies? Agricultural Intensification, Economic Development and the Environment,* Lee, D.R, Barrett, C.B. (eds), CABI, Wallingford, UK: 35–56.

Bilsborrow, R.E. et al.
2004 "Changes in population and land use over time in the Ecuadorian Amazon", *Acta Amazonica,* 34(4): 635–647.

Black, R.
2001 *Environmental Refugees: Myth or Reality?* Working Paper 34, United Nations High Commissioner for Refugees (UNHCR), Geneva.

Blaikie, P., Brookfield, H.
1986 *Land Degradation and Society*, Methuen, London.

Blalock, H.
1985 "Cross-level Analysis." In J. Casterline, /The Collection and Analysis of Community-level Data. /WFS Seminar on Collection and Analysis of Data on Community and Institutional Factors, 20-23 June 1983. London: International Statistical Institute, World Fertility Survey, pp. 187-206.

Boserup, E.
1965 *The Conditions of Agricultural Growth: The Economics of Agrarian Change under Population Pressure*, Aldine Publishing Company, Chicago.

Boyle, P. & Shen, J.
1997 "Public housing and migration: a multi-level modelling approach", *International Journal of Population Geography, 3:* 227–42.

Browder, J.
1995 "Redemptive communities: Indigenous knowledge, colonist farming systems, and conservation of tropical forests", *Agriculture and Human Values,* 12(1): 17–30.

Brown, O.
2007 "Climate Change and Forced Migration: Observations, Projections and Implications", United Nations Development Programme (UNDP), New York.

Burnett, C. & Blaschke, T.
2003 "A multi-scale segmentation/object relationship modelling methodology for landscape analysis", *Ecological Modelling,* 168: 233–249.

Carr, D.
2008 "Migration to the Maya Biosphere Reserve, Guatemala: Why place matters", *Human Organization*, 67: 37–48.

Cassels, S. et al.
2005 "Do migrants degrade coastal environments? Migration, natural resource extraction and poverty in North Sulawesi, Indonesia", *Human Ecology,* 33(3): 329–363.

Charnley, S.
1997 "Environmentally-displaced peoples and the cascade effect: lessons from Tanzania", *Human Ecology,* 25(4): 593–618.

Chomitz, K. & Gray, D.
1996 "Roads, land use, and deforestation: a spatial model applied to Belize", *World Bank Economic Review,* 10(3): 487–512.

Chopra, K. & Gulati, S.
2001 *Migration, common property resources and environmental degradation: interlinkages in India's arid and semi-arid regions,* Sage Publications, Thousand Oaks, CA.

Cincotta, R. & Engelman, R.
2000 *Nature's Place: Human Population and the Future of Biological Diversity*, Population Action International, Washington, DC.

Cline-Cole R.A. et al.

1990 "On fuelwood consumption, population dynamics and deforestation in Africa", *World Development,* 18(4): 513–27.

Cochrane, M. et al.

1999 "Positive feedbacks in the fire dynamic of closed canopy tropical forests", *Science* 284 (5421): 1832–1835.

Collins, J.

1986 "Smallholder settlement of tropical South America: the social causes of ecological destruction", *Human Organization,* 45(1): 1–10.

Conelly, W.T.

1994 "Population pressure, labour availability, and agricultural disintensification: the decline of farming on Rusinga Island, Kenya", *Human Ecology,* 22(2): 145–170.

Cruz, M.C.

1997 "Effects of population pressure and poverty on biodiversity conservation in the Philippines", in *Population, Environment and Development,* Pachauri, R.K. & Qureshy, (Eds), Tata Energy Institute, New Delhi: 69–94.

Dasgupta, S. et al.

2009 "The impact of sea level rise on developing countries: a comparative analysis", *Climatic Change,* 93(3): 379–388.

DaVanzo, J.

1981 "Microeconomic approaches to studying migration decisions", in *Migration Decision Making: Multidisciplinary Approaches to Microlevel Studies in Developed and Developing Countries,* De Jong, G. & Gardner, R. (Eds), Pergamon Press, Oxford.

Davis, K.

1963 "The Theory of Change and Response in Modern Demographic History", *Population Index*, 294: 345–366.

de Janvry, A. et al.

1997 *Determinants of Mexico–U.S. Migration: The Role of Household Assets and Environmental Factors*, report prepared for the Natural Heritage Institute (NHI), San Francisco.

De Jong, G. & Gardner, R. (Eds)
1981 *Migration Decision Making: Multidisciplinary Approaches to Microlevel Studies in Developed and Developing Countries*, Pergamon Press, Oxford.

De Walt, B.
1985 "Microcosmic and macrocosmic processes of agrarian change in Southern Honduras: The cattle are eating the forest", in *Micro and Macro Levels of Analysis in Anthropology: Issues in Theory and Research,* DeWalt, B., Pelto, P. (Eds), Westview Press, Boulder, CO.

El-Hinnawi, E.
1985 *Environmental Refugees,* United Nations Environment Programme (UNEP), Nairobi.

Ellis, F.
2000 *Rural Livelihoods and Diversity in Developing Countries,* Oxford University Press.

Entwisle, B. & Stern, P. (Eds)
2005 *Population, Land Use, and the Environment*, National Research Council, National Academy Press, Washington, DC.

Ezra, M.
2003 "Environmental vulnerability, rural poverty, and migration in Ethiopia: a contextual analysis", *Genus,* 59(2): 63–91.

FAO
2005 *State of the World's Forests 2005,* Food and Agricultural Organization of the United Nations (FAO), Rome.

Faulkingham, R.H. & Thorbahn, P.F.
1967 "Population dynamics and drought: a village in Niger", *Population Studies,* 29(3): 463–477.

Findley, S.
1988 "From seasonal migration to international migration: an analysis of factors affecting the choices of families in the Senegal River Valley", in *Proceedings of the Second Annual African Population Conference, Dakar*, vol. 2, International Union for the Scientific Study of Population (IUSSP), Paris.

Findley, S.
1994 "Does drought increase migration? A study of migration from rural Mali during the 1983–1985 drought", *International Migration Review,* 28(3): 539–553.

Findley, S. et al.
1987 "An interactive contextual model of migration in Ilocos Norte, the Philippines", *Demography,* 24(2): 163–190.

Fox, J. et al. (Eds)
2003 *People and the Environment: Approaches for Linking Household and Community Surveys to Remote Sensing and GIS*, Kluwer Academic Publishers, Boston/Dordrecht/London.

Geist, H. J. & Lambin, E.F.
2001 *What drives tropical deforestation? A meta-analysis of proximate and underlying causes of deforestation based on sub-national case study evidence*, LUCC International Project Office, Louvain-la-Neuve, Belgium: 116.

Geist, H.J. & Lambin, E.F.
2002 "Proximate causes and underlying driving forces of tropical deforestation", *Bioscience,* 52(2): 143–150.

Geoghegan, J. et al.
2001 "Modeling tropical deforestation in the Southern Yucatan peninsular region: comparing survey and satellite data", *Agriculture, Ecosystems and Environment,* 85: 25–46.

Godoy, R.H. et al.
2002 "Local financial benefits of rain forests: comparative evidence from Amerindian societies in Bolivia and Honduras", *Ecological Economics,* 40(3): 397–409.

Grandia, L. et al.
2001 *Petén: Salud, Migración y Recursos Naturales: Resultados del Módulo Ambiental en la Encuesta de Salud materno Infantil, 1999*, Instituto Nacional de Estadística, Guatemala City; USAID; and Measure/DHS Macro International.

Gray, C.
2009 "Environment, land and rural out-migration in the Southern Ecuadorian Andes", *World Development,* 37(2): 445–468.

Groenewold, G. & Bilsborrow, R.E.
2008 "Design of samples for international migration surveys: methodological considerations and lessons learned from a multi-country study in Africa and Europe", in *International Migration in Europe: New Trends and Methods of Analysis, Bonifazi, C. et al. (Eds),* Amsterdam University Press: 293–312.

Gutmann, M. et al.
2005 *"Two population–environment regimes in the Great Plains of the United States, 1930–1990", Population and Environment, 27(2): 191–225.*

Hecht, S.B.
1985 "Environment, development and politics: capital accumulation and the livestock sector in eastern Amazonia", *World Development,* 13(6): 663–684.

Heckandon, M.S. & McKay, A. (Eds)
1984 *Colonización y Destrucción de Bosques en Panamá: Ensayos sobre un Grave Problema Ecológico*, Asociación Panameña de Antropología, Panamá.

Heilig, G.K.
1997 "Anthropogenic factors in land-use change in China", *Population and Development Review*, 23(1): 139–168.

Henry, S. et al.
2004 "The impact of rainfall on the first out-migration: a multi-level event-history analysis in Burkina Faso", *Population and Environment,* 25(5): 423–460.

Henry, S. & Bilsborrow, R.E.
2006 "How migrants choose their destination in Burkina Faso", presented at European Population Conference, Liverpool, UK, June 21–24.

Higgins, G. et al.
1983 *Potential Population Supporting Capacity of Lands in the Developing World*, Food and Agricultural Organization of the United Nations (FAO), Rome.

Hugo, G.
1981 "Village-community ties, village norms, and ethnic and social networks: a review of evidence from the Third World", in *Migration Decision Making: Multidisciplinary Approaches to Microlevel Studies in Developed and Developing Countries,* De Jong, G. & Gardner, R. (Eds), Pergamon Press, Oxford.

Hugo, G.
1996 "Environmental concerns and international migration", *International Migration Review,* 30(1): 105–131.

Humphries, S.
1998 "Milk cows, migrants, and land markets: unraveling the complexities of forest-to-pasture conversion in Northern Honduras", *Economic Development and Cultural Change,* 47(1): 95–124.

Hunter, L.
2005 "Migration and environmental hazards", *Population and Environment,* 26(4): 273–302.

Ibrahim, F. & Ruppert, H.
1991 "The role of rural–rural migration as a survival strategy in the Sahelian zone of the Sudan – a case study in Burush, N. Darfur", *GeoJournal,* 25(1): 31–38.

Jensen, J.
2000 *Remote Sensing of Environment: An Earth Resource Perspective,* Prentice Hall, New Jersey.

Jiang, L.
2005 "Water resources, land exploration and population dynamics in arid areas: the case of the Tarim River Basin in Xinjiang China", *Population and Environment, 26(6): 471–503.*

Jokisch, B.
2002 "Migration and agricultural change: the case of smallholder agriculture in highland Ecuador", *Human Ecology,* 30(4): 523–550.

Joly, L.G.
1989 "The conversion of rain forests to pastures in Panama," in *The Human Ecology of Tropical Land Settlement in Latin America,* Schumann, D. & Partridge, W. (Eds), Westview Press, Boulder, CO: 89–110.

Kaimowitz, D. et al.
1999 "The effects of structural adjustment on deforestation and forest degradation in lowland Bolivia, *World Development,* 27(3): 505–520.

Kalipeni, E.
1999 "Demographic responses to environmental pressure in Malawi", *Population and Environment,* 17(4).

Kayastha, S. & Yadava, R.
1985 "Flood induced population migration in India: a case study of Ghaghara Zone", in *Population Redistribution and Development in South Asia*, Kosinski, L., Elahi, K. (Eds), D. Reidel, Dordrecht, Netherlands: 79–88.

Kulu, H. & Billari, F.
2004 "Multilevel analysis of internal migration in a transitional country: the case of Estonia", *Regional Studies,* 35(6): 185–198.

Lambdin, E.F. et al.
2001 "The causes of land-use and land-cover change: moving beyond the myths", *Global Environmental Change,* 11(4): 261–269.

Laurian, L. et al.
1998 "Migration decisions among settler families in the Ecuadorian Amazon: the second generation", *Research in Rural Sociology and Development,* 7: 169–195.

Lee, E.
1966 "A Theory of Migration", *Demography,* 3(1): 47–57.

Leinbach, T.R. & Smith, A.
1994 "Off-farm employment, land, and life cycle: transmigrant households in South Sumatra, Indonesia", *Economic Geography,* 70(3): 273–296.

Liu, J. et al.
2003 "Human impacts on land cover and panda habitat in Wolong Nature Reserve: linking ecological, socioeconomic, demographic and behavioral Data", in Fox, J. et al. (Eds), op cit.: 241–263.

Liverman, D. (Ed.)
1998 *People and Pixels: Linking Remote Sensing and Social Science*, National Research Council, National Academy Press, Washington, DC.

Lonergan, S.
1998 "The role of environmental degradation in population displacement", *Environmental Change and Security Report,* 4: 5–15.

Malthus, T.R.
1960 *On Population (First Essay on Population, 1798, and Second Essay on Population, 1803),* Modern Library, for Random House, New York.

Marquette, C.
1998 "Land use patterns among small farmer settlers in the northeastern Ecuadorian Amazon", *Human Ecology,* 26(4): 573–598.

Marquette, C. & Bilsborrow, R.E.
1998 "Population and environment relationships in developing countries: recent approaches and methods", in *People and Their Planet: Searching for Balance,* Baudot, B. & Moomaw, W. (Eds), St Martin's Press, New York: 29–44.

Massey, D.
1990 "Social structure, household strategies, and the cumulative causation of migration", *Population Index,* 56(1): 3–26.

Massey, D. et al.
1993 "Theories of international migration: a review and appraisal", *Population & Development Review,* 19(3): 431–466.

Massey, D. & Espinosa, K.
1997 "What's driving Mexico–U.S. migration? A theoretical, empirical, and policy review", *American Journal of Sociology,* 102(4): 939–999.

Massey, D. et al.
2007 *Environmental Change and Out-Migration: Evidence from Nepal*, Population Studies Center Research Report 07-615, University of Michigan, Ann Arbor, MI.

McCracken, S. et al.
1999 "Remote sensing and GIS at farm property level: demography and deforestation in the Brazilian Amazon", *Photogrammetric Engineering and Remote Sensing,* 65(11): 1311–1320.

McCracken, S. et al.
2002 "Deforestation trajectories in a frontier region of the Brazilian Amazon", in *Linking People, Place, and Policy: A GIScience Approach*, Walsh, S., Crews-Meyer, K. (Eds), Kluwer Academic Publishers, Boston/Dordrecht/London: 215–234.

Mena, C.F. et al.
2006 "Socioeconomic drivers of deforestation in the Northern Ecuadorian Amazon", *Environmental Management*, 34(10): 1831–1849.

Mertens, B. et al.
2001 "Prediction of the impact of logging activities on forest cover: A case-study in the East province of Cameroon", *Journal of Environmental Management* 62(1): 21–36.

Meze-Hausken, E.
2000 "Migration caused by climate change: How vulnerable are people in dryland areas? A case study in Northern Ethiopia", *Mitigation and Adaptation Strategies for Global Change, 5: 379–406.*

Mincer, J.
1978 "Family migration decisions", *The Journal of Political Economy,* 86(5): 749–773.

Moran, E.
1994 "Integrating Amazonian vegetation, land use and satellite data", *BioScience,* 44(5): 329–338.

Moriniere, L.
2009 "Tracing the footprint of 'environmental migrants' through 50 years of literature", in *Linking Environmental Change, Migration and Social Vulnerability*, Oliver-Smith, A. & Shen, X. (Eds), Publication Series 12, United Nations University Institute for Environment and Human Security: 22–29.

Myers, N.
1993 "Environmental refugees in a globally warmed world", *Bioscience,* 43(11): 752–761.

Myers, N.
1997 "Environmental refugees", *Population and Environment,* 19(2): 167–182.

Myers, N.
2002 "Environmental refugees: a growing phenomenon of the 21st Century", *Philosophical Transactions of the Royal Society of London Series B – Biological Sciences*, 357(1420): 609–613.

Natural Heritage Institute
1997 *Environmental Degradation and Migration: The U.S.–Mexico Case Study*, Natural Heritage Institute (NHI), San Francisco.

Nelson, G.C. & Hellerstein, D.
1997 "Do roads cause deforestation? Using satellite images in econometric analyses of land use", *American Journal of Agricultural Economics,* 79: 80–88.

Nepstad, D. et al.
1999 "Large-scale impoverishment of Amazonian forests by logging and fire", *Nature* 398 (6727): 505–508.

Oglethorpe, J. et al.
2007 *People on the Move: Reducing the Impacts of Human Migration on Biodiversity,* Conservation International and World Wildlife Fund, Washington, DC.

Oliver-Smith, A. & Shen, X. (Eds)
2009 *Linking Environmental Change, Migration and Social Vulnerability*, Publication Series 12, United Nations University Institute for Environment and Human Security, Bonn.

Pan, W.K. & Bilsborrow, R.E.
2005 "A multilevel study of fragmentation of plots and land use in the Ecuadorian Amazon", *Global and Planetary Change,* 47: 232–252.

Pan, W. et al.
2004 "Farm-level models of spatial patterns of land use and land cover dynamics in the Ecuadorian Amazon", *Agriculture, Ecosystems and Environment,* 101: 117–134.

Panayotou, T. & Sungsuwan, S.
1994 "An econometric analysis of the causes of tropical deforestation: The case of northeast Thailand", in *The Causes of Tropical Deforestation,* Brown, K. & Pearce, D. (Eds). University College of London Press, London: 192–209.

Perz, S.
2003 "Social determinants and land use correlates of agricultural technology adoption in a forest frontier: A case study in the Brazilian Amazon", *Human Ecology,* 31(1): 133–165.

Perz, S.G. & Skole, D.
2003 "Secondary forest expansion in the Brazilian Amazon and the refinement of forest transition theory", *Society and Natural Resources,* 16(4): 277–294.

Pfaff, A. et al.
2007 "Road investments, spatial intensification and deforestation in the Brazilian Amazon", *Journal of Regional Science,* 47(1): 109–123.

Pichon, F.
1997 "Settler households and land-use patterns in the Amazon frontier: Farm-level evidence from Ecuador", *World Development,* 25(1): 67–91.

Pichón, F. & Bilsborrow, R.E.
1999 "Land tenure and land use systems, deforestation, and associated demographic factors: Farm-level evidence from Ecuador", in *Population and Deforestation in the Humid Tropics*, Bilsborrow, R.E., Hogan, D. (Eds), International Union for the Scientific Study of Population (IUSSP), Liège, Belgium: 175–207.

Preston, D. et al.
1997 "Fewer people, less erosion: the twentieth century in southern Bolivia", *The Geographical Journal,* 163(2): 198–205.

Radel, C. & Schmook, B.
2008 "Male transnational migration and its linkages to land-use change in a southern Campeche *ejido*", *Journal of Latin American Geography,* 7(2).

Ravenstein, E.
1885 "The laws of migration", *Journal of the Statistical Society,* 48: 167–227.

Ravenstein, E.
1889 "The laws of migration", *Journal of the Statistical Society,* 52: 214–301.

Read, J. & Lam, N.
2002 "Spatial methods for characterizing land cover and detecting land-cover change for the tropics", *International Journal of Remote Sensing,* 23(12): 2457–2474.

Reardon, T. et al.
2001 "Rural nonfarm employment and incomes in Latin America: overview and policy implications", *World Development,* 29(3): 395–409.

Reardon, T. & Taylor, J.
1996 "Agroclimatic shock, income inequality, and poverty: Evidence from Burkina Faso", *World Development,* 24(5): 901–914.

Rindfuss, R. et al.
2008 "Panel studies and migration", *Social Science Research,* 36: 374–403.

Ritchey, P.
1976 "Explanations of migration", *Annual Review of Sociology,* 2: 363–404.

Roncoli, C. et al.
2001 "The costs and risks of coping with drought: livelihood impacts and farmers' responses in Burkina Faso", *Climate Research,* 19 (2): 119–132.

Rudel, T.K.
1983 "Roads, speculators, and colonization in the Ecuadorian Amazon", *Human Ecology,* 11(4).

Rudel T. et al
2002 "A tropical forest transition? Agricultural change, out-migration, and secondary forests in the Ecuadorian Amazon", *Annals of the Association of American Geographers,* 92 (1): 87–102.

Ruitenbeek, H.J.
1996 "Distribution of ecological entitlements: Implications for economic security and population movement", *Ecological Economics,* 17(1): 49–64.

Sader, S.A. et al.
1997 "Human migration and agricultural expansion: an impending threat to the Maya Biosphere Reserve", *Journal of Forestry,* 95(12): 27–32.

Schmidle, N.
2009 "Wanted: a new home for my country", *New York Times*, 10 May.

Schmidt-Verkerk, K.
2009 "The potential influence of changing precipitation and temperature patterns on migratory behavior in the State of Zacatecas, Mexico", in Oliver-Smith and Shen (Eds), op. cit.

Schmink, M., Wood, C.
1992 *Contested Frontiers in Amazonia*, Columbia University Press, New York.

Schmook, B. & Radel, C.
2008 "International labour migration from a tropical development frontier: globalizing households and an incipient forest transition. The Southern Yucatan Case", *Human Ecology,* 36: 891–908.

Schoorl, J. et al.
2000 *Push and Pull Factors of International Migration: A Comparative Report,* Eurostat, European Commission, Luxembourg.

Schwartz, M.L. & Notini, J.
1994 "Desertification and migration", in *Governing Global Desertification: Linking Environmental Degradation, Poverty and Participation,* US Commission on Immigration Reform (Ed.), Natural Heritage Institute (NHI), San Francisco.

Sessay, R. & Mohamed, F.
1997 "Forced migration, environmental change, and wood fuel issues in the Senegal River Valley", *Environmental Conservation,* 24(3): 251–260.

Seto, K.C.
2002 "Measuring landuse change with Landsat TM: evidence from Pearl River Delta", *International Journal of Remote Sensing*, 23: 1985–2003.

Shaw, R.
1974 "Land tenure and the rural exodus in Latin America", *Economic Development and Cultural Change,* 23(1): 123–132.

Shrestha, N.R.
1990 *Landlessness and Migration in Nepal*, Westview Press, Boulder, CO.

Shrestha, S. & Bhandari, P.
2007 "Environmental security and labour migration in Nepal, *Population and Environment,* 29: 25–38.

Siegel, J. & Swanson, D. (Eds)
2004 *The Methods and Materials of Demography,* Second edition, Elsevier Academic Press, Amsterdam.

Sierra, R. & Stallings, J.
1998 "The dynamics and social organization of tropical deforestation in Northwest Ecuador, 1983-1995", *Human Ecology*, 26(1): 135–161.

Sjaastad, L.
1962 "The costs and returns of human migration", *The Journal of Political Economy,* 70: 80–93.

Som, R.
1973 *Recall Lapse in Demographic Enquiries*, Asia Publishing House, New York.

Song, C. & Woodcock, C.
2003 "Monitoring forest succession with multitemporal Landsat images: factors of uncertainty", *IEEE Transactions on Geoscience and Remote Sensing*, 41(11): 2557–2567.

Song, C. et al.
2008 "Empirical evidence for impacts of internal migration on vegetation dynamics in China from 1982 to 2000", Sensors, 8: 5069–5080.

Stark, O. & Bloom, D.
1985 "The new economics of labour migration", *American Economic Review,* 75: 173–178.

Stark, O. & Lucas, R.
1988 "Migration, remittances, and the family", *Economic Development and Cultural Change,* 36(3): 465–482.

Stark, O. & Taylor, J.
1991 "Relative deprivation and migration: theory, evidence, and policy implications", in *Determinants of Emigration from Mexico, Central America, and the Caribbean*, Diaz-Briquets, S., Weintraub, S. (Eds), Westview Press, Boulder, CO.

Talbot, L.M.
1986 "Demographic factors in resource depletion and environmental degradation in East African rangeland", *Population and Development Review*, 12(3): 441–51.

Todaro, M.
1969 "A model of labour migration and urban unemployment in less-developed countries", *The American Economic Review,* 59: 138–48.

Tucker, C. et al.
2005 "An extended AVHRR 8-km NDVI data set compatible with MODIS and SPOT vegetation NDVI data", *International Journal of Remote Sensing*, 26: 4485–4498.

United Nations
2001a "Migration, population change and the rural environment", *World Population Monitoring 2001: Population, Environment and Development*, United Nations Population Division, New York: 41–53.

United Nations
2001b *World Population Monitoring 2001: Population, Environment and Development*, United Nations Population Division, New York.

United Nations
2008 *World Urbanization Prospects: The 2007 Revision*, United Nations Population Division, New York.

Unruh, J. et al.
2005 *Environmental Change and its Implications for Population Migration,* Kluwer Academic Publishers, New York.

Walker, R. et al.
2000 "Deforestation and cattle ranching in the Brazilian Amazon: external capital and household processes", *World Development,* 28(4): 683–699.

Walsh, S. et al.
2003 "Integration of longitudinal surveys, remote sensing time-series, and spatial analyses: approaches for linking people and place", in *Linking Household and Remotely Sensed Data: Methodological and Practical Problems,* Fox, J. et al. (Eds), Kluwer Academic Publishers: 91–130.

Walsh, S.J. et al.
2001 "A multiscale analysis of LULC and NDVI variation in Nang Rong District, Northeast Thailand", *Agriculture Ecosystems and Environment,* 85(1–3): 47–64.

Walsh, S. & Crews-Meyer, K. (Eds)
2002 *Linking People, Place, and Policy: A GIScience Approach*, Kluwer Academic Publishers, Boston, MA.

Walsh, S. et al.
1999 "Scale dependent relationships between population and environment in northeastern Thailand", *Photogrammetric Engineering and Remote Sensing,* 65(1): 97–105.

Warner, K. et al.
2009 *In Search of Shelter: Mapping the Effects of Climate Change on Human Migration and Displacement*, Columbia University, New York; Center for International Earth Science Information Network(CIESIN); UN University; and Care International. Available at: http://www.ciesin.columbia.edu/documents/clim-migr-report-june09_final.pdf.

White, M.
2005 "Internal migration", in *Handbook of Population,* Poston, D., Micklin, M. (Eds), Kluwer Academic Publishers/Plenum Publishers: 311–346.

Whitten, A.J.
1987 "Indonesia's transmigration program and its role in the loss of tropical rain forests", *Conservation Biology*, 1(3): 239–246.

Wolpert, J.
1965 "Behavioral aspects of the decision to migrate", *Papers and Proceedings of the Regional Science Association,* 15: 159–169.

Wood, C.
1982 "Equilibrium and historical-structural perspectives on migration", *International Migration Review,* 16(2): 298–319.

Wood, C. & Skole, D.
1998 "Linking satellite, census and survey data to study deforestation in the Brazilian Amazon", in Liverman (Ed.), op. cit.: 70–93.

Wood, C. & Porro, R. (Eds)
2002 *Deforestation and Land Use in the Amazon*, University Press of Florida, Gainesville, FL.

World Bank
1991 *Forestry,* Sector Report, The World Bank, Washington, DC.

World Bank
2005 *International Migration in Latin America: Brain Drain, Remittances and Development*, Wodon, Q. (Ed.), The World Bank, Washington, DC.

Zhu, J.
1998 "Rural out-migration in China: a multilevel model", in Bilsborrow, R.E. (Ed.), op. cit.: 157–186.

Zimmerer, K
1993 "Soil erosion and labour shortages in the Andes with special reference to Bolivia, 1953–1991: implications for 'conservation-with-development", *World Development*, 21(10): 1659–1675.

Zweifler, M. et al.
1994 "Land-use evolution in hill regions of the Dominican Republic", *Professional Geographer,* 46(1): 39–53.

Researching environmental change and migration: evaluation of EACH-FOR methodology and application in 23 case studies worldwide[1]

Dr. Koko Warner[2], Dr. Tamer Afifi[3], Mr. Marc Stal[4], Ms. Olivia Dun[5]

[1] The authors would like to thank Frank Laczko, Christine Aghazarm, Aleksandra Koluvija, Humaira Daniel, Thomas Loster, Janos Bogardi, Andras Vag, Jill Jäger, colleagues from the EACH-FOR Consortium, and the participants in the expert workshop on environment and migration held in Munich on 16–18 April 2008, for their useful comments and support in preparing this document.

[2] Head of Section on Environmental Migration, Social Vulnerability, and Adaptation; E-mail: warner@ehs.unu.edu

[3] Associate Academic Officer; E-mail: Afifi@ehs.unu.edu

[4] Research Associate; Email: stal@ehs.unu.edu

[5] Research Associate; Email: dun@ehs.unu.edu

1. Introduction

This chapter evaluates the fieldwork methodology of the EACH-FOR Project, identifying some of the challenges and opportunities of conducting research about the linkages between the environment and migration.

In recent years, policy makers and scientists have become interested in the dynamic links between migration and environmental change (Döös, 1997; Adger et al., 2001; Gunderson et al., 2002; Scoones et al., 2007; Galaz et al., 2008). Yet only recently have empirical observations begun to be gathered to underpin discussions about the increasing role environmental change is expected to play in decisions relating to migration. To contribute to the base of knowledge about the links between environmental change and migration, the European Commission co-sponsored the Environmental Change and Forced Scenarios (EACH-FOR) project to assess the impact of environmental change on migration at the local, national, regional and international level.[6]

This chapter has three purposes. First, the authors explore the methodology design in the first global survey of environmental change and migration. Section 2 describes how the nature of the research topic posed many challenges and trade-offs for research design, and how the project attempted to create a method that would produce comparable results. The second purpose of this chapter is to examine how field researchers implemented and used this methodology in the EACH-FOR Project. The chapter focuses on an analysis of the fieldwork methodology applied in investigating the 23 EACH-FOR Project case studies. The third purpose of the chapter is to explore the direction of

[6] The Environmental Change and Forced Migration Scenarios Project was a two-year-long research project within the Sixth Framework Programme (Policy-oriented research) of the European Commission (EC). Findings, case study reports, policy briefings and materials from the 2008 Bonn conference on environment and migration (EFMSV) can be found at the project website: www.each-for.eu.

future research, building on lessons learned from this initial fieldwork experience.

Motivation to develop a project to investigate environmental change and migration

In 2006, the European Commission funded this research consortium to support comparative research of factors underlying migration and refugee flows, including illegal immigration and trafficking in human beings. In response, seven organizations formed a consortium to investigate whether and how environmental change affects migration. The overarching purpose of the EACH-FOR Project was to provide a greater understanding of the role of environmental change in causing forced migration (both internal and international) and its related societal consequences.

The EACH-FOR Project was conceptualized from the beginning as a multidisciplinary study that aimed to undertake original empirical research through case study research complemented by desk studies, statistical and other information sources. The research consortium had the opportunity to create a set of comparable studies using a unified methodological research approach. It was hoped that using one methodological approach across almost two dozen case studies would create a set of internally valid results, with some degree of external validity, as well. This type of investigation had not yet been done, so researchers first conducted an analysis of literature and methodologies (see Vag et al., 2007). This survey revealed a diversity of individual case studies with some mention of migration or environmental change (ibid.). Yet a lack of consistent and comparable data on migration related to environmental change (Castles, 2002; Black, 2001) suggested that a qualitative approach to first describe and then analyse the diversity of migration patterns would be useful, given the parameters of the project that was in development in the autumn of 2006. Subsequent synthesis reports supported this approach (Boano et al., 2007; Brown, 2008; Kniveton et al., 2008; Piguet, 2008; McLeman et al., 2006).

Researchers undertook 23 case studies in the seven major regions of the world under diverse local conditions, looking at different types of environmental change, and in widely different cultural contexts. Section 3 discusses some of the practical considerations and shortcomings of the method in practice, and illustrates how local researchers in three of the case study areas (Mozambique, Niger and Viet Nam) managed the challenges of their complex assignment. Section 4 draws lessons from the EACH-FOR Project experience and considers how future empirical research attempts could build and improve on these experiences. Section 5 concludes the chapter with the hope that this discussion will facilitate future research efforts on the topic.

From the outset, the research team faced pragmatic and scientific constraints. It was clear that the project needed to encompass a scope of different academic disciplines including migration and demography, environmental sciences, geography, economics, history, sociology, etc. Migration experts and the literature supported the idea that multiple factors contribute to migration; that it is difficult to isolate these contributing factors from one another (i.e., to determine which factor had the more important role in contributing to migration (Black, 2001; Castles, 2002; Biermann, 2007; Boano et al., 2008). Environmental experts on the team, supported by the literature, emphasized that environmental change is not monolithic: diverse patterns and complex processes are involved in changing environments. It was also clear from the project budget and two-year lifespan that long-term observations (to assess environmental change processes, and how these might affect migration through time) were not possible.

These facts posed a frame of limitations within which the EACH-FOR Project needed to work to produce results. The project consortium was therefore looking for a set of methodologies that would allow for an ex-post observation (observation after the fact) of whether or not environmental variables affected migration and, if so, the processes through which this occurred. One implication of this was that an important part of the EACH-FOR Project´s work was to gather information about how people perceive the influence of environmental factors on their decision to migrate. The EACH-FOR Project was fortunate to have key resources in the form of data and partners, which helped it address some of the practical limitations of completing an ambitious two-year project on a complex research topic. The project had access to good statistical data sources and geo-information, as well as a partnership with the International Organization for Migration (IOM), and local institutions such as the Public Policy Research Center Almaty, the Institute of Demography Tajikistan, the Agricultural University of China, and the University of Auckland and Otago (New Zealand).

2. EACH-FOR methodology

The research design process of the EACH-FOR Project research followed the process described in table 1, and encountered methodological design issues at each step of the process. Section 2 describes how the project addressed the challenges involved in investigating links between environmental change and migration.

Table 1: The EACH-FOR Project research steps and design issues

Step	Description	Design issue
1. Hypothesis	• Discernable environmental signal in migration today. Null hypothesis: no discernible environmental signal in migration today.	• How to establish whether the environmental signal is discernible in migration patterns? • How to assess or measure environmental signals?
2. Variables	• Independent variable of interest: environmental change • Dependent variable: migration	• How to isolate the independent variable of environmental change? • How to determine that presence of independent variable caused dependent variable?
3. Intervention group and control group	• Intervention group is made up of people that will experience environmental change • Control group is made up of people that will not experience environmental change	• How to isolate control group that does not experience environmental change (independent variable)?
4. Introduce intervention	• Environmental change	• Impossible to control environmental change; need to carefully select case study countries.
5. Measure dependent variables in intervention group and control group	• Did migration occur when environment changed?	• How to prove that migration would not have occurred in the absence of environmental change?

Initial methodology design for global comparable fieldwork

The EACH-FOR Project began its methodology design by forming a hypothesis that its desk and field research would test: There is a discernible environmental signal in migration patterns today. The project considered its general hypothesis to hold true if fieldwork found empirical qualitative and quantitative evidence that migration occurred, in part, due to environmental factors. The failure to find migrants in whose mobility pattern environmental causes played no role or were negligible would negate the central hypothesis. The weakness of this hypothesis lies in the difficulty of determining a measure for 'discernible' and environmental 'signals'.

The project took the approach of asking experts, migrants and non-migrants about their perceptions of environmental factors and whether these factors had anything to do with the decision to migrate or not to migrate. Where participants answered positively, this was considered as evidence that environmental factors were perceived as having played a role in migration (discernible). The project's case studies investigated a wide range of environmental change variables, from natural hazards of a sudden and gradual nature, to longer-term processes. The hypothesis testing did not attempt to quantify the strength of the environmental variables, but rather attempted to determine whether they played a discernible role or not. The causes of migration are manifold and the project did not expect to find cases where environmental reasons were the sole driving factor behind migration (with the possible exception of extreme natural hazards). Instead of trying to find a 'pure' type of environmentally induced migrant, the project aimed to test a hypothesis of whether the environment is a factor in migration (particularly when the environment changes in ways that worsen the welfare of those dependent upon it). The set of resulting case studies provides insights into this phenomenon worldwide and contributes to more rigorous hypothesis building and testing in future work.

Variables: Environmental change and migration

The EACH-FOR Project´s dependent variable was migration (including a range from internal to international migration) and the independent variable was a set of environmental stressors (including a range of complex phenomena from sudden to gradual and creeping processes).

Lonergan described migration as "an extremely varied and complex manifestation and component of equally complex economic, social, cultural, demographic, and political processes operating at the local, regional, national, and international levels" (Lonergan, 1998). The project recognized the difficulty of attempting to explain the patterns and trends of migration, both international and internal, using only one approach or academic discipline. For example, push factors were frequently mentioned in fieldwork, as migrants sometimes mentioned declining livelihoods from farming at home, due to land degradation or erosion, and the sense that a combination of environmental and economic factors contributed to migration. It was recognized from the beginning that it might be difficult to interpret research results, as the literature has established that migration outcomes have multiple causal factors. Also, the existing data on migration are inconsistent, with much of the information based on international migration figures from census data that do not necessarily capture temporal or geographic dynamics of human movement (Kniveton, 2009; Afifi & Warner, 2008). This presented a situation in which it would be challenging to measure any difference between migration in the absence of the independent variable and migration in the presence of the independent variable (environmental change).

The EACH-FOR Project treated the independent variable as certain types of environmental change and employed a multi-case study approach to examine major types of environmental change and how they might affect migration. Taking environmental change as an independent variable was associated with at least three issues, from the outset. First, the set of environmental variables that make up 'environmental change' are difficult, if not impossible, to isolate from other factors driving migration. Similarly, it is not possible to control the independent variable in this kind of research attempt: there are likely few cases of migration where it is possible to fully exclude the environmental variables of interest. Environmental processes are ongoing and omnipresent in all migration or non-migration situations, making it quite challenging to devise a methodology that can accurately test the impact of environmental change on migration. The project relied on the relative importance that interviewees placed on environmental factors to begin to isolate the relevance of environmental change variables in the migration choice. Second, 'environmental change' is comprised of many different phenomena, spanning different geographical and temporal scales. Third, because the EACH-FOR Project placed such importance on an ex-post methodology, the ability of subjects to perceive change required them to be at a boundary where change could be observed – either a physical boundary, such as the desert noticeably advancing onto a subject´s field, for example, or a noticeable time boundary, such as a violent storm or an exceptionally dry period (in a time scale relevant to human memory).

In light of these methodological limitations, some scholars recommend not attempting this kind of research, or at least not this 'driver-focused' framing of environmental factors and migration (Black, 2001). However, the EACH-FOR consortium recognized the need to address the knowledge gaps and was given the opportunity to gather information from the field and report its findings back to a wider academic and policy-centred community. The EACH-FOR Project accepted these limitations and shaped itself as a scoping study that would contribute to the building of a basis upon which more rigorous studies could be undertaken.

Considerations in the research design

Aside from the issues discussed in defining hypotheses, central research questions and variables, the project faced several considerations about how to design the research approach. These considerations stemmed from the multidisciplinary nature of the research question, but especially the omnipresence and characteristics of the independent variable environmental change. This section discusses some of the key design issues for the project, such as how to address intervention and control groups, controlling the intervention, and measuring the dependent variable after the independent variable had been introduced.

Challenge of defining a control group

Both the literature and experts consulted in the methodology design phase emphasized that it would not be surprising to find that the environment was one of many factors that contributed to migration (Faist, 2007; Castles, 2002). The project struggled to find ways to isolate the independent variable(s) in order to create a research design that would develop comparable, internally valid results. The ideal design would have allowed for the isolation of intervention groups and control groups in every case study area, so that the hypothesis could be established or rejected in each case. Two practical difficulties arose. First, individuals could not be randomly assigned into two groups. Second, it was unclear as to how to isolate a control group that did not experience environmental change, in each field study. Some case studies offered conditions in which some parts of the country experienced a particular kind of environmental change, while other areas remained intact. Examples of these will be offered below in the context of specific case studies. A pre-test – an assessment of the migration situation before environmental change was introduced – was not possible, due to limited time and budget. More fundamentally, a pre-test was not possible under the fieldwork circumstances, where it was impossible to control the independent variable (environmental change).

This left the project with a significant design issue: Could the EACH-FOR consortium attempt to define a control group against which the intervention group could be compared? Without a control group in each case, how would the project know whether migration would or would not have happened, even in the absence of environmental change without a control group? In other words, without a control group that was sure not to have experienced environmental change, how could the project establish whether there was an environmental signal in migration patterns?

Faced with the challenge of defining meaningful control groups, the project instead defined eight central questions to guide interviews during fieldwork. It was hoped that the answers to these questions would aid researchers in determining the validity of the hypothesis in the absence of true control groups.

Central questions for fieldwork

A set of questions helped test the central hypothesis of the project, and guided the collection of data in desk study and fieldwork activities. These questions were formulated in a way that would avoid drawing a deterministic relationship between environmental degradation and migration, which was considered inappropriate for the topic area. The guiding questions were intended to identify cases where environment plays an important role as a contributor to population movement. The following questions served as the basis for all research efforts in the global survey, and were intended to create a comparable set of descriptions of how environmental factors interact with migration pressures in the 23 field studies.

1. Who is migrating away from situations of environmental degradation/change?
2. Where are environmentally induced migrants coming from and where are they going to?
3. Why have people migrated (i.e., what role has environmental degradation or change played)?
4. How does environmental degradation interplay with other social, economic and political factors in decisions relating to migration?

5. What might prevent people from migrating when they are faced with environmental degradation (i.e., what assistance was needed, what was lacking)?

6. Why do some people remain in areas of environmental degradation/change while others migrate (i.e., what are their coping/adaptation strategies and capacities)?

7. How does environmentally induced migration occur (e.g., choice of destination, networks used)?

8. What is the role of people's perception of environmental degradation in triggering them to move?

These eight questions provided a basis upon which individual case studies could build additional falsifiable hypotheses about the particular relationships between environmental factors in specific areas and migration trends there. EACH-FOR was conceived as an initial study upon which further extensive research would be built. Its case studies were intended to provide insights into the many possible hypotheses that could subsequently be formulated and tested.

Intervention and case study selection

The project aimed to study environmental change affects on migration, but could not in any way manipulate the independent variable in fieldwork. To address this challenge, the project carefully selected case study countries to ensure the presence of several different types of environmental processes and migration processes. The project design aimed to observe cases where both independent and dependent variables were present, in order to determine whether there was a discernible environmental signal in migration patterns (i.e., whether the independent variable affected migration).

Case areas were selected to create a snapshot of environmental processes and their possible interactions with migration. For example, case study areas with documented environmental problems of one or more of the following types were selected:

- extreme flooding
- desertification
- land degradation
- water shortages and drought

- the potential of sea-level rise
- industrial pollution.

This approach allowed the project to identify 'hotspot' countries with potentially high descriptive value, but it was noted that multiple environmental processes, as well as complex migration processes, could be going on in each country. The fieldwork was not able to cover the entire country exhaustively and some processes were cross-border. Some areas with underreporting of migration (especially internal migration), or areas with environmental degradation of a creeping nature that is not reflected in international databases, were possibly passed over in the case country selection.

Measuring the intervention and control group

EACH-FOR researchers tried to find ways to establish whether migration would not have occurred in the absence of environmental change. To test whether there was indeed an impact on migration when the environment became less hospitable, the project had a three-step procedure: first, desk research was undertaken to examine historical patterns of both environmental change and migration; second, expert interviews were conducted to help capture the dynamics of environmental change and how this might have affected human mobility in the past and current situation in a given case study; and, finally, a questionnaire was given to migrants and non-migrants who had stayed behind in areas with documented cases of environmental degradation.

This latter comparison of migrants and non-migrants was hoped to reveal answers to the central question of the project: what role has environmental degradation or change played in people's decision to migrate or not migrate? For those individuals that remained behind, the project was keen to understand what factors intervened to keep people from migrating, even when they faced environmental problems. It was hoped that this set of answers would help researchers verify or reject the hypothesis, and also help measure the environmental signal in migration patterns (especially in relation to other factors).

Confounding factors in fieldwork

The project managed several confounding factors that could threaten the validity of project findings. The general purpose of the methodology design activity in EACH-FOR was to provide guidelines that would help produce comparable data from fieldwork, while taking into account the diverse field conditions in the 23 case study areas. It was considered advantageous for the project to define the eight research questions, and then use a few standard methods complemented by methods tailored to local conditions.

The EACH-FOR Project involved multiple case studies, research teams and field workers, and it created a comparable questionnaire for both migrants and non-migrants, as well as guidelines for semi-structured expert interviews.[7] The questionnaires were pre-tested in an early case study and then adjusted and revised before all other case study work began. All investigators received field guidelines about how to work with participants/interviewees, record their results, and interpret the results (to ensure interpreter reliability). Yet the reality of fieldwork inevitably led some researchers to change wording to meet local conditions. Language translation further exacerbated the instrumentation confound, and complicated the interpretation and comparison of results from one case study to another. This created locally specific and useful case studies with a moderate degree of comparability.

Non-probability sampling technique

One of the most significant confounding factors was the possibility of selection bias (i.e., no control group, no random assignment and no control over assignment of participants to groups). The project specifications did not allow for a random sample of a large population of people. Field researchers were looking for people exposed to environmental problems in order to ask whether those problems affected the participants´ decisions about whether or not to migrate. In most cases, researchers were only able to interview people from a limited number of areas, due to time and budget constraints. Expert interviews and desk studies were used to help balance the sampling biases that would emerge in migrant and non-migrant questionnaires. A non-probability sampling method was chosen because it was suitable for research during which the population of interest is not fully visible and where accurately defining the population of interest

[7] The migrant and non-migrant questionnaires can be found online at: http://www.each-for.eu/index.php?module=project_outline.

is problematic. This sampling method is used frequently in sociological studies in hidden populations involved in sensitive issues, and in the study of human systems where factors with the most influence in a system are not necessarily those whose exact characteristics are known. This method fitted the nature of the problem: to better understand the impact of a little-understood variable (environmental change) on decisions relating to migration.

The snowball, or chain-referral, sampling method was used in the project. Researchers identified an initial set of relevant respondents in pre-fieldwork preparations. During field interviews, researchers requested that participants suggest other potential subjects who shared similar characteristics or had relevance in some way to the object of study. This second set of subjects was then interviewed, and also requested to supply names of other potential interview subjects. The process continued until the individual researcher felt that the sample was large enough for the purposes of the study (a minimum of 15 expert interviews and 30 migrant and 30 non-migrant interviews was performed in each case study). The limited amount of time for each case study – an average of seven weeks – prevented an exhaustive sampling.

The researcher for each case study was directly involved in developing and managing the initiation and progress of the sample. Each researcher sought to ensure that the chain of referrals remained within boundaries that were relevant to the study. Researchers were instructed to ensure that the initial set of respondents was sufficiently diverse so that the sample was not skewed excessively in any one particular direction (Tansey, 2006:12).

Controlling for threats to validity of project findings

The project considered several designs to help control for threats to validity and increase the internal and external validity of results. For example, the two-group post-test design was explored. In this design, the researcher has no control over assignment of participants. This leaves the static-group comparison design open to irresolvable validity threat. There is no way of telling whether the two groups were comparable at time 1, before the intervention, even with a comparison of observations 1 and 3. The researchers can only guess as to whether the intervention caused any differences in the groups in time 2. The short nature of the project (24 months) weighed against the nature of the independent variable (environmental change, which may happen abruptly or gradually or not at all, in a discernible way, in 24 months) and did not offer significant advantages over the ex-post-facto research design that was chosen.

An 'ex-post-facto' design and a 'static group comparison design' were particularly considered and, in the end, the 'ex-post facto' design was chosen. The reasons for this choice are discussed below.

The ex-post-facto design is used when a single group of people is measured on some dependent variable (migration) after an intervention (environmental change) has taken place. This research design is often used when it is impossible to manipulate the dependent variable (migration). The researcher tries to evaluate the experiment by interviewing people (observation) and assessing the impact of the intervention. This design involves no pre-test or control group – two characteristics that fit the situation of the EACH-FOR Project. Yet this research design makes it difficult to be sure that the observations from fieldwork are the result of some particular intervention (environmental change). In spite of this weakness, the ex-post-facto design has the potential to produce useful results based on numerous migrant responses about environmental factors that contributed to household migration decisions.

3. Field experiences: Viet Nam, Mozambique and Niger

Following the design stage, the EACH-FOR Project moved to the fieldwork stage to attempt to apply the methodology in a way that would produce a set of globally comparable case studies. This section examines some of the fieldwork experiences using this methodology in the case study countries portrayed in figure 1 below.[8]

Figure 1: EACH-FOR case study locations

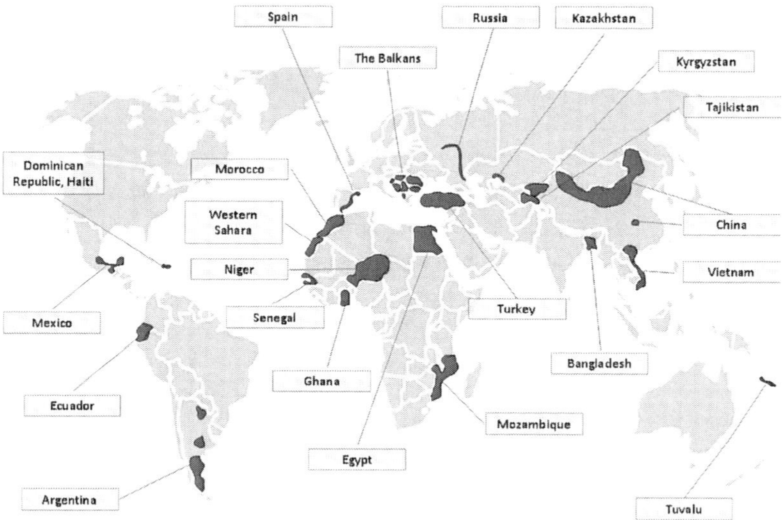

Experiences in Mozambique, Niger and Viet Nam highlight a number of complexities in achieving the project's goal of establishing whether linkages exist between environment and migration.

[8] For more detailed information about the 23 different case studies, see appendix, table 2: EACH-FOR case studies overview (regions/countries; environmental issues addressed; case study sites).

Case study descriptions: Mozambique, Niger and Viet Nam

Mozambique

The south-eastern African country Mozambique is one of the poorest countries in the world. With flooding as a focus, the Mozambique case study presents sudden disasters as a cause of displacement, with a minor focus on slow-onset environmental degradation within the EACH-FOR Project case studies.

In the last decade, several floods and tropical cyclones have devastated Mozambique and its low-lying deltas. These floods and tropical cyclones caused a large amount of population displacement in the country. People's livelihoods in Mozambique are particularly dependent on the natural environment, and a large majority of the population are subsistence farmers and fishers, especially those living along the Zambezi River where the impacts of floods and cyclone in 2001, 2007 and 2008 were felt.

The Mozambicans who lost their houses due to tropical cyclones in both 2000 and 2007 managed to stay in their places of origin and rebuild their houses with basic improvements in construction for storm resistance. On the other hand, the floods in Mozambique resulted in the displacement of thousands of people who had been living in the low-lying river areas. These low-lying river areas are not only high-risk areas for flooding, but are also the most fertile areas for agriculture. People not only lost their houses and belongings during the flooding; they also lost their harvest and, therefore, their means of livelihood.

The inhabitants of the Zambezi River Valley are displaced on a temporary basis, generally during the flood emergency period. Following recurring flooding events, people tend to be relocated on a permanent or semi-permanent basis. After the recurring flooding events in the last decade, the majority of the inhabitants of these low-lying river areas have resettled in government resettlement centres, where they attempt to stay. However, such temporary mass displacement has been observed to take on the characteristics of permanence.

The Government of Mozambique is trying to develop rural areas by providing the essential infrastructure and giving people incentives to produce more solid houses within the resettlement process. Nevertheless, resettlement does not seem to be the best option for dealing with the existing and upcoming impacts of environmental change in Mozambique. Resettlement is causing additional problems.

It is not solving the problems of the people who are, even after the resettlement process, still dependent on government and international aid and remain very vulnerable to further flooding.

The floods of 2007 and 2008 highlight the argument that government and international aid is indispensable. The flood-affected inhabitants are not able to survive on their own, as floods destroyed their living basis (their crops) for two consecutive years. If the flooding trend continues, people may be forced to move again if no government or international aid is available. In the future, it is uncertain whether people living in flood-affected areas would be able to survive without external support, especially in light of predictions of heavily increasing precipitation and extreme events in Mozambique.

The research process in Mozambique involved data collection via semi-structured expert interviews and semi-structured migrant interviews at resettlement centres in the Zambezi River Valley. Topic-related data, such as reports, images and relevant literature, were also gathered.[9]

Niger

The West African country Niger offered the opportunity of researching the linkages between slow environmental change processes and migration. Niger has been suffering from frequent and strong droughts over the past century. The strongest droughts that hit the country occurred in 1973 and 1984. Although more than three decades have passed since then, people are still suffering from the consequences of these droughts, which had a negative impact on soil quality.

The ongoing droughts harm the livelihoods of people, especially farmers and cattle herders, who must then seek new means of generating income. Consequently, people have been cutting down, chopping and selling trees as firewood. This causes further deterioration of soil quality and contributes to sand intrusion and siltation of local waterways. In particular, siltation and water pollution, which hinder fish reproduction, are problematic for fishermen who make their livelihoods along the Niger River.

[9] For further information about the Mozambique case study, please refer to Stal, 2009.

The shrinking of Lake Chad was another environmental problem investigated within the Niger case study. Lake Chad, which formerly extended into Niger, no longer exists within that country's territory. This environmental change has led in different ways to the migration of the people whose livelihoods mainly depend on the environment.

The main region visited in Niger was Tilabéri, in the south of the country, which also includes the capital, Niamey. The researcher personally interviewed migrants who had moved to the region of Tilabéri from other regions within Niger. Migrants who had left Niger because of the shrinking of Lake Chad and were temporarily visiting Niamey were also interviewed in person. The shrinking of Lake Chad also prompted people to move to other regions of Niger, and some of these migrants were interviewed by telephone. Semi-structured interviews with experts were conducted in the capital of Niamey.[10] A general overview of the field research outcomes is given in the following paragraphs.

Those who migrate are usually young men, while those left behind are, in most cases, women, children and the elderly. Women rely on small businesses to feed their family members and sometimes even work in other farmers' fields, as a last resort – a sign of desperation. In many cases, women attempt to restore the environment by planting new trees and fixing sand dunes. However, taking care of the children and the elderly limits their efforts. In general, women use their knowledge of edible plants to sustain themselves and their families in the deteriorating environment.

The link between environmental degradation and migration in Niger is, in most cases, indirect. The migrants usually name economic factors, such as decreasing income and unemployment, as reasons for migration. However, the root causes are environmental problems that lead to migration through these economic mechanisms. Most of the environmentally induced migrants interviewed did not want to give up farming, cattle herding or fishing. Instead, migration occurs in a step-by-step process as environmental degradation processes drive people short distances in search of places to pursue their livelihoods. In cases of more advanced environmental degradation, they may leave the region. In the most severe cases, they leave the country for other

[10] For further information about the Niger case study, please refer to Afifi, 2009.

countries in the South, where they practise similar livelihood patterns. Others change their work and take random jobs in the capital or move to Libya and mainly work in the construction sector before returning to Niger.

Environmentally induced migration in Niger is often a survival strategy following the loss of livelihood due to the deteriorating environment in the villages/regions of origin.

Viet Nam

The South-East Asian case study of Viet Nam focused on the linkages between regular flooding events, migration and resettlement in the Mekong Delta region. The research for the Viet Nam case study evolved distinctly through the field research phase to become a study focused on the linkages between regular anticipated flooding events and migration/displacement, as opposed to a study between flash flooding events and migration/displacement.

The Mekong Delta is one of the most densely populated areas on earth (MRC, 2005), and the slow and regular flooding of the Mekong River is considered an integral part of the livelihoods of the Vietnamese population living in the Delta (Be et al., 2007). The enriched flood plain constitutes approximately 40 per cent of the cultivated land in Viet Nam and is known as the 'rice bowl' of the country. Regular flooding of the Mekong River occurs annually between July and November and affects 40–50 per cent of the land area across nine provinces. Over the past four decades, the frequency of major floods in the Mekong River, which would normally only have a return period of 50 years, has been a major concern (Lettenmaier, 2000 in White, 2002: 11) and flood patterns for the Mekong Delta show a worsening trend (Be et al., 2007).

The research for this case study was conducted in four areas: An Giang Province, Ho Chi Minh City and Hanoi within Viet Nam as well as Phnom Penh, Cambodia. Expert interviews were conducted in all four locations, while interviews and questionnaires with migrants and non-migrants took place in An Giang Province, Ho Chi Minh City Province and Phnom Penh only. An Giang Province was selected as the site of possible migrant origin within the Mekong Delta because it is the province that experiences the highest level of flooding during the annual wet season. Ho Chi Minh Province and Phnom Penh, as major urban centres, were selected as destinations for migrants from the Mekong Delta.[11]

Where flooding was an underlying trigger in people's decision to migrate, the research identified different types of people on the move, including children and poorer households seeking refuge in cities after the annual flood, youth (especially girls) vulnerable to human trafficking, single household members carrying out seasonal work in different locations or whole households seeking alternative livelihoods in a new location. It was not possible to identify conclusive patterns that revealed where environmentally induced migrants were coming from or where they were going, due to the lack of systematic sampling or review of census data. However, the research did reveal that initial trends indicated that those affected by flooding tended to move from rural locations to urban areas or peri-urban areas, where there were perceived to be greater employment and livelihood diversification opportunities or social ties indicating that pull factors also play a role in determining the destination of environmentally induced migrants.
In the case of Viet Nam, the anecdotal research revealed that those who had moved because of flooding had generally experiencing repeated flood events as opposed to a one-off flood event. In general, it was the negative impact of repeated flooding, leading to livelihood threats, loss and/or debt that triggered the decision to migrate. People migrated when their main source of livelihood was destroyed by flood – for example, the loss of a rice crop, leading to loss of income and debt – and when disaster aid relief expired. This indicated that alternative livelihood and income options, crops that could withstand flooding and its impacts, and disaster relief aid may have prevented people from migrating. Overall, this indicated that the socio-economic conditions and political context played an important role in influencing the migration outcome.

It was revealed generally that those who stayed in areas of environmental degradation possessed land, had lived there for generations and generally wanted to stay in their usual place of residence and were able to find a means of surviving there. There were also poorer people who remained in areas of environmental degradation because they did not have the means or resources to move, despite expressing a

[11] For further information about the Viet Nam case study, please refer to Dun, 2009.

desire to do so. Overall, the research in Viet Nam was preliminary but revealed the need to make a more systematic comparison between households within the same village, affected by the same environmental change, in order to understand why some choose to migrate and others do not. This would provide greater insight into the role of factors such as level of income, assets, access to resources, information, social networks and capital in preventing or prompting migration in the Mekong Delta area.

EACH-FOR methodology in practice, lessons learned

In all three case studies (Mozambique, Niger and Viet Nam), IOM local in-country offices were engaged as partners prior to the fieldwork commencing. They played a crucial role as locally-based partners in identifying and establishing initial contact with relevant experts, arranging logistics and assisting with translation and implementation of the EACH-FOR questionnaire. A key contact point within IOM was established as the main interpreter and assistant for the duration of the field research in all three locations.

Multiple field objectives and tasks

While there was one overall field research goal (that of investigating whether there were linkages between the environment and migration), there were several field research objectives, as mentioned above in section 2. The focus on whether there was a link between environment and migration meant that the field research did not strike a balance in terms of exploring the range of other reasons why people migrated. The project attempted to address each objective with the questionnaire, semi-structured migrant and non-migrant interviews and interviews with experts. Yet each of the field objectives might require a specific type of methodology.

The fieldwork was a scoping exercise, but one that did not allow for repeat visits to follow up on information gathered. Researchers also had the difficult task of conducting both expert interviews, migrant and non-migrant interviews and questionnaires within the field research phase. There was little time for analysis of expert interviews before the researchers interviewed migrants and non-migrants. At times, for example, the researchers obtained crucial pieces of information from experts, following interviews with migrants – information that would have been useful to know before interviewing migrants.

Differing perceptions and definitions of 'environment'

In conducting this research, it was important to clearly and distinctly define what was meant by the term 'environment'. The EACH-FOR Project researcher and the local IOM research partner needed to have a common understanding and focus for the research.

It was also important for the experts, migrants and non-migrants being interviewed to have a common understanding of the term so that they could provide relevant information. As researchers, this created a dilemma, because providing an explanation of the term 'environment' influenced the respondents' answers. For instance, in the case of Niger, the local partner had to provide clear examples for environmental problems, such as droughts, water pollution and sand intrusion. This might have oriented the interviewees towards discussion about these problems, and they may have neglected to highlight other factors influencing their migration decision.

As for the expert interviews in the case of Viet Nam, it became important to define not only what environment meant but also what was meant by the term 'flooding' in the particular research locality. For example, flooding in northern and central Viet Nam is commonly flash flooding, with little warning, while in southern Viet Nam the type of flooding experienced occurs slowly and regularly. Since many of the experts interviewed were located in the capital city, Hanoi (in northern Viet Nam), they often had a different concept of flooding to that of the researcher, especially those from humanitarian relief agencies. Additionally, in the research analysis phase, it was important to take account of the background and perception of the experts interviewed.

In Viet Nam, it was mainly those that indicated they had suffered negative consequences from environmental degradation that were identified as having migrated due to underlying environmental factors. Since our focus was on identifying people who had been triggered to move because of environmental degradation, research naturally tended to focus on, and be framed around, the negative impacts of environmental degradation.

Differing perceptions and definitions of 'migration'

The questionnaire, semi-structured migrant interviews and expert interviews were useful for addressing the field research question of why people had migrated (i.e., what role environmental degradation or change had played). These qualitative techniques allowed for spontaneous responses from those interviewed, with a view to identifying the broader range of reasons why people had migrated. When those interviewed were specifically asked, in the questionnaire, why environmental problems had played a role in their decision to move, their answers helped to shed light on the various ways in which the environment played a role in the decision to move.

The EACH-FOR researchers were interested in all types of movement linked to environmental issues, such as seasonal or permanent migration as well as internal and cross-border movement. During some interviews in Viet Nam and Niger, experts often assumed that the researchers were examining the links between environmental issues and international migration. The researchers had to explain that they were interested in all types of human mobility linked to the environmental issues under investigation. It was also necessary for researchers to be aware of different interpretations of the term 'migration' in the research analysis phase for all three case studies.

The literature review performed by EACH-FOR found no clear distinctions of environmentally-induced migration and displacement. In the early stages of the EACH-FOR Project, working definitions of 'environmentally displaced person'[12] were compiled. These definitions contributed towards the conceptual context of the EACH-FOR case study research and helped to establish consensus among EACH-FOR Project members regarding definitions for people who migrate or are displaced by environmental factors.

In some cases, researchers investigated the issue of resettlement due to environmental factors – for example, in Mozambique and Viet Nam. In other case studies, the researchers interviewed those who had migrated independently of any formal programme. Lacking clear definitions of environmentally induced migration reduced the comparability of case studies. However, using the working definition of 'environmentally induced migration' also allowed for the broader spectrum of issues relating to human mobility and environmental change to emerge.

In Mozambique and Viet Nam, where the phenomenon of migration related to disaster-type environmental problems (more commonly,

[12] The working definitions for the EACH-FOR Project can be found online at: http://www.each-for.eu/documents/Environmentally_Displaced_Persons_-_Working_Definitions.pdf

sudden events) was investigated, it was important to distinguish between temporary evacuation during the 'environmental' event (such as temporary local evacuation due to high flood water levels) and longer-term movement after the 'environmental' event. This was not the case in Niger, since the environmental changes under examination were slow and progressive.

Identifying experts to interview

According to criteria outlined by the EACH-FOR researcher, IOM was requested to identify experts who worked in the migration, environment and humanitarian sectors, with a view to interviewing them. Organizations such as the International Committee of the Red Cross, Oxfam and CARE International, as well as government agencies, national universities and research institutes, helped identify additional experts or migrants for interview. This proved to be very beneficial, allowing for rapid access to experts and reducing logistical boundaries given the short time available for the field research.

In most case study areas, it was rare to find specific experts in the area of 'environmental migration'. Some experts were reluctant to be interviewed when the researchers approached them explaining that the research was about environment and migration linkages, since the experts felt this was not their area of expertise. However, once the researchers explained that they were interested in knowing about their particular expertise, the experts agreed to be interviewed.

One issue that emerged as a result of relying on the IOM office of Viet Nam as the initial contact point was that not so many experts were known to them in the environment sector (since migration is their main area of expertise). This could be seen as a disadvantage but it also became advantageous as it provided a point of departure for discussion of what exactly was meant by the term 'environment', which helped to refine the research. For example, this occurred after the first expert interview was conducted with an IOM contact in the environment sector. This contact was very knowledgeable about a variety of environmental issues but was in charge of Waste Management Services within the Department of Natural Resources and Environment of Ho Chi Minh City. The discussion in this interview revolved more around pollution of water ways rather than flooding. What became evident after this interview was that a clearer definition of 'environment' needed to be provided to interviewees.

Fortunately, the IOM office in Maputo, the capital of Mozambique, had been actively involved in the evacuation and resettlement process along the Zambezi River Valley during the 2001 and 2007 flooding events. They were therefore familiar with the environmental context and provided several relevant contacts with environmental expertise. In the case of Niger, the staff at the Niamey IOM office recognizes environmental causes as factor in migration. This is because environmental problems in Niger are numerous, present and dominant. Moreover, there is a historical tradition of nomadic and farmer migration.

Finding migrants: source–destination dilemma

A common question when one is trying to conduct research on environment–migration linkages is whether to begin researching in the source location of potential migrants (i.e., the location of environmental impact) or the destination location of migrants. In the three EACH-FOR case studies discussed here, there was a tendency to first head for the zones where there were environmental problems, rather than conducting research in locations where people migrated. This was because of the limited time available for conducting field research, which meant that less time was spent concentrating on interviewing migrants – for example, people in urban areas who had migrated from the countryside.

Approaching the research in this manner facilitated the identification of individuals who had been affected by changes in the environment. Thus, the researchers faced fewer challenges engaging in discussions about people who had migrated or been resettled due to environmental factors. On the other hand, by spending less time conducting interviews with migrants in localities at a distance away from the environmental problem under investigation (for example, in urban centres), the researchers were not able to cover *all* of the reasons why people migrate, and thus faced limitations in assessing the role of environmental factors in migration decisions.

In 12 semi-structured interviews with migrants who had moved from the Mekong Delta to Can Gio District, Ho Chi Minh Province, only one migrant interviewed revealed there was a connection between flooding and his decision to migrate. Migrants interviewed were selected by the Ho Chi Minh City Department of Agriculture and Rural Development (DARD), based on the researcher's request to speak with migrants who had moved the Mekong Delta to Can Gio District within the previous ten years. The majority of those interviewed had migrated from the Delta to Ho Chi Minh Province for economic and social reasons. However, when the researcher was conducting research in the localities of the Mekong Delta where flooding occurred, there were clearer indications that people were moving or being relocated due to the impacts of flooding. This suggested that geographic proximity to the environmental problem under investigation played a role.

This experience was repeated in Mozambique. In order to find migrants in Mozambique, a field trip to one source of the flooding events, the Zambezi River Valley was carried out. The field trip included visits to 13 resettlement centres, where interviews with displaced people were conducted. The visit started in the Chinde District, Zambezi Province, in the delta of the Zambezi River Valley. Here, several resettlement centres were visited, expert interviews with non-governmental organization associates were conducted and informal conversations with government representatives were held. The field trip carried on along the Zambezi River Valley further inland to the Mutarara district. On the way from the Zambezi River delta to the Mutarara District, several resettlement centres were visited and interviews with displaced persons were conducted ad hoc along the way. Most of the interviewees responded similarly and all of them mentioned the flooding as the main cause of their resettlement.

In Niger, some people moved within rural areas, but when the environmental problems threatened their livelihoods, they decided to migrate to the capital city and completely changed their activities from farming to unskilled work – for example, selling mobile-phone cards. Other people from Niger affected by environmental problems even moved to Libya. However, people who insisted on continuing their farming and fishing activities eventually moved to neighbouring countries, such as Benin, Chad, Mali and Nigeria. The main challenge here was that the researcher was not able to trace the migrants to all the countries. It was only possible to know about them from their friends and relatives who stayed in the country. The only migrants who crossed the borders and were personally interviewed by the researcher were those who returned as part of the re-integration programme of IOM (returnees from Libya) and some others who left the Lake Chad Region in Niger for Nigeria and came for short visits to Niamey. Telephone interviews were also conducted.

Selecting migrants for interview

Due to the time limitations and exploratory nature of this research project, the migrant selection was not systematic. Migrants were interviewed when opportunities were created as a result of expert interviews or by random selection of individuals in the field sites. This proved to be both beneficial and restrictive; in all three cases, interviews conducted with randomly selected individuals in the field locations proved to be useful in terms of highlighting issues that would not have been anticipated by a more systematic selection of migrants. In the case of Niger, the researcher attempted to divide the migrants into the following categories: migrants who left the country and did not return (according to their friends or relatives or to conversations with said migrants during their short visits to the country); migrants who left for Libya and returned; and migrants who left one village/region and moved to another. There was a fourth category of non-migrants. It is difficult to ascertain whether the samples were entirely representative, due to the small number of interviewees in each case, which is an important limitation of the study. Nevertheless, judging from the similarity of answers of each group within one sample, one can tell that the samples were reliable at this stage of the study.

Contacts provided by experts facilitated rapid access to the potential target community under study. In Viet Nam, for example, some migrants interviewed were those to whom the researcher was granted access through the Department of Agriculture and Rural Development (DARD). The researcher was able to ask DARD to identify migrants who had moved from the Mekong Delta within the last decade, with a view to interviewing them. This provided an organized approach to selecting migrants but was not sufficient to determine systematically whether or not the environment was a significant push factor for migration. This is where time and budget limitations became a restricting factor for the researcher in Viet Nam. Had there been more time, the researcher would have chosen to focus on interviewing a more representative sample of migrants by selecting those who had moved from a particular district and province of the Mekong Delta affected by flooding.

Random selection of migrants in Viet Nam was useful in identifying a broad spectrum of people on the move who were affected by flooding – for example, prior to the field research, the researcher had not considered children in child-care centres as a potential target research group for examining environment–migration linkages. However, due to the selection of migrants being random, it was difficult to know if the categories of people identified above were an anomaly or representative of people prompted to move due to underlying environmental factors.

IOM Niamey provided an excellent opportunity for the researcher to interview people who left Niger (especially those who went to Libya) and came back as part of a re-integration programme. The researcher was concerned about a possible lack of trust on the part of the interviewees; this category of returned migrants is financially supported by IOM, on the condition that they stay in Niger and re-integrate into its society. Therefore, during the interview, they might, for example, have denied that they would be leaving the country for environmental reasons, in order not to be deprived of the financial support they were expecting from IOM. As a result, this sample was taken as an indicator of the reasons behind the decision to migrate for those who had already migrated and returned home, rather than as an indicator of their future plans, which cannot be determined in advance.

A different category of people that the researcher met via the IOM office were those who left the Lake Chad region within the territory of Niger for other countries, such as Nigeria, and made short visits to Niamey. These people passed by the IOM office, either because they had good contacts there or because they were phoned by the researcher's assistant and asked for an interview. Without the IOM contact, it would have been hard for the researcher to interview this category, since the field visit was limited to Tilabéri and not the Lake Chad region or other countries sharing the lake with Niger.

In Mozambique, selecting appropriate migrants for interviews was less complicated, because there was an organized resettlement programme for people displaced by floods. Therefore, mainly resettled people along the Zambezi River Valley were selected for migrant interviews. This was beneficial, as the linkage between flooding and displacement was clear, but other migrants who had moved for various reasons or who had moved further distances could not be identified, given the limited amount of time the researcher spent in the field. With more time and budget, comparative research of people displaced by flooding events in other parts of the country could potentially provide a deeper understanding of different coping strategies.

Timing issues: Gap(s) between environmental events and site visits

Researchers found that the more recently a particular locality had experienced the environmental problem or issue under investigation, the more people in that locality were aware of the situation. Therefore, they were able to discuss their recent experience as a factor leading to livelihood impacts and possible migration or resettlement. For example, in the EACH-FOR Mozambique case study, communities in flood-prone areas had experienced the 2007 flooding event six months prior to the researcher's visit. On the other hand, in the Viet Nam case study, the last major disaster flood in the Mekong Delta occurred in 2000. Considering that the research was conducted seven years later, it was more difficult to pinpoint the exact migratory impacts of that particular event, despite the fact that people could clearly remember the event.

In addition, the personal experience of a migrant or non-migrant with the environmental event in the past also played a significant role in shaping research findings. For example, in Mozambique, interviews were carried out with people who had experienced multiple flooding events. This revealed a change in attitude towards migration or resettlement from temporary evacuation to permanent resettlement. The affected people accepted the fact that they should not move back to their places of origin. Similarly, in Viet Nam, the researcher found that the impacts of repeated flooding (as opposed to a unique flooding event) could play a crucial role in prompting people to migrate.

Since slow-onset environmental change is a constant part of life in Niger, the time gap between environmental events and the site visit did not play a significant role in influencing the results of migrant interviews. Even in the cases of the severe droughts of 1973 and 1984, farmers and cattle herders have continued to suffer from the cumulative impacts of those events.

Another timing issue that had to be taken into account was the daytime and seasonal timing, especially for rapid research, for interviewing migrants/non-migrants in the field location. For example, the researchers in Mozambique and Niger found that the farmers and cattle herders were only available in the villages at certain times of the day and the year. Likewise, when the farmers living along the Zambezi River Valley were busy seeding their almost inaccessible fields, it was mainly the elders and the children who were interviewed. In the case of Niger, the researcher visited the country in the dry season (January/February 2008), during which many farmers wander with their cattle to other regions that were not covered in the field visit. Therefore, the researcher had to rely on interviews with people who stayed in the Tilabéri region, as well as people in the capital.

4. Recommendations for future environment–migration research

EACH-FOR gathered empirical observations and contributed to the building of a research agenda for investigating the complex relationships between environmental factors and migration (Warner & Laczko, 2008a). Future research will certainly improve on these efforts and increase the level and quality of information available about these relationships. Some considerations may help guide the design of research methodologies for further investigation.

Emerging toolkit and methods

A spectrum of emerging tools for further research is available – from macro-level to micro-level approaches. We have identified several possible approaches from recent literature, although many more exist. To gain a global or regional overview, Perch-Nielson (2008) and others illustrate the possibility of linking climate and environmental models with migration models. Other researchers have begun introducing environmental variables into geographic regression models (Neumayer, 2005; Barrios et al., 2006; Afifi & Warner, 2008), including a multi-level approach that simultaneously uses area and individual data (Henry, 2004). A particularly promising area appears to be simulation with agent-based modelling (Kniveton & Schmidt-Verkerk, 2008).

To gain more comparable and geographically specific insights, longitudinal research needs to be undertaken with panel studies of the evolution of the environment and of the migration behaviour (Massey, 2007). Looking back, historical analogues can provide insights into coping mechanisms and tipping points beyond which coupled human–ecological systems began to break down (Piguet, 2008; Mc Leman, 2006). Other authors have, with reasonable success, developed indices of vulnerability to localize 'hotspots' of environmental change and migration (Erhart et al., 2008; Dasgupta, 2007). Finally, empirical fieldwork such as that performed by EACH-FOR, particularly using a combination of survey and ethnographic study methods, will provide detail and help build hypotheses about the relationships between the multiple factors affecting migration.

Methods and project design are rarely free from considerations of funding. Ideally, funding will become available for longer-term longitudinal studies that would allow for the establishment of a globally comparable set of data on environmental change and migration. Case studies could continue to provide rich localized insights that may be aggregated and compared in meta-analyses.

Research questions and hypotheses

There is a need for methods and approaches that isolate the role of environmental variables in the decision to migrate. For example, it is widely accepted in the literature that environmental factors usually accompany other factors (particularly economic) that contribute to the ability of poor people to sustain their livelihoods. Environmental problems often accompany economic factors that affect migration, but economic factors do not necessarily accompany environmental problems in affecting migration. The results from the EACH-FOR Project point in one direction. Future research may find effective ways to begin isolating environmental variables in certain cases from others, particularly economic. A case study in which environmental variables were present, but economic variables were not clearly present as drivers of migration, would be a starting point in developing methods for better isolating the role of environmental variables in migration. This type of research would also strengthen the ability to create falsifiable hypotheses.

Another way to help isolate environmental variables in the migration decision would be to test different typologies of environmentally induced migration (see, for example, Renaud et al., 2007 and 2009). It would be useful to have a clearer understanding of the role of sudden and gradual or creeping environmental processes in migration processes. Defining the parameters for environmental change (natural catastrophes, output of the land, desertification, etc.) would increase the comparative strength of empirical research. For instance, a research project that looked at comparable areas with desertification and without desertification could form hypotheses about whether or not migration would occur in both locations to the same extent, and then test the hypothesis through empirical study. Use of objective information sources, such as satellite imagery, may provide an additional means of improving the validity of research results from subjective sources, such as migrant interviews.

Empirical research, promising approaches

The review of the EACH-FOR Project methodology and actual use of the method in conducting research into the linkages between environment and migration reveal some lessons relevant for future research. Four core components should be defined in the design of future research efforts.

It is crucial to have a **clear definition** of the terms 'environment' and 'migration'. It is important to have a clear understanding of how the research subjects (such as migrants or expert interviewees) will **perceive** these definitions and concepts. Time and distance have been important factors in the perception of these definitions. The scale at which the research is conducted, in terms of time and geographical distance, must also be considered.

By '**time**' we mean that a researcher must consider whether, and how recently, the migrant (person being interviewed) has experienced the environmental problem under study. (The interviewer might, for example, ask the migrant when he/she last experienced the flooding event or the impacts of land degradation and what the magnitude of that event was.) Cognitive dissonance has been noted in studies of people's perception of their risk due to natural hazards. If the event of interest has not occurred for some time, awareness about the potential risk tends to be low. Participants tend to forget the role that environmental factors may play in influencing migration decisions, if those environmental factors have not been perceived for some time (in other words, they are temporarily forgotten). Future research could uncover useful additional information about the role of perception in the decision to migrate in the presence of environmental problems.

By '**geographical distance**' we mean that the methodologies engaged by a researcher will vary depending on how far away the migrant being interviewed is from the environmental phenomenon under study. Participants´ perceptions of environmental issues may be more pronounced if they are in closer physical proximity (or if some other factor has raised awareness of the issues). The closer people are to an area impacted by the environmental event, the more likely they are to understand the environmental phenomenon a researcher is referring to during an interview. A researcher interviewing people at a geographic distance from the environmental event under study will have to carefully consider ways of both assessing whether interviewees perceived environmental change, and identifying whether the environmental factor played a role in the decision to migrate.

5. Conclusions

Interest in the environmental drivers of migration is growing, yet empirical research remains limited. The EACH-FOR Project represented the first major, global attempt to explore a set of hypotheses in fieldwork, and provided a valuable point of departure for further research. Some of the most significant results of the project were that it created an overview of patterns of environmental change and migration in different types of ecosystems worldwide – from drylands and small island developing states, to deltas, mountain areas, and flood-prone areas (Warner et al., 2009). The 23 EACH-FOR case studies provided insights about ways that environmental factors affect human mobility – from sudden natural hazards, such as flooding and storms, to slow-onset phenomena, such as desertification, sea-level rise and other forms of land and water degradation that often affect migration as a result of livelihoods being affected.

EACH-FOR confirmed the observation that environmentally induced migration has the potential to become a phenomenon of a scale and scope never experienced in human memory. Given an inconceivable scale of environmental change, migration may be an adjustment mechanism of first resort, or a survival mechanism of last resort (Laczko, 2008; Bogardi & Warner, 2008). Migration may be an adaptation mechanism for those with the resources to move early and far enough away from danger. Alternatively, in extreme cases and for those with fewer means to move, migration may be an expression of failed adaptation (Renaud et al., 2007; Afifi & Warner, 2008).

The debate about environmental change and migration is still emerging; it is not yet clearly framed. As the dialogue on the topic increases, however, there will be a struggle to define the debate in terms of security, humanitarian action, development, disaster management and climate change. While the parameters are still undefined more empirical research and data will be needed in order to better understand the interactions between environmental factors and human mobility, as well as the intervening factors that affect migration or non-migration when environmental changes occur. These questions are of paramount policy importance in the current context of climate change, and will continue to grow in importance in coming decades. Empirical research has a role to play in helping to support and shape effective policy.

6. References

Adger, W.N. et al.
2001 "Environment, society and precipitous change", in *Living with Environmental Change: Social vulnerability, adaptation and resilience in Vietnam*. Routledge, London.

Afifi, T. & Warner, K.
2008 *The Impact of Environmental Degradation on Migration Flows across Countries*, Working Paper No. 5/2008. UNU-EHS, Bonn. Available at: http://www.ehs.unu.edu/article:476?menu=94

Afifi, T.
2009 Niger Case Study Report for the Environmental Change and Forced Migration Scenarios Project. See: www.each-for.eu

Barrios, S. et al.
2006 "Climatic change and rural–urban migration: The case of sub-Saharan Africa". *Journal of Urban Economics,* 60(3): 357–371.

Barnett, J. & Adger, N.
2007 "Climate Change, human security and violent conflict", in *Political Geography*, 26: 639–655.

Bates, D.C.
2002 "Environmental refugees? Classifying human migrations caused by environmental change", in *Population and Environment 23(5)*: 465–477.

Be, T.T. et al. (Eds)
2007 *Literature Analysis. Challenges to Sustainable Development in the Mekong Delta: Regional and National Policy Issues and Research Needs*. The Sustainable Mekong Research Network (Sumernet), Bangkok.

Biermann, F.
2007 "Earth system governance as a crosscutting theme of global change research". *Global Environmental Change*, 17: 326–337.

Black, R.
2001 "Environmental r*efugees: myth or reality?" In* New Issues in Refugee Research. Working Paper No. 34, University of Sussex, Brighton.

Boano, C. et al.
2008 *Environmentally displaced people: Understanding the linkages between environmental change, livelihoods and forced migration. Forced Migration Policy Briefing 1*. Refugee Studies Centre, Oxford Department of International Development, University of Oxford.

Brown, O.
2008 *Migration and Climate Change*. IOM Migration Research Series No. 31. International Organization for Migration, Geneva.

Castles, S.
2002 "Environmental change and forced migration: making sense of the debate". In *New Issues in Refugee Research*. Working Paper No. 70. United Nations High Commissioner for Refugees (UNHCR), Geneva.

Conisbee, M. & Simms, A.
2003 *Environmental Refugees*. The Case for Recognition. New Economics Foundation, London.

Döös, B.
1997 "Can large-scale environmental migrations be predicted?" in *Global Environmental Change*, 7(1): 41–61.

Dun, O.
2009 *Linkages between Flooding, Migration and Resettlement, Vietnam Case Study Report for the Environmental Change and Forced Migration Scenarios Project*. See: www.each-for.eu

El-Hinnawi, E.
1985 *Environmental refugees.* United Nations Environment Programme, Nairobi.

Erhart, C. et al.
2008 Humanitarian implications of climate change. Mapping emerging trends and risk hotspots: Hazards associated with climate change: using GIS to identify areas of high humanitarian risk, CARE International, Geneva.

Faist, T.
2007 Personal communication on the validity of EACH-FOR methodology and field questionnaire. 14 August 2007.

Flintan, F.
2001 *"Environmental refugees – a misnomer or a reality?"* A contribution to the Wilton Park Conference Report on Environmental Security and Conflict Prevention, 1–3 March 2001.

Galaz, V. et al.
2008 *Ecosystems under pressure. A policy brief for the International Commission on Climate Change and Development.* Stockholm Resilience Centre, Stockholm Environment Institute, United Nations University Institute for Environment and Human Security. Stockholm.

Gunderson, L. H. & Holling, C.S.
2002 *Panarchy – Understanding Transformations in Systems of Humans and Nature*, Island Press, Washington, DC.

Homer-Dixon, T.F.
1994 "Environmental scarcities and violent conflict", I*nternational Security*, 19: 1–40.

Hussein, K. & Nelson, J.
1998 *Sustainable livelihoods and livelihood desertification.* Working Paper 69, Institute of Development Studies, Brighton.

IOM
2007 *Discussion Note: Migration and the Environment* (MC/INF/288 – 1 November 2007, 94th Session), International Organization for Migration (IOM), Geneva.

Kniveton, D. et al.
2008 *Climate change and migration: Improving methodologies to estimate flows.* IOM Migration Research Series, No. 33, International Organization for Migration (IOM), Geneva.

Lonergan, S.
1998 The Role of Environmental Degradation in Population Displacement. Global Environmental Change and Security Project Report, Research Report 1, July 1998 (2nd edition), University of Victoria, Canada.

McLeman, R. & Smit, B.
2006 "Migration as an adaptation to climate change", *Climatic Change* 76: 31–53.

MRC
2005 *About the Mekong – Special Places*, Mekong River Commission (MRC) Secretariat, Vientiane. Available at: http://www.mrcmekong.org/about_mekong/special_place.htm_

Neumayer, E.
2005 "Bogus Refugees? The determinants of asylum migration to Western Europe", *International Studies Quarterly*, 49: 389–410.

Perch-Nielsen, S.L. et al.
2008 "Exploring the link between climate change and migration". *Climatic Change*, 91: 375–393.

Piguet, E.
2008 *Climate change and forced migration: How can international policy respond to climate-induced displacement?* UNHCR Evaluation and Policy Analysis Unit, Geneva.

Renaud, F.G. et al.
2009 "Deciphering the Importance of Environmental Factors in Human Migration". Paper submitted to *International Migration*.

Renaud, F.G. et al.
2007 *Control, Adapt or Flee: How to Face Environmental Migration?* United Nations University Institute for Environment and Human Security, Bonn.

Scoones, I. et al.
2007 *Dynamic Systems and the Challenge of Sustainability*, STEPS Working Paper 1, STEPS Centre, Brighton.

Stal, M.
2009 Mozambique Case Study Report for the Environmental Change and Forced Migration Scenarios Project. See: www.each-for.eu.

Tansey, O.
2006 Process Tracing and Elite Interviewing: A Case for Non-probability Sampling. Available at: http://www.nuff.ox.ac.uk/politics/papers/2006/tansey.pdf

Vag, A. et al.
2007 *Research Guidelines Book 1, Version 3*. Working document, EACH-FOR Work Package 3: Methodology & Synthesis. See: www.each-for.eu

Warner, K. & Laczko, F.
2008a "A global research agenda. Climate change and displacement", *Forced Migration Review*. Issue 31, October 2008.

Warner, K. & Laczko, F.
2008b "Migration, environment and development: New directions for research", in *International Migration and Development, Continuing the Dialogue: Legal and Policy Perspectives*, Chamie, J. & Dall´Oglio, L. (Eds), International Organization for Migration (IOM), Geneva and Center for Migration Studies (CMS), New York.

Warner, K. et al.
2008 "Field observations and empirical research. Climate change and displacement", *Forced Migration Review* 31: 13–14.

Warner, K. et al.
2009 *In Search of Shelter: Mapping the Effects of Climate Change on Human Migration and Displacement*, a policy paper prepared for the 2009 Climate Negotiations, Bonn, Germany: United Nations University, CARE, and CIESIN-Columbia University, in close collaboration with the European Commission Environmental Change and Forced Migration Scenarios Project, the UNHCR and the World Bank. Available at: http://www.reliefweb.int/rw/lib.nsf/db900sid/ASAZ-7SVCRB/$file/CARE_May2009.pdf?openelement

White, I.
2002 *Water Management in the Mekong Delta: Changes, Conflicts and Opportunities*. IHP-VI Technical Papers in Hydrology No. 61, UNESCO, Paris.

7. Appendix

Table 2: EACH-FOR case studies overview (regions/countries; environmental issues addressed and case study sites)

Case study region/ countries	Environmental issues addressed:	Case study sites:
Asia-Pacific		
Bangladesh	Sea-level rise; cyclones	Coastal Regions of Bangladesh (South West), chars (moving islands) on Jamuna River (North West), Dhaka
China	(a) Dam construction (b) Desertification	(a) Three Gorges Dam affected regions: Shangdong Province, Jiangsu Province, Chongming Island of Shanghai, Zhejiang Province (b) Erenhot, Inner Mongolia Autonomous Region
Tuvalu	Sea-level rise, erosion, waste disposal, water stress	Funafuti Atoll, Tuvalu and Auckland, New Zealand
Viet Nam	Flooding	Mekong Delta, particularly An Giang Province
Central Asia		
Kazakhstan	(a) Desertification and water stress (b) Nuclear testing	(a) Aral Sea region, Almaty (b) Semipalatinsk
Kyrgyzstan	Soil pollution, waste disposal, landslides, earthquakes	Whole country, with particular focus on the Ferghana Valley
Tajikistan	Soil pollution, degradation and erosion, mud flow, landslides, floods, earthquakes	Whole country, with particular focus on the Ferghana Valley
Europe and Russia		
Spain	Water shortage and desertification	South-eastern regions of Spain – Murcia and Almeria

Turkey	(a) Dam construction (b) Water destruction	(a) South-east Turkey (Adiyaman – Samsat District; Urfa city centre); West of Turkey (Didim-Yalikoy village; Izmir – Torbali) (b) South-east Turkey (Urfa – Suruc District); Istanbul
Balkans	Unavailable at time of printing	Danube Basin
Russia	Unavailable at time of printing	Unavailable at time of printing
Latin America and Caribbean		
Argentina	(a) Floods, increase of rain - water excesses with periods of abnormal droughts (b) Droughts - decrease in water availability, melting of glaciers (c) Droughts - decrease in water availability.	(a) Pampa Arenosa and Depresión del Salado north-west of the Province of Buenos Aires (b) Pre-Andean region (Comahue and the city of Jáchal, San Juan) (c) Yungas in the Salta Province
Dominican Republic, Haiti	Deforestation (and its consequences during tropical storms)	Province of Independencia, Dominican Republic; Port-au-Prince, Haiti
Ecuador	Water quality and availability; soil degradation; climate issues (ENSO and its consequences)	Guayas, El Oro, Pichincha, Manabi, Imbabura, Bolívar, Tunguruha, Azuay eta Quevedo provinces
Mexico	(a) Tropical storms, landslides, flooding; (b) Desertification, soil degradation	(a) Soconusco/Chiapas, south-eastern Mexico; (b) Western Tlaxcala (approx 60km east of Mexico City)
Middle East and North Africa		
Egypt	Water shortage	Newly reclaimed desert lands (Western Cairo), Cairo slums, Nile Valley and Nile Delta, Upper Egypt (Southern Egypt).
Morocco	Water shortage, desertification and the impact of other environmental challenges on rural villages in arid areas	Desert fringe villages in south-east Morocco: the two most southern Oases of the Draâ river valley: Mhamid and Tagounite (Province of Zagora)
Western Sahara	Desertification and water shortage	Algeria: Interviews with refugees from Western Sahara in refugee camps in Algeria (Tindouf region) under the control of the *Frente Polisario* government-in-exile

Sub-Saharan Africa		
Ghana	Unreliable rainfall, poor soil fertility	Source area: Upper West Region; Destination area: Brong Ahafo Region
Mozambique	Flooding, roughts	Central Mozambique - Zambezi River Valley
Niger	Droughts, deforestation, overgrazing, land degradation, Niger River problems and Lake Chad drying out	Niamey, Tilabéri
Senegal	Desertification, drought and water management	Fatick and Kaolack (the Peanut Basin) in Central Senegal and the Sénégal River Valley in Northern Senegal

Migration and Natural Disasters

Asmita Naik[1]

[1] Consultant; E-mail: asmitanaik.consultancy@gmail.com

1. Introduction

Background

Human migration patterns have long been a reflection of climatic changes and natural disasters. Recent catastrophes, such as the Indian Ocean tsunami and Hurricane Katrina in the United States of America (USA), bring home the potential of such disasters to uproot human life. It is anticipated that climate change will result in higher temperatures, more intense rainfall and more extreme weather events, such as droughts, storms and floods (Brown, 2008a:16). These, in turn, will likely prompt further population movements. The purpose of this chapter is to find out what history can teach us about migration[2] and natural disasters – whether natural disasters have led to migration and, if so, what numbers and scale of population movements have been involved, how people have coped with disaster, and what were the patterns ofmovement out of disaster-affected areas, as well as into those areas by new migrants and returnees.

A study of previous natural disasters can provide invaluable insights for future preparedness if, as anticipated, increased catastrophes result from climate change. Although the connections seem evident, there is surprisingly little reference, to date, on natural disasters in the literature on climate change and displacement (Ferris, 2007: 8). If natural disasters are expected to increase, there is a need for policy action to strengthen both systems of risk-mitigation and response mechanisms. There is significant evidence that the negative effects of natural disasters can be diminished through planning, foresight

[2] The terms 'migration' and 'displacement' are used interchangeably in this paper to refer to all movements away from the usual place of residence post-disaster, whether these are internal or international movements, permanent or temporary, forced or voluntary. The paper may also use more specific terms such as 'forced migration' when it is citing the work of authors writing on this subject. As the section on 'Definitions' in the introduction will show, there is a lack of consensus on terminology to describe movements post-disaster.

and a commitment of resources. For example, the Government of Bangladesh has made progress in protecting populations by building cyclone shelters (Ferris, 2007: 10). If human displacement is expected to increase in the wake of climate change, there are policy actions that can be taken in preparation.

Climate change, in itself, does not directly displace people but it produces environmental effects that make it difficult for people to survive where they are. There are several ways in which climate change and displacement may interlink. Ferris argues that there is empirical backing to support the claim that climate change causes an increase in the number and severity of sudden natural disasters that displace people but that other models linking climate change and displacement through long-term environmental effects (such as desertification), or conflict over resources, inadequate governmental responses or long-term development projects, all require further research (Ferris, 2007: 4–5). The UN International Strategy for Disaster Reduction (UNISDR) says disasters – storms, floods and droughts – have increased threefold over the past 30 years. It is also generally accepted that global warming will cause a rise in sea levels that will, in turn, displace people; according to a 2007 World Bank study, sea levels rising a single metre would displace 56 million people in 84 developing countries (Ferris, 2008). Thus, while there are considerable differences of opinion about the impact of climate change on displacement, there is, at least, consensus on this one area – that the number and severity of sudden-onset natural disasters will increase and that they will, in turn, lead to displacement (Ferris, 2007: 7–8). This brings natural disasters to the very core of the climate change debate.

While an examination of past natural disasters is useful, a word of caution is also necessary. The use of information on previous natural disasters as a predictor of migration caused by climate change does have some limitations, given that the scale and pattern of future disaster events are uncertain and may far exceed what has been recorded in human history. Moreover, debates over causality and the difficulty of establishing evidence-based linkages between migration and climate events continue. Recognizing the gaps in knowledge raised by this issue, this chapter aims to produce a baseline analysis of the literature on migration and natural disasters, with a focus on the following:

(a) An analysis of the numbers of people estimated as displaced by environmental disasters (disaggregated by types of disasters). How are they estimated?

(b) Migration patterns in response to environmental disasters. What factors shape the choices of destination and settlement of the environmentally displaced? What are the implications of displacement by environmental disaster for internal and international migration patterns? What is the social composition of those who migrate?

(c) The potential responses to sudden climatic events and disasters. Is mass displacement a typical response? To what extent is migration in disaster-prone areas a proactive prevention response?

(d) The return of migrants to post-disaster areas. To what extent do the environmentally displaced return to their homes? Can patterns be identified?

(e) The role migration can play in aiding the reconstruction of disaster-affected areas. What measures need to be put in place to ensure that such efforts are sustainable?

Methodology

This paper is based on a literature review carried out through Internet, library and journal research. The search involved using various database terms connected with migration and natural disasters in a variety of combinations. These terms included 'natural disasters', 'migration' and 'displacement'. The terms 'migration' and 'displacement' were then coupled with references to particular types of disaster, such as 'earthquakes', 'tsunamis', 'floods', 'volcanoes', 'storms', 'hurricanes', 'droughts' and 'famines'. The research so carried out is by no means exhaustive but it is comprehensive enough to identify the main works in this field. The emphasis in this literature review is to bring together existing theoretical and empirical academic and policy-oriented work on this issue, rather than provide new analysis based on primary data. The paper identifies gaps in the existing literature and makes recommendations for the way forward. The questions listed above are covered in the following three sections of this chapter:

- Migration statistics (section 2) provides an analysis of the numbers of those displaced by natural disasters (point a);
- Migratory movements out of disaster-affected areas (section 3) explores migratory patterns following natural disasters, as well as responses by affected populations that do not involve migration (points b and c);

- Migratory movements into disaster-affected areas (section 4) covers new migrants coming to disaster-struck areas for reconstruction work as well as the return of those affected (points d and e).

Definitions

Both 'natural disaster' and 'migration/displacement' are terms used to describe a wide range of environmental and social processes. The starting point of this paper is to note the parameters within which it is working. Firstly, what is meant by 'natural disasters' and where do they sit within the realm of 'environmental disasters'? 'Natural disasters' can be seen as a subcategory of 'environmental disasters', which have been classified by theorists in different ways, according to their cause and temporal nature. The causes of displacement are difficult to disentangle, given the political, economic, demographic and environmental factors at play, and various frameworks for categorizing disasters have been put forward. Table 1 summarizes a selection of these typologies.

Table 1: Selection of frameworks for categorizing disasters

Author	Framework
Bates	Three-prong classification based on: disasters (unintended catastrophic events), expropriations (wilful destruction of environment, making it unfit for human habitation), and deterioration (incremental deterioration of environment) (Bates, 2002: 475).
Keane	Five categories: natural disasters, long-term environmental degradation, development, industrial accidents and remnants of war (Keane, 2004).
Richmond	Categorizes environmentally related disasters into: naturally induced disasters; technologically induced disasters; economically induced disasters; politically induced disasters; and socially induced disasters. Naturally induced disasters includes hurricanes, earthquakes, floods, fires, droughts, tornadoes, volcanic eruptions, electric storms, famines, whirlwinds, avalanches, hail and snow storms, lightning and plagues (Hugo, 2009).

These typologies show that there is no standard accepted way of classifying disasters. All the frameworks seem to operate along a spectrum of disasters that moves from the natural to the man-made. But there is much scope for debate about what is truly 'natural', especially within the context of discussions about the role of mankind in climate change. Although discussions about causation abound, there does appear to be some consensus, at least, as to what constitutes a natural disaster – sudden catastrophes such as floods, storms, earthquakes, tsunamis, volcanic eruptions, hurricanes and cyclones. This chapter thus uses this definition of natural disasters and focuses on migration flows attributed to them. 'Droughts' and 'famines' do not neatly fall within this definition as they are not sudden events but tend to occur more slowly and are not as 'natural', in the sense that there is more scope for arguing human causation. Nonetheless, 'droughts' are included in this paper, since much of the literature focuses on this aspect and also because they provide an illuminating counterpoint to sudden disasters in terms of their impacts on migration (for further discussion of droughts as well as desertification please see Leighton Chapter 6 in this volume).

Secondly, what is meant by '*migration*'? The complexity of migration movements means that there is a range of concepts and terminology used to describe and analyse the different types of migration experience. In short, migration means the 'movement of a person to a new place of residence'. The terms most commonly used to describe movements post-disaster are 'forced migration', which denotes the non-voluntary movement of persons to escape, inter alia, natural disasters, and 'displacement', which refers to people who are forced to leave their habitual residence spontaneously in order to flee, inter alia, natural disasters. Although displacement often refers to movement to an internal location, there are no agreed definitions and, in fact, both 'forced migration' and 'displacement' are used to refer to both internal and international/cross-border movements. Box 1 provides a selection of IOM definitions of terms relevant to this text. This paper uses the terminology of migration or displacement interchangeably to refer to all movements away from the usual place of residence, whether these are internal or international, permanent or temporary, forced or voluntary. It uses more specific terminology, such as 'forced migration', at times when it is citing the work of a particular author and in order to stay true to the referred text.

Box 1: Migration terminology

Forced migration is the non-voluntary movement of a person in order to escape armed conflict, a situation of violence, violation of his or her rights, a natural disaster or a man-made disaster. This term applies to refugee movements and forced exchanges of populations between states.
Return migration is the movement of a person returning to his or her country of origin, or of habitual residence, after spending at least one year in another country.
Trafficking in persons is defined in the *Protocol to the UN Convention Against Transnational Organized Crime* as the recruitment, transportation, transfer, harbouring or receipt of persons, by means of the threat or use of force or other forms of coercion, of abduction, of fraud, of deception, of the abuse of power or of a position of vulnerability or of the giving or receiving of payments or benefits to achieve the consent of a person having control over another person, for the purpose of exploitation. Exploitation includes, at the minimum, the exploitation of the prostitution of others or other forms of sexual exploitation, forced labour or services, slavery, or practices similar to slavery, servitude or the removal of organs.
Internally displaced person is defined in the *Guiding Principles on Internal Displacement* to mean a person forced to leave his or her habitual residence spontaneously, in order to flee an armed conflict, situations of widespread violence or systematic human rights violations, or to escape natural or man-made disasters or their effects. This term also covers persons displaced within the borders of their country of origin, who are not covered by the 1951 *Convention relating to the Status of Refugees* because they did not cross an internationally recognized border.
Refugee is a person who, pursuant to the 1951 *Convention relating to the Status of Refugees*, owing to well founded fear of being persecuted for reasons of race, religion, nationality, membership of a particular social group, or political opinion, is outside the country of his or her nationality and is unable, or owing to such fear, unwilling to avail himself or herself of the protection of that country.

Source: International Organization for Migration (IOM), 2009

Added to this complexity is a third question about what people who flee their homes for environmental reasons should be called. This debate has occurred within the context of climate change rather than natural disasters per se, but is worth recounting. Academics and campaigners have proposed a variety of terms: 'environmental refugees' (El-Hinnawi, 1985), 'climate refugees' and 'climate migrants' (Brown, 2008a: 10). There is no consensus regarding these terms. Brown says that the word 'refugees' has fewer negative connotations but is not accurate under international law, and that there would be resistance to expanding the definition (and, hence, protection) of political refugees to include climate refugees (who are also mostly internally displaced). 'Climate evacuee' sounds too short term. 'Climate migrant' emphasizes the pull from the destination rather than the push towards it. He concludes that forced climate migrants fall through the cracks of international refugee and immigration policy: *"So far, there is no 'home' for climate migrants in the international community, both literally and figuratively"* (Brown, 2008a: 10) (also see Martin Chapter 7, Zetter Chapter 8 in this volume).

The lack of an agreed definition is due to the difficulty of isolating environmental factors from other drivers; debates over forced versus voluntary migration; scholarly tension over the desire to establish this as a specific field in the context of migration; and disputes over numbers. Underlying this is a tension between 'alarmists' who isolate environmental factors and 'sceptics' who insist on the complexity of migration process (Dun & Gemenne, 2008: 10). The lack of an interdisciplinary approach to the migration–environment nexus adds to this problem (Hugo, 2009). This paper uses the working definition proposed by IOM for *'environmental migrants'*: *"persons or groups of persons who, for compelling reasons of sudden or progressive changes in the environment that adversely affect their lives or living conditions, are obliged to leave their homes or choose to do so, either temporarily or permanently, and who move either within their country or abroad"* (IOM, 2008a).

2. Migration and natural disaster statistics

Lack of statistics

There are no global statistics on migratory movements prompted by natural disasters. At best, there are estimates and indications that can be derived from displacement data relating to particular crises or other data on general trends. Agencies and governments do collect location-specific data, indicators and geographic information necessary to plan assistance and reconstruction, some of which is aggregated into global trends. The *World Disasters Report* published annually by the International Federation of the Red Cross and Red Crescent Societies (IFRC), for example, regularly publishes extensive data on natural and man-made disasters, including numbers of disasters (by continent and year); people killed and affected (by continent and year); estimated damage in dollars; numbers of refugees and asylum seekers. This report sometimes has figures for internally displaced persons – for example, the 2004 *World Disasters Report* included a table of significant populations of internally displaced people (IFRC, 2004) but references to internal displacement do not usually distinguish between those displaced by natural disasters and those displaced for other reasons – violence, human rights violations etc. Nor is there any indication of those displaced internationally by natural disaster. Thus, while there are some data on migration and displacement caused by specific natural disasters, they are not systematically compiled or analysed at a global level; the overall picture of migration trends following natural disaster is, therefore, piecemeal.

Reasons for lack of statistics

The lack of data is largely due to the absence of an adequate definition to cover migrants affected by natural disasters under international law and the fact that responsibility for this group has not been designated to any particular organization (Brown, 2008a: 15). The scientific categorization of natural disasters is imprecise; causal relationships between disaster and displacement are difficult to define and often depend on a great variation in individual decision-making (Brown, 2007: 2, 6). Hurricane Katrina demonstrated the difficulty of disaggregating the role of weather events from other environmental, economic and social factors, as it affected different parts of the population in different ways (Brown, 2008a: 12).

There are wider challenges with collecting data on natural disasters. Investment in data collection on natural disasters is relatively recent; until the funding of the International Emergencies Disaster Database (EM-DAT – see later in this section for details) and other such initiatives, data gathering on disasters was carried out on an ad hoc basis (Guha-Sapir et al., 2004: 13). Even today, *"data on disaster occurrence, their effect upon people and cost to countries remain, at best, patchy"* (Guha-Sapir et al., 2004: 15). No single institution has the role of prime provider and the original information (such as reports from governments, insurance companies, press agencies, aid agencies etc.) upon which the various databases (including EM-DAT) rely is not specifically gathered for statistical purposes. The lack of a standardized system of data collection, methodologies, definitions, verification, protocols and storage is a weakness (ibid). Further challenges in collecting data on natural disasters are noted in box 2.

Box 2: Additional challenges in collecting data on disasters

- **Lack of comprehensive data** - There are data on certain aspects, such as deaths, but not on others, such as economic losses, migration or the fact that data collection is for limited objectives (e.g., insurance companies may not collect data in poorer parts of the world that are not as relevant for the insurance market).
- **Ambiguous definitions of disaster types** - There is, for instance, a lack of agreed temporal and spatial definition of when a drought starts and ends. This problem is compounded by 'cascading' disasters whereby an initial hazard (e.g., earthquake, flood) triggers a secondary event (e.g., landslides). With no common methodology for the local reporting of disaster-related losses, impacts might be associated with either event.
- **Political and organizational interests** - Data may be manipulated to minimize or exaggerate the impact of an event.
- **Geo-referencing** - Disasters may transcend national boundaries or occur in parts of a country leading to double-counting or missed counts (Guha-Sapir et al., 2004: 15–19).
- **Standardizing processes** - Data-collection processes are severely challenged by shifting populations, remoteness of locations and in situations of physical danger and chaos, where survival (not data collection) is the main priority (Reed et al., 1998: 2).
- **Populations who 'self-settle'** - Those who do not rely on agencies for support or who move to urban areas are particularly difficult to track.

- A lack of staff knowledge and training, bureaucratic structures and organizational priorities can all be impediments to data collection (Reed et al., 1998: 7–13).
- Moreover, national census data which are often the foundation for assessing populations at risk are inadequately maintained and disaggregated by age, gender, race etc (BESR et al., 2007: 5–6).

The lack of capacity at both national and international levels inhibits the effective collection of data and the capacity that exists is focused on cross-border migration, particularly refugee flows, rather than internal displacement (the most likely option for natural disaster victims). Consequently, data collection by the United Nations High Commissioner for Refugees (UNHCR) on refugee movements has enjoyed considerable success and is based on statistics collected from governments and sometimes other sources or registration in camps (Reed et al., 1998: 7). Hovy of the United Nations (Population Division), reflects that it is not possible to replicate this success for data on victims of natural disaster – firstly, because refugees cross an international border and the receiving country has an interest in counting their numbers (whereas most victims of natural disaster are internally displaced and also return to their homes more quickly) and, secondly, because refugees are in need of international protection and have an international statute and agency dedicated to this task (as compared to victims of natural disasters, who still fall under the protection of their own governments) (Hovy, 2009).

Initiatives to collect statistics

There are several initiatives to collect statistics on natural disasters but, as noted, none collects data on migration. The Emergency Disasters Database (EM-DAT), managed by the World Health Organization's Brussels-based Collaborating Centre for Research on the Epidemiology of Disasters (CRED), is the most complete, publicly accessible international database (http://www.emdat.be/) on estimates of human and economic losses. It was established in 1988 with the main objective of serving the purposes of humanitarian action and contains core data on the occurrence and effects of over 16,000 mass disasters worldwide from 1900 to the present day. Data are organized by: disaster number; country, disaster groups (i.e., natural, technological, complex emergencies); disaster type; date; persons killed; persons injured; and persons made homeless. It also includes a category of 'persons affected' –*"those requiring immediate assistance during a period of emergency; it can also include displaced or evacuated people"* (EM-

DAT, 2009). Although the database does not explicitly cover migrants, numbers affected should incorporate numbers displaced and thus give a top-level indication of possible migrants. That said, EM-DAT's analysis does not tend to refer to the issue of migration much; for instance, its 30-year review of natural disasters between 1974 and 2003 only mentions migration as a risk factor – i.e., increasing migration to urban and hazardous areas (Guha-Sapir et al., 2004: 36). CE-DAT, the Complex Emergency Database, (http://www.cedat.be/) is a linked portal that was created in 2003 to provide information on conflicts and complex humanitarian emergencies. It collects data on indicators concerning crude mortality, global acute malnutrition, and under-five mortality. It does not refer to migration.

The Disaster Database Project (http://learning.richmond.edu/disaster/index.cfm) operated by the University of Richmond (USA) is an independent publicly accessible database on natural and man-made disasters. Its main sources of information are government reports, newspapers and scholarly texts. It collects information on location, date and time, class and intensity of event, people missing and injured, animals killed, buildings destroyed, area covered, and the cost of the disaster. Interestingly, this database includes displaced persons but its design makes it difficult to obtain an aggregate figure as it is necessary to be very specific and to search by disaster. The inclusion criteria may mean that some disasters are missed and the dependence on public sources may also affect reliability of statistics. Box 3 gives information on other initiatives aimed at collecting data on natural disasters, which may cover migration, to some extent.

Box 3: Other data-collection initiatives on natural disasters

- **Disaster Management Information System (DMIS)** - This is an information system of the International Federation of the Red Cross and Red Crescent Societies (IFRC). It is a comprehensive dataset drawing on regular data from Red Cross national societies or governments. In addition to statistical data, it also links to substantive reports, updates and appeals.
- **The National Hazards Assessment Network** (NatHan) (http://mrnathan.munichre.com/) and **Sigma** (http://www.swissre.com/) databases are managed by Munich Re and Swiss Re respectively, two of the world's largest reinsurance companies. These databases are not accessible to the public and focus on data of interest to insurers (e.g., victims, insured and uninsured losses). They do not cover migration.
- **Global Disaster Identifier Number** (GLIDE) (http://www.glidenumber.net/) is a project of the Asian Disaster Reduction Centre (ADRC), in collaboration with a range of international agencies. This database is limited to Asia and includes information on dates, disaster type, countries, duration of events, locations, magnitude, information

source, human and economic loss but it does not systematically cover migration.

- **Dartmouth Floods Observatory** - This is a USA database that tracks global flooding events.
- **British Association for Immediate Care** (BASICS) (http://www.basedn.freeserve.co.uk/) is a UK-based charity with a database on natural and technological disasters, which appears not to have been updated for several years.
- **Other sources include**: the Pan American Disaster Response Unit (**PADRU**) – a region-specific database created by the IFRC; the National Oceanic and Atmospheric Administration (**NOAA**) – an American Government database covering events in the Americas; the United Nations Office for the Coordination of Humanitarian Affairs (**OCHA**)'s **Reliefweb**; and international news agencies such as Reuters AlertNet, Associated Press, Xinhua, Agence France-Presse, Deutsche Press Agentur, BBC and CNN.

Statistics on natural disasters

As a result of these initiatives, and despite the challenges involved, data collection on certain aspects of natural disasters has improved in recent years and there are credible statistics on natural disasters in general terms. These confirm a number of important trends that can be assumed to impact on displacement levels, even if data proving causal connections are absent. These trends include the following:

- *Increase in natural disasters*: Severe weather events, especially hydro-metereological events (i.e. floods and wave surges, storms, droughts and related disasters, extreme temperatures and forest/scrub fires, and landslides and avalanches) are on the rise, as shown by graph 1, below (Ferris, 2008: 7). Over last 30 years, there has been an increase in the reporting of natural disasters, from 100 in 1974 to 400 in 2003. The increase is, no doubt, partly due to improved methods of reporting and data collection, making it difficult to say that the number of natural disasters has actually increased (Guha-Sapir et al., 2004: 23–24).

Graph 1: Trend of increasing reports of natural disasters

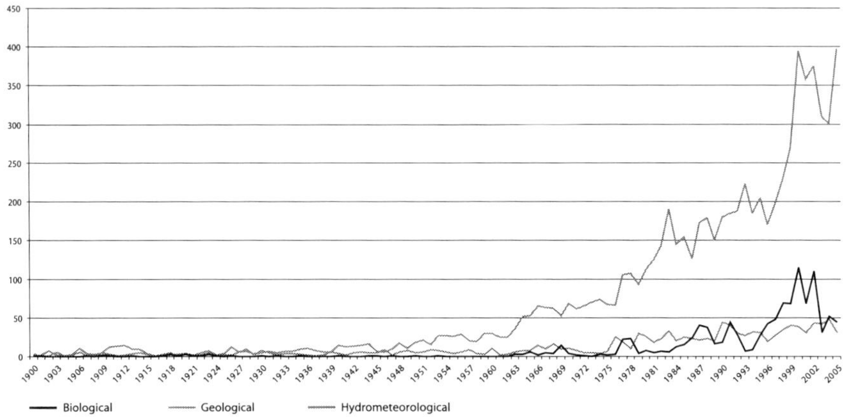

Source: EM-DAT table from Ferris, 2007:7

- *Increase in persons affected by natural disasters*: Although the numbers of deaths caused by natural disasters are decreasing, the numbers affected are increasing. More than 255 million people were affected by natural disasters globally each year, on average, between 1994 and 2003, with a range of 68 million to 618 million (Guha-Sapir et al., 2004: 13). This increase in victims is a trend that is likely to continue because of environmental factors and population pressures, despite improvements in mitigation measures and the effectiveness of responses. Although excess mortality related to acute and chronic disasters has decreased, there is an increase in victims who survive with physical injuries or experience severe social and economic devastation. EM-DAT notes that mega-disasters are mega not because of an objective measurement scale such as Richter but because of the size of population they affect. "*Therefore, densely populated areas tend to have more 'disasters' from relatively insignificant events because large numbers of people are affected... Conversely, disasters occurring in sparsely populated areas affect fewer people and are less likely to qualify as a major humanitarian crisis*" (Guha-Sapir et al., 2004: 53). The 2000 IFRC World Disasters Report attributed this increase in persons affected to 'hydro-meteorological events' (Brown, 2007: 11), whereas UNHCR puts this down to rising vulnerability rather than an increase in the frequency of hazards per se (Hugo, 2009).

- *Disproportionate impacts on developing countries*: Another established fact is that developing countries suffer the worst impacts – for example, of the 262 million people affected by climate disasters annually from 2000 to 2004, over 98 per cent of them were in the developing world (United Nations Development Programme (UNDP), 2008: 8). Natural disasters disproportionately affect South and East Asia, sub-Saharan Africa and small island states, with Asia accounting for almost 70 per cent of all lives lost through natural hazards (IOM, 2008a). Bangladesh, China and India were among the top ten worst affected countries between 1974 and 2003, in terms of numbers of people affected (Guha-Sapir et al., 2004: 29). The prevalence of floods and droughts in these places mean that numbers quickly reach hundreds of thousands and, in some cases, millions (Guha-Sapir et al., 2004: 30–33).

Available statistics on migration and natural disasters

There are credible statistics on specific recent disasters but global statistics are missing. In terms of specific disasters, Hurricane Katrina resulted in rich demographic data and analysis; it caused 1.5 million to be displaced temporarily and 500,000 permanently (IRIN, 2009) and US demographers were able to track these flows in considerable depth, using census data and population sampling techniques (Rosenberg, 2006; Johnson et al., 2008: 9). Estimates for the tsunami were less precise and ranged from 1 to 2 million displaced across 12 affected countries (Hedman, 2005; Hugo, 2009).

For global statistics on migrants and natural disasters, there are only best estimates. As discussed below, current estimates for displacement caused by natural disasters range from 10 million through to 25 million, and up to 255 million a year if the broader category of 'persons affected' is used. This paper takes a cautious approach to re-producing any numbers, given the difficulty of finding reliable data.

One figure worth further discussion is the often quoted 25 million 'environmental refugees' – defined as people forced from their homes, both internationally and internally, by a range of serious man-made and natural environmental pressures, including dam construction, pollution (the degradation of the Niger delta due to oil spills), land degradation, droughts (Sahel), desertification (e.g., in China, Libyan Arab Jamahiriya, Morocco and Tunisia), and soil erosion (Turkey) (Brown, 2008a: 11; Ratha, 2007). The 2001 IFRC *World Disasters*

Report repeated the estimate of 25 million current 'environmental refugees' (Brown, 2008a: 11). This figure comes from a paper by Norman Myers in 1995; he admits that estimates such as these are "*no more, and no less, than a first-cut assessment*" and that they veer towards the cautious side (Myers, 1995:1). Myers arrives at this figure by adding up known estimates of 'environmental refugees', such as the 7 million in sub-Saharan Africa, 6 million in China, 5 million in the Sahel, 4 million in the Horn of Africa, 2 million in Mexico, and 1 million in other locations. It is important to note is that the definition and estimates used by Myers include people displaced by man-made disasters and development projects and not solely by natural climate events, which are the focus of this paper. There are other estimates, too; in 2002, UNHCR came to a similar number of 24 million for people fleeing floods, famine or other environmental disasters (Hugo, 2009). Keane estimates that there are 10 million 'environmental refugees' worldwide (Jacobsen, 2008; Keane, 2004).

These figures are only estimates and not based on gathered statistics. It is arguable, therefore, that EM-DAT is the most credible source at the current time for data on natural disasters. The database does not, however, specifically isolate 'migrant/displaced populations', including them instead in overall figures of populations 'affected' by natural disasters. As EM-DAT collects and analyses these data systematically and, given that its definition of 'affected' (see above) includes those who are displaced, these statistics can be taken as a very top-end estimate of those who might have suffered migration/displacement of any kind, following a natural disaster. Table 2, based on EM-DAT's analysis, shows the numbers killed and affected by different categories of natural disasters between 1979 and 2008.

Table 2: Numbers killed and affected by certain types of natural disasters between 1979 and 2008

Disaster type	Number of events	Numbers killed	Numbers affected
Earthquakes	734	387,129	134 million
Droughts	427	558,554	1.6 billion
Floods	3,005	198,390	2.8 billion
Volcanoes	145	25,474	4.2 million
Storms	2,458	430,865	718 million

Source: EM-DAT http://www.emdat.be; data accessed 29.04.09.

While the data do not disaggregate the numbers of persons displaced, they do indicate that relatively slow-onset natural disasters (those with some warning), such as droughts and floods, appear to affect far greater numbers than do sudden events. For example, the numbers in table 2 show that 2.8 billion and 1.6 billion were affected by floods and droughts respectively, compared to far lesser numbers for volcanoes (4.2 million), earthquakes (134 million) and storms (718 million). This highlights a role for disaster preparedness and planning in limiting migration flows or finding alternatives for affected populations. Although 'affected' is far from synonymous with 'displacement' or 'migration' and should not be taken as such, the data at least give a very top-level (maximum) indication of what these numbers might be.

Statistics on internal displacement could be a useful indicator, given that most population movements following natural disasters are internal and not cross-border. However, organizations such as the Internal Displacement Monitoring Centre (IDMC) were set up to track statistics on those who are internally displaced as a result of conflict or human rights violations and not those displaced by natural disasters, once again highlighting the lacunae in monitoring how people are affected by natural disasters.

IDMC, in collaboration with OCHA, has recently started to look at displacement caused by natural disasters and taken a significant step towards addressing these gaps. A new report on natural disasters and forced displacement by OCHA -IDMC includes a statistical analysis of EM-DAT data on natural disasters. This found that, of 207,631,038 people affected by natural disasters in 2008, a total of 36,062,843 were found to be displaced. This figure comprised 28,650,653 people who were displaced due to complete destruction of their homes and 7,412,190 who had to be evacuated, either as a preventive measure or after the disaster, as their homes had been rendered temporarily uninhabitable. Seventeen per cent of people reported to have been affected by natural disasters were found to have been displaced. This preliminary study does not disaggregate internal and cross-border migration (OCHA–IDMC, 2009) but shows important progress in tackling these gaps in information.

3. Migratory movements out of disaster-affected areas

Typical response

Migration, whether permanent or temporary, is a typical survival strategy used by people confronting natural disasters (Hugo, 1996). It is an expected response, given pressing issues of safety and security, lack of material assistance and protection, employment and income, and the scale of damage wrought. The capacity of communities to cope will be determining factors in their response (Overseas Development Institute (ODI)/UNDP, 2005: 11). Academic works by Blaikie et al. 1994; Boon and Paul, 2003; Cannon, 1994; Lovell, 1994; Parker et al., 1997; and Smith and Ward, 1998 confirm that natural disasters generate migration due to the loss of livelihoods or because of fear of the event itself (Paul, 2005: 373). Yang is quoted as saying, *"At a very high level, migration correlates with natural disasters in developing countries ... Shocks stimulate people to migrate to escape the negative effects of disasters"* (Murray, 2009).

There are numerous historical examples of out-migration post-disaster, such as population movements caused by drought in East Africa in the 1980s and 90s, following Hurricane Mitch in Central America in the late 1990s, and due to perennial flooding in South Asia (Naik et al., 2007: 39). Of the 23 case studies comprising the EACH-FOR Project[3], several provide evidence of migration out–for example, post floods and landslides in the Ferghana Valley, Kyrgyzstan (Warner et al. Chapter 4 in this volume; Gemenne & Reuchlin, 2008: 15), post droughts and floods

[3] Environmental Change and Forced Migration Scenarios (EACH-FOR) is a two-year research project co-financed by the European Commission and being carried out by several research institutes, including the United Nations University. The objectives of EACH-FOR are (1) to discover and describe the causes of forced migration in relation to environmental degradation/change and their association with other social, political and economic phenomena in Europe and in the main countries of migration origin; and (2) to provide plausible future scenarios of environmentally induced forced migration. The list of methodological tools contains all traditional elements of research, from primary data collection (semi-structured expert interviews, semi-structured field interviews, migrant questionnaires, non-migrant questionnaires) via statistical analysis and environmental evaluation, to modelling. There are 22 case studies from the following regions: Europe and Russia, Newly Independent States (NIS) and Central Asia, Asia, sub-Saharan Africa and Ghana, Middle East and Northern Africa, and Latin America. Further information is available at http://www.each-for.eu/index.php?module=main.

caused by El-Niño in Ecuador (Alvarez Gila, 2009), due to tropical storms in the Chiapas region of Mexico (Alscher et al., 2009), and flooding in Mozambique (Stal, 2009) and Viet Nam (Dun, 2009). The question is not whether migration post-disaster happens but what its scope and form are; these, in turn, depend on a variety of factors including the scale of the disaster and destruction, aid response, likelihood of recurrence, loss of livelihood, and poverty levels, remittances and reconstruction, as well as opportunities in destination areas.

Migration–environment literature

Before moving on to consider literature relating to the specific questions posed by this study, it is worth setting the background in terms of how environmental disasters fit with migration theory (see also Kniveton et al. Chapter 2 in this volume). Hunter provides a synthesis on migration and environmental hazards and notes that classical theoretical perspectives acknowledge that environmental conditions influence migration but that they rarely emphasize it. Hunter says that theorists such as Wolpert (1966) ('stress-threshold'), Speare (1974) ('threshold of dissatisfaction') and De Jong and Fawcett (1981) ('value-expectancy') suggest models that take a micro-level approach, looking at individual factors affecting migration motivation. Others, such as Petersen (1958 and 1975) and Gardner (1981), emphasize the macro social, economic and geographic context in influencing migration (Hunter, 2005). Bates, in 2002, argued that environment is an indirect causal factor: "*environmental changes affect migration decisions only after being filtered through the local context*" (Bates, 2002: 475). This is supported by other theorists who argue that natural disasters play a causal role but not necessarily a direct and immediate one. Green et al. (1983), Tobyn & Montz (1997) and Ward (1978) quoted in Paul (2005: 373) categorize the direct and indirect impacts of natural disaster, where direct effects are death, injury, disruption of socio-economic activities and property damage. These effects can be further subdivided into tangible and intangible effects. Paul (ibid) also refers to Parker et al. (1997) and Smith & Ward (1998) further who divide tangible and intangible effects into primary and secondary categories, whereby primary, for example, would be crop destruction, whereas out-migration would be a secondary, intangible, indirect impact (Paul, 2005: 373).

Diversity of migration patterns

The theory supports what is visible on the ground – that disasters induce a variety of responses and that different types of disaster lead to a range of migratory movements, from permanent to temporary displacement, cyclical to linear etc, to different scales of displacement and destinations, and conceal different motives (Warner & Laczko, 2008:60). Brown says that sudden natural disasters displace large numbers for relatively short periods of time but that slow-onset drivers are likely to displace many more people permanently (Brown, 2008a: 1; Hugo, 2009). The motivation for migration due to slow-onset disasters is also seen as more complex, requiring not only a 'push' but also a 'pull' in terms of socio-economic prospects, access to money, family, networks and contacts (Brown, 2008a: 19; Warner & Laczko, 2008: 60). McLeman and Smit argue that climate-related migration can function in numerous ways, and patterns might include: (1) repetitive migration as part of ongoing adaptive responses to variation (e.g., East Africa); (2) short-term movements in response to particular climatic stimulus (e.g., Hurricane Mitch); (3) large-scale movements that build slowly to gain momentum as adverse climatic conditions coincide with other adverse socio-economic conditions. *"The point to be made here is that there is no simple 'if A then B' formula to explain migration response to climatic stimuli. This highlights the importance of assessing the possible migration responses to climate change in the context of exposure to risk and adaptive capacity in the particular community and country"* (McLeman & Smit, 2004).

The EACH-FOR case studies provide tentative support for this notion of diversity in response patterns, though it should be noted that the project itself considers that, although linkages between the environment and migration have been identified, the distance and the magnitude of the environment as a push factor for migration still needs further investigation. The Bangladesh case study shows how households have adopted different strategies to deal with floods and cyclones. In some cases, fathers temporarily migrate to urban centres in order to work and send remittances; others migrate in a seasonal fashion and move where the work is; some migrate with the whole family, some are forced into human trafficking and others do not to migrate at all, due to a lack of resources to enable them to do so (Poncelet, 2009). The Viet Nam case study found a similarly diverse outcome to flooding events: seasonal labour migration and movement towards urban centres during the flooding season (both internal and international (regional) migration); individuals or entire households migrating if successive flooding events led to a destruction of crops (on more than one occasion); increased risk of, and vulnerability to, human trafficking in situations of disaster; families or children seeking protection in child shelters in Ho Chi Minh City after each annual flooding season, once disaster relief aid is exhausted; and a planned resettlement of people in vulnerable zones through a government programme. Initial findings suggest that the environment is an indirect, rather than a direct, cause of migration. *"Rather, environmental change (flooding, in this case study) is shown to be a trigger for independent migration decisions when livelihoods are negatively affected – e.g., crops are lost, generally on more than one occasion. In such cases, livelihood stress is the direct cause of migration and environmental factors act as the trigger. This supports the argument put forward by Black (2001) and Kibreab (1997) that migration is multi-faceted"* (Dun, 2008: 12–13; Warner et al., 2008: 14).

Migration–environment researchers have sought to put a framework around these different types of movements and created typologies linking disasters and migration. A selection of these is given in table 3, which summarizes typologies that distinguish different types of environmental disasters, leading to different forms of migration.

Table 3: Selection of typologies linking disasters and migration

El-Hinnawi	Categorization based on likelihood of return: those displaced temporarily who can return once homes are repaired; those who are permanently displaced and resettle elsewhere; and those who migrate in search of a better quality of life due to degradation of their original habitat (Keane, 2004).
Hugo	'Environmental migration' to encompass migration induced by environmental disasters; environmental degradation; migration and climate change; and migration forced by environmental change caused by large-scale projects (Hugo, 2009: 2, 15).
IASC	Typology based on four types of events/causes of movement linked to different types of movement – i.e., hydro-meteorological extreme hazard events; environmental degradation and/or slow-onset extreme events; significant permanent losses in state territory as a result of sea-level rise etc.; armed conflict/violence over shrinking natural resources (IASC, 2008: 2–3).
McLeman	Two climate drivers of migration from a meteorological perspective: climate processes, such as sea-level rise, salinization of agricultural land, desertification and water scarcity, which cumulatively erode livelihoods and make it difficult to stay in one place; and climate events that are sudden and dramatic hazards, such as flooding, storms and glacial lake outburst floods and force people off land more dramatically and quickly (Brown, 2008a: 17–18).

Oliver-Smith argues that all demographic movements vary along a number of scales or continua:

- proactive–reactive (Richmond, 1993)
- voluntary–forced
- temporary–permanent
- physical danger–economic danger
- administrated–non-administrated.

He says these five pairs are best viewed as poles on a series of continua rather than closed or opposing categories (Oliver-Smith, 2006). Even these continua are not exhaustive and others can be added – for example, a spatial continuum (internal–international) based on how far people move from their homes, and even a spectrum of social vulnerability (vulnerability–resilience), which captures how socio-economic status plays into decisions to migrate. This paper seeks to capture the complexity of migratory patterns by organizing the literature under the following headings:

- voluntary–forced
- temporary–permanent
- internal–international
- vulnerability–resilience.

Voluntary–forced migration

Migration is often described as a continuum ranging from totally voluntary migration to totally forced migration – a non-voluntary movement of persons in order to escape armed conflict, violence, violation of rights or natural or man-made disasters. At the extreme end, evacuation in the face of life-threatening climate events is one form of forced migration. Disasters can also make people vulnerable to coercion. Trafficking, though not normally categorized as 'forced migration' but rather 'irregular migration', has long been recognized as a risk in such situations (UNHCR, 1995). The EACH-FOR case studies in both Bangladesh and Viet Nam uncovered trafficking as an extreme consequence, for some individuals (Warner et al., 2008: 14). Natural disasters in Bangladesh reinforced pre-existing trafficking trends, with widows and children being sent to India, and men taken away to sweatshops (Poncelet, 2009). The preventive measures taken by governments and international agencies following the Asian tsunami in 2004 (alerts at airports, directives, registration of children etc.) may have averted the occurrence of trafficking in that situation (Naik et al.,

2007: 43–44). At the other end of the spectrum from forced migration, some victims of natural disaster may appear to make voluntary choices, particularly in the case of insidious, slow-onset natural catastrophes where the threat is not immediately apparent.

Researchers generally agree that environmental migration and, particularly, migration due to sudden-onset natural disasters, constitute forced migration (Hugo, 2009; Dun & Gemenne, 2008: 10). But the nature of migration may change over time. Hugo (1996), when talking of the continuum of voluntary-to-forced migration, said that, in case of hurricanes and other disasters, "*initial migration is indeed forced, though the decision to return to the place of origin may become a more individualistic cost-benefit analysis as time progresses*" (Myers et al., 2008:274). As such, although migration stemming from natural disasters tends towards the forced end of the spectrum, the issue would benefit from more empirical study to test the nuances in the voluntary–forced character of such movements.

Temporary–permanent migration

Another area of enquiry is how long displacement induced by natural disasters continues and whether it can be defined as 'temporary' or 'permanent'. There is no common understanding of what constitutes 'temporary' or 'permanent' migration. Definitions range from six months away from the place of origin with no plans to return for 'permanent' migration (Findley, 1994 citing Hugo, 1980; Standing, 1985; Prothero & Chapman, 1985), to movement from the usual place of residence to another country for a period of at least three months but less than a year for 'short-term migration' (IOM, 2004: 60). Some governments have simply decided that displacement ends when a decree is signed to this effect – for example, in Angola. The Inter-Agency Standing Committee *Framework for Durable Solutions* argues that the ending of displacement is a process through which the need for specialized assistance and protection diminishes (Ferris, 2008). In fact, knowing whether movements are permanent or temporary following natural disasters would require accurate demographical measurements spanning several years (Smith & McCarty, 1996: 265, 294).

It is argued that natural disasters tend to displace persons temporarily rather than permanently (Keane, 2004). Certainly, examples from around the world show that temporary displacement does occur. There are examples of wholesale permanent community relocation

– for example, the village of Valmeyer, Illinois (USA) was completely relocated following the 1993 Mississippi floods – but this type of relocation is a relatively rare occurrence, especially in developing nations where migration is less organized (Myers et al., 2008: 284–5). One common thread in the literature is the assertion that the permanency of migration depends on the type of disaster in question. In other words, slow-onset disasters lead to permanent migration and sudden events lead to short-term displacement – for example, drought leads to the highest number of people considering permanent moves; floods lead to localized temporary out-migration; and cyclones lead to temporary displacement (Raleigh et al., 2009: 6–7). It is estimated that only a small share of migrants (up to 30%) choose to relocate permanently (op.cit.: 37).

Although it seems likely that sudden disasters at least lead to more temporary rather than permanent displacement, it is difficult to say this with authority, on the basis of the current literature, as the boundaries between the two are often blurred in practice. The EACH-FOR Mozambique study is a case in point. Central Mozambique is an area of recurring flooding events and it experienced a cyclone in 2007. People are displaced during flood emergency and following recurring flooding events, and are relocated on a permanent or semi-permanent basis. Temporary mass displacement along the Zambezi River is taking on the characteristics of permanency (Warner, 2008: 13).

Moreover, circular migration patterns are common in disaster-prone areas. A study of India's Ghaghara floodplain found a mixture of movements: permanent moves, when areas are so severely hit by floods as to cause loss of crops and houses, and periodic movements, when people migrate mostly to higher ground for shelter and temporary jobs. Unskilled workers prefer temporary jobs as they can return home for the harvest (Kayastha & Yadava, 1985: 85), thus perpetuating circular patterns of migration.

In fact, temporary circular migration emerges as an adaption strategy in numerous developing countries. A survey on floods in Bangladesh showed that 64 per cent of households reported being displaced once, and by a short distance only, due to a lack of resources, the presence of kin and a belief that the flooded land would recover (Myers et al., 2008: 285). Other forms of this type of household coping strategy include situations where particular household members make a permanent move and others stay behind. In western Sudan, older male members are sent to Khartoum to seek wage labour when drought conditions occur (McLeman & Smit, 2004 quoting research by Afolayan & Adelekan, 1998). The same happens in West Africa's Sahel region where young people are sent away to earn money (Brown, 2008a: 22). Ezra's study on drought-prone rural areas of Ethiopia found out-migration to be highest during the 1984 famine (Ezra, 2001: 763), with families using it as a coping mechanism; out-migration was seen as a way of assisting relatives and also reducing the burden on the household (op. cit.: 767).

There are also questions of causation and the notion that natural disasters cause temporary displacement, but not authentic 'permanent' migration, in the sense that it is other factors that come into play post-disaster that cause the displacement to become permanent. A study of migrants following the 1970 Peruvian earthquake found that most were compelled to seek employment elsewhere as the disaster intensified poverty in their villages but that only a handful indicated that the earthquake itself was the primary motive for moving (Myers et al., 2008: 286). Some writers have argued that permanent migration occurring as a result of a natural disaster is due to deficient responses of weak or corrupt states or other push and pull factors, rather than the hazard itself. Black says that the emphasis on environmental factors is a distraction from the central issues of development, inequality and conflict resolution (Castles, 2004) that actually determine the length and quality of the migration experience.

Internal–international migration

A clearer picture emerges from the literature as to how far victims of natural disaster are likely to go. Firstly, displacement post-disaster is presumed to be not only internal but also local, to the nearest safe destination (Raleigh et al., 2008: 20, 23). This conclusion is supported by studies from both the developing and developed world. A regression analysis of population data following Hurricane Katrina showed out-migration from areas experiencing the most damage but the greatest in-migration to nearby counties that were not damaged and able to give refuge (Myers et al., 2008: 286–7). Kayastha and Yadava's 1985 study of the Indian Ghaghara floodplain found that few moved permanently and even those that did remained within the floodplain area (Kayastha & Yadava, 1985: 85). The same was found in studies in Bangladesh (Raleigh et al., 2008: 24, quoting Zaman, 1991).

The evidence on whether natural disasters promote further rural-to-urban migration is conflicting. Hugo's 1996 analysis of Asian environmental migrants from 1976 to1994 shows migration to urban areas due to floods in Bangladesh and China (Hugo, 1996; and researchers have found greater migration to cities following earthquakes in El Salvador and India (Gujarat) in 2001, floods in Pakistan in 1974 (Paul, 2005: 372), and recent droughts in Kenya (Adow, 2008: 34). Enarson's study of migration flows in Gujarat in 2001 found a more nuanced picture of migration patterns that vary according to types of disaster. In this research, villages reported migration in from cities following the 2001 earthquake as city people dealt with quake losses by returning to their extended families in rural areas. But when the region was experiencing drought, this caused migration away from villages to cities. Family size was seen to increase more in the wake of the earthquake (with 14 villages, or 38 per cent, reporting an increase in family size) than as a result of the drought (after which 12 villages, or 32 per cent, reported an increase in family size) (Enarson, 2001). Ezra's study on Ethiopia finds rural–rural migration to be the most prevalent (Ezra, 2001:767). Despite general trends towards urbanization in the developing world, the available literature shows that people make pragmatic and disaster-specific choices about where best to migrate. In the developed world, too, behaviour can challenge assumptions about population movements: following the 1999 floods in North Carolina caused by Hurricanes Floyd, Dennis and Irene, homeowners used government relocation funds to move from urban to rural areas, rather than the other way round (de Vries, 2006).

There is strong consensus that those displaced by natural disasters remain within their own borders (Ferris, 2007: 9; Hugo, 2009). This is in marked contrast to refugees fleeing persecution and armed conflict, as the number of people who cross national borders because of natural disasters seems to be much lower than those displaced internally; in many cases, conflicts force people to leave not only their communities, but also their countries (Ferris, 2008). Although this is a widely accepted belief, it is not always backed by empirical data but rather proven by the negative and the absence of evidence showing international migration post-disaster. The 2004 tsunami experience indicates that the displaced stayed in the region and did not go to OECD countries;[4] there was no evidence of international migration from Indonesia–for example, to Malaysia – or India (Naik et al., 2007:8). Although there was

[4] Countries that have signed the Convention on the Organisation for Economic Co-operation and Development: http://www.oecd.org/document/58/0,3343,en_2649_201185_1889402_1_1_1_1,00.html

documented out-migration from tsunami-affected districts in Sri Lanka (Gallina, 2007: 15), there is little evidence that this resulted in moves abroad (Naik et al., 2007: 8). Likewise, with Hurricanes Rita and Katrina in the USA, evacuees did not go across the (nearby) border but moved to their families around the country (Brown, 2008a: 23). The EACH-FOR Mozambique study did not find large-scale international migration or significant rural–urban migration and confirms that movements were localized (Warner et al., 2008: 13). Adow's work on Kenya finds that most affected persons are living on the urban periphery as internally displaced persons, since borders are closed (Adow, 2008: 34).

Interestingly, study findings suggest that natural disasters may actually inhibit long-distance and international migration. Findley's study on droughts in Mali between 1982 and 1989 found changes in migration patterns in times of hardship. The Sahel traditionally has two forms of labour migration: short-cycle migration, whereby younger males, especially those from poorer families, go to nearby cities for poorly paid jobs and women go as domestics; and long-cycle migration, which is of a longer duration than long-distance migration to locations including foreign places such as France, Gabon and Senegal (these people could actually be considered long-cycle circulators as they stay abroad for two-to-three years, have home visits and send remittances). Long-cycle migration can require years of planning and is thus an impractical spontaneous response to drought-induced crop failure (Findley, 1994: 541). In the 1983–85 drought, the majority of families depended on migration to get them through the crisis; some depended on remittances from prior migrants (including long-cycle migrants), and other families sent some of their members to become new migrants. The actual rate of migration did not increase (perhaps because it was already at capacity) but there was a shift to short-cycle migration to destinations in Mali (Findley, 1994: 549).

Where migrants do go abroad, they tend to travel along pre-existing paths where they have family ties or old colonial relationships (for example, Bangladeshis go to India, Indonesians to Sumatra etc.) (Brown, 2008a: 24). The only example of large-scale international migration due to natural disasters in recent years is Hurricane Mitch in Central America. Mitch is estimated to have caused 20,000 deaths and 13,000 injuries, to have left 1.5 million homeless, and to have affected another 2 million in other ways (Kugler, 2006: 5). Emigration increased after Hurricane Mitch in 1998 (and following two earthquakes in El Salvador in 2001, and Hurricane Stan in 2005) (Davy, 2006) both to the USA and within the region (Mahler & Ugrina, 2006). Emigration from

Honduras almost tripled and emigration from Nicaragua increased by about 40 per cent (Kugler, 2006: 6). These migrants tended to go towards the American southern states of California, Florida and Texas – areas that had traditionally received this group before the disasters. As a formal response to the migration generated by Hurricane Mitch, on 30 December 1998, the Immigration and Naturalization Service (INS) designated Temporary Protected Status (TPS) for a period of 18 months, whereby Hondurans and Nicaraguans who had entered the country would not be subject to removal and would be eligible for permission to work in the USA (Kugler, 2006: 6). This example shows that intra-regional migration is an important phenomenon in the developing world. Studies from other regions show similar trends – for example, there is a dense network of disaster-spawned migration in West Africa (Raleigh et al., 2008: 21; Findley, 1994), and regional migration in South-East Asia post-floods (Dun, 2009).

Vulnerability–resilience migration

Natural disasters as geophysical events do not solely generate risk; rather, it is the state of human development that shapes vulnerability to natural hazards and exacerbates their effects and consequences. The terms 'vulnerability' and 'resilience' as used in this chapter refer to the structural factors that render certain groups 'vulnerable' or 'resilient' in the face of natural disasters; it does not refer to their internal capacities, emotional strengths or personal qualities.

The fact that developing countries are more vulnerable than developed countries is already well established. The section on migration and natural disaster statistics provided data to support this claim. Further evidence that national wealth determines vulnerability is demonstrated by the fact that, between 1994 and 2003, natural disasters in countries of high human development killed an average of 44 people per event, while disasters in countries of low human development killed an average of 300 people each (Brown, 2008a: 43). In another statistic, between 1991 and 2005, developing and least developed countries suffered 884,845 deaths, compared to OECD countries, which suffered 61,918 deaths due to natural disasters (Basher, 2008: 35). Indirect consequences – population changes, brain drain, political instability, health impacts and gender impacts – can serve to further undermine national development.

There is also a variation in vulnerability within countries. Many natural disasters are partly 'man-made'; a natural hazard may only become a disaster if a community is particularly vulnerable – for example, due to a lack of early warning, poorly built houses, or if people don't know how to protect themselves. Government policy, population growth and community-level resilience all contribute to people's vulnerability, which may be mediated by class, race, ethnicity, gender or age (Oliver-Smith, 2006). Castles gives the example of a mudslide that buries a Brazilian shantytown, which may be labelled by the media as a natural disaster but, on closer examination, might be seen as the result of land speculation, unplanned urban growth and a lack of government accountability (Castles, 2004). According to Brown, *"A community's vulnerability, then, is a function of its exposure to climatic conditions (such as a coastal location) and the community's adaptive capacity (the capacity of a particular community to weather the worst of the storm and recover after it"* (Brown, 2008a: 18). Thus, except in extreme disaster events, the actual displacement is due to a complex pattern of factors including political, social, economic as well as environmental forces (Castles, 2004).

Studies in the developing and developed world show how socially disadvantaged or marginalized groups are disproportionately susceptible to displacement by disasters. Those at the lower socio-economic levels are known to be worst affected, due to poor housing and vulnerable locations, and fewer prevention and response capabilities (Naik, 2007: 20). Out-migration following natural disaster is a rational response for certain groups and the theory suggests that vulnerability of place is negatively related to net migration–in other words, more socially vulnerable places will have incurred greater population loss (Myers et al., 2008: 275). There are a number of empirical studies that consider how different variables interact with vulnerability and migration. The factors that re-emerge in the literature are discussed here:[5]

- **Race/ethnicity**

  A study in the USA by Morrow-Jones (1991) identified racial minorities, as one of the groups especially likely to migrate following disasters (Morrow-Jones, 1991). This supports findings by other authors such as Belcher-Bates in 1983, Enarson in 1998, Fordham in 1999, Haas el in 1977 (Myers et al., 2008: 274).

[5] As most studies use multiple variables, they are organized under specific criteria purely to illustrate the point in question but may also be relevant to other variables.

Some studies suggest that it is a combination of race and class that is the key. Koerber's study of Hurricane Katrina finds that a larger percentage of African-Americans than white Americans moved out, perhaps because the latter owned houses and retained their jobs after the hurricane (Koerber, 2006: 10). This supports previous research on the disproportionate displacement of low-income African-Americans post-disaster (Myerset al., 2008: 274, 286–7, citing Frey & Singer, 2006 et al.).

Elliot's research on Hurricane Katrina found counter evidence on class and race, as, according to that study, low-income African-Americans remained in New Orleans and did not evacuate. The authors attribute this to a variety of factors, including a lack of adequate personal transport, limited spatial networks, lack of hotel reservations, and the need to remain in situ to collect social security payments. A belief that the storm would not be as bad as predicted prevented all races/classes from leaving early but the better off (i.e., the white population) had the resources to eventually leave at the last minute under their own steam (Elliot & Pais, 2006: 309).

- Wealth

The evidence of the role of poverty in stimulating migration post-disaster is surprisingly conflicting. Morrow-Jones (1991), Myers et al. (regression analysis of Hurricane Katrina data from the US Census Bureau) (2008: 286–7) and Koerber (2006) all support the notion that poorer, more disadvantaged groups are more likely to migrate post-disaster. Other connected factors may mean that poor people are more likely to migrate. Studies from the USA (Hurricane Katrina) and developing countries show that the poor tend to live in vulnerable areas, on marginal land and in coastal zones, from where they are more likely to have to move. Mass displacement is also increased when disasters strike highly populated areas (Myers et al., 2008: 275) – again, more likely to be inhabited by poorer groups.

A number of significant studies postulate the opposite position and suggest that better-off groups migrate. They find that people who migrate do not necessarily share the demographic characteristics of populations most at risk – for example, migration in East Africa after the droughts is undertaken by young males; migration from Mali is by those with the economic resources, education and experience to undertake ambitious and potentially more lucrative migration projects to foreign climes (Findley, 1994: 549); migration from flood-prone areas in Peninsular Malaysia is only an option for wealthier households, while poorer families are restricted by poverty and, even if they do relocate, they just end up in different floodplains (Myers et al., 2008: 285–286, quoting Chan, 1995); and Elliot's research cited above on Hurricane Katrina (2006: 309). Research on the USA Dust Bowl years of the 1930s found that migrants to California were intact nuclear families of above-average education, from a range of occupational backgrounds and with extended family in California (McLeman & Smit, quoting the work of Gregory, 1989) – that is, atypical of those most adversely affected in the source area, who were landless people, the rural poor, the sick, the elderly and those with little support. *"Those most vulnerable are not necessarily the most likely potential climate change migrants. To undertake migration, particularly over long distances, requires access to money, family networks and/or some other form of assistance or capital. In the 1930s shantytowns sprawled on the outskirts of urban centres in the southwestern plains states, this appears to be where persons adversely affected but lacking the ability to migrate away from the region ended up..."*(McLeman & Smit, 2004).

- **Home ownership**

  This emerges as an important factor, and consistent findings from both the developing and developed world show that homeowners are less likely to migrate than non homeowners. A study of the effects of the tsunami in Sri Lanka found that migration was less likely among those with land/home ownership (Grote et al., 2006: 1). Migrations during the USA Dust Bowl years of the 1930s mainly comprised tenant farmers without ties (Brown, 2008a: 22, based on the works of Gregory). Koerber's study post-Katrina found that renters were more likely to migrate (2006). Research from Florida (USA) shows that lower-income groups are more likely to rent, have mobile homes, and live in lower-quality constructions that are less able to withstand disasters – for example, they do not have hurricane-resistant windows or sufficient insurance (Peacock & Girard, 1977; Myers et al., 2008: 274).

- **Education**

  The role of education is inconclusive. Some studies suggest that those with lower levels of education are likely to migrate (Morrow-Jones USA study, 1991 and Grote's study in post-tsunami Sri Lanka,

2006) but there is also evidence to the contrary – for example, in Mexico, where weather-related disasters are estimated to be responsible for 80 per cent of economic losses between 1980 and 2005, the segment of the population most likely to emigrate are those with higher levels of education but living far from trading centres (Saldana-Zorrilla & Sandberg, 2009). The USA Dust Bowl migration also comprised better-educated people (quoted in Brown, 2008a: 22).

- **Age**

  The limited analysis of age as a variable comes from studies in the USA. Koerber (2006) tracked migration changes following Hurricane Katrina and discovered that younger people were more likely to leave the area. Morrow-Jones (1991) found the opposite – that older people are more likely to migrate (Morrow-Jones, 1991). It is not possible to generalize on the basis of this information.

- **Gender**

  Natural disasters have distinct gendered effects: women respond to natural disasters by mobilizing social networks, whereas men adopt strategies that take them away from their families and communities–for example, seeking external employment, temporary or permanent migration and, sometimes, abandoning families (Delaney & Shrader, 2000: 15). Research findings reveal a consistent pattern of men leaving and women staying behind – for example, following Hurricane Mitch in Honduras and Nicaragua (Delaney & Shrader, 2000: 25) and in environmental migration in Niger (Afifi, 2009). In fact, increased male migration has the effect of restricting female migration. Halliday's research on Central American earthquakes in 2001 found that men were freed up to migrate to the USA because women stepped in to cover production at home (Halliday, 2007:18). The consequences of male migration can be twofold. Women becoming de facto heads of households can facilitate their empowerment, on the one hand (Delaney & Shrader, 2000: 5). On the other, male migration can impede women's recovery from disaster (Brown, 2008a: 35), by leaving them overburdened and less able to rebuild home-based businesses, re-plant fields or relocate (Delaney & Shrader, 2000: 21–22).

  Other studies confirm that gender roles and societal norms determine whether women are able to migrate. One study from the USA found that female-headed households (i.e., where men were absent) were actually more likely to migrate post-disaster (Morrow-Jones, 1991). A study from Mali shows that women and children were sent away during droughts on pretexts of family visits or health care, as economic motives were considered dishonourable in that context (Findley, 1994: 542). When women do migrate post-disaster, they may be exposed to additional risks, hazardous jobs, gender discrimination and trafficking. In a study on drought in India, Enarson concluded that migration, remarriage or sex work were the only options for Indian women made destitute by gendered vulnerability to drought (Enarson, 2000).

The literature discussed above shows that various factors enhance or diminish a person's vulnerability to natural disaster and the likelihood that people will migrate. The studies use a range of similar but not identical variables, so comparisons between these studies is inevitably flawed. Table 4 below presents the key characteristics provided by the literature, to date, regarding the influence of various individual characteristics on migration in response to natural disasters.

Table 4: A summary of the influence of select variables on migration post-natural disaster

Characteristics	Influence on migration*
Race/ethnicity	Racial minorities are more likely to migrate – Morrow - Jones (USA), Myers (USA), Koerber (USA) Racial minorities are less likely to migrate – Elliot (USA)
Wealth	Poor are more likely to migrate – Morrow-Jones (USA), Myers (USA), Koerber (USA). Connected factors: living in more vulnerable areas that were densely populated - Myers (USA); suffered more housing damage – Myers (USA); did not have jobs – Koerber (USA). Better-off are more likely to migrate – Findley (Mali), Elliot (USA), Chan (Malaysia), Gregory (USA Dust Bowl).
Home ownership	Homeowners less likely to migrate – Grote et al. (Sri Lanka tsunami). Non-homeowners are more likely to migrate – Grote (Sri Lanka), Chan (Malaysia), Gregory (USA Dust Bowl), Koerber (USA). Lower-income groups are more likely to rent – Peacock et al. (USA).
Education	Less educated are more likely to migrate – Morrow-Jones (USA) More educated are more likely to migrate – Grote (Sri Lanka), Chan (Malaysia), Gregory (USA Dust Bowl), Saldana (Mexico), Findley (Mali).

Age	Older people are more likely to migrate – Morrow-Jones (USA), Afolayan &Adelekan (Sudan). Younger people are more likely to migrate – Koerber (USA); McLeman (USA), Findley (Mali)).
Gender	Women are more likely to migrate – Morrow-Jones (USA). Men are more likely to migrate – Delaney (Hurricane Mitch), EACH-FOR Project (Niger), Halliday (Central America), Delaney (Andrew).

*Details in brackets refer to area, country or climatic event studied.

Admittedly, this table presents a crude analysis, given the limited number of studies available, and the problems of making like-to-like comparisons. Nonetheless, some observations can be made. Probably the most surprising finding is on poverty: it is often assumed that poorer groups are more likely to migrate, that they are likely to live in more vulnerable, densely populated areas that have suffered more damage, and that they are less likely to own homes or have jobs to keep them in situ. This is supported by studies that also link poverty to other social characteristics (such as race, ethnicity, gender and age), thus accounting for why certain groups move and others don't.

However, there is significant evidence to the contrary – that, in fact, relatively better-off groups migrate. The different conclusions arrived at by these studies can perhaps be reconciled in two ways. Firstly, the studies that show that better-off people migrate seem to refer to slow-onset or repetitive disasters, such as droughts and floods, rather than sudden emergencies. These are situations where people with the means to escape can make considered life choices to move away from disaster-prone areas. There appears to be a difference in the sustainability of these migratory movements. Secondly, they also highlight what is already known – that it is not always the absolute poorest or the most vulnerable groups in society that migrate. Except for situations of immediate flight, the ability to migrate requires a modicum of financial and social resource (Brown, 2008a: 9).

In terms of other variables, the findings on race/ethnicity, mainly from the USA, highlight the interplay between racial minority, low income levels and the likelihood of migration post-disaster. The findings on gender are generally consistent: men migrate post-disaster and women stay behind. It would be interesting to see how socio-economic status affects these patterns as the studies cited tend to refer to poorer groups that use migration as a family coping strategy. Home ownership is a decisive factor, and those who own their homes experience a strong compulsion and obligation to remain in the disaster-affected area. By definition, they are likely to be of middle-income levels, not wealthy enough to abandon their homes completely but not poor enough to have to leave. The studies do not say anything conclusive about age or education. Presumably, however, it can be assumed that older people may lack mobility and are thus less likely to move. Education, conversely, would facilitate mobility and well planned migration.

Aside from these factors, there are always deeply personal and individual experiences that influence decisions on whether to stay or go. The post-tsunami study in Sri Lanka, mentioned above, highlighted emotional and personal experiences of the tsunami itself: households that had been more negatively affected by the tsunami (in terms of number of dead, missing or seriously injured members), or who had had bad experiences with the sea before the tsunami, were more likely to migrate elsewhere (Grote et al., 2006: 1).

Macro-level factors also come into play. Local economic conditions shape disaster vulnerability and migration–for example, pre-disaster depressed economies and those that are overly dependent on a particular sector are more vulnerable (Myers et al., 2008: 275). Jeffery argues that large-scale agricultural production in the Dominican Republic created vulnerability by reducing the ability of local people to recover from disasters and led to urban migration (Jeffrey, 1982). A study on Ethiopia found that community vulnerability to food crisis had a significant positive effect on out-migration (Ezra, 2001: 763). In the USA, too, the government's decision not to reconstruct Homestead Air Force Base, a major regional employer, after Hurricane Andrew, may have accounted for some of the permanent migration that ensued (Oliver-Smith, 1992). Social networks in the destination area can encourage migration, as shown by Gregory's work on the USA Dust Bowl migration (McLeman, 2004). A study in Bangladesh found that kin groups, lineages or even entire villages shifted their home to nearby urban areas as a survival strategy during floods. Social networks in destination areas were an important draw and people from the poorest quartiles had to migrate without having such social networks in place (Rayhan & Grote, 2007: 94).

Out-migration does not always take place

There is compelling evidence that migration post-disaster does not always take place, even where it might be expected. This might be because the initial fear and terror caused by the natural event diminish and people feel psychologically able to stay or because, as discussed above, people lack reasonable alternatives and have no option but to remain in high-risk areas. In addition, there may be 'pull' factors that keep people in the disaster-affected location. Paul's study provides empirical evidence that out-migration did not occur following the 2004 tornado in Bangladesh. The availability of aid, the effectiveness of distribution and the limited area affected all served to stem outflows. Notwithstanding arguments about aid discouraging self-sufficiency, the study shows how effective aid delivery can help people to successfully remain in affected areas (Paul, 2005: 381–2). This confirms the experience of developed countries, where governments discourage post-disaster migration by giving economic incentives such as subsidies, low-interest loans and tax credits. In 2003, for example, the USA leaders of five tornado-affected cities gave financial incentives to residents to rebuild or relocate nearby (op. cit.: 373–4).

Where disaster-affected populations do not migrate out, they may use other adaptation strategies to enable them to remain in their homes. In the developed world, it is not so much exposure to events but changes in adaptive capacity that decrease outflows post-disaster. For example, the US drought of the 1930s caused widespread population displacement but subsequent droughts have not, and this can be attributed to enhanced adaptive capacity, such as changes to land tenure, new technology, crop insurances, subsidies and financial relief. Adaptive capacity has increased so much that migration has not been an adaptive option in the USA since 1930s. In less developed countries such as Ethiopia and Sudan, where governments cannot provide assistance, migration will continue to be an adaptive strategy for as long as remaining in situ brings a risk of starvation (McLeman & Smit, 2004).

In developing countries, diversification of livelihoods is a typical response by vulnerable communities in areas prone to natural disasters as a way of spreading risks and mitigating damage (Naik et al., 2007:39). Typical diversification strategies include non-farm work, animal holdings, sending of remittances, diversifying income, social support networks, dispersed grazing, changes in planting practices, collecting goods, inter-household transfers and loans, use of credit, food rationing, sale of assets, commodity trading, and reliance on relief aid (Raleigh et al., 2008: 18). Studies also indicate that there are differences in strategies adopted in response to slow-onset disasters as compared to sudden catastrophes. The most critical strategies in slow-onset disasters are diversification of livelihood, consolidation of savings into incontestable forms, and social investment (which may include planned migration), whereas sudden crises require liquidation of savings, service labour and movement (which may amount to forced migration) (Raleigh et al., 2008: 19, quoting Shipton, 1990).

Meze-Hausken's study on sedentary farmers in Ethiopia shows that stratification of coping strategies can delay migration post-disaster but that it cannot avert it altogether. The study found that not all households were equally vulnerable at the beginning of the drought, as some had other coping strategies, but that, after a number of months of critical food and water shortages, differences in vulnerability between households disappeared and families were forced to migrate (Meze-Hausken, 2000: 23). Migration becomes a critical option for disaster-affected peoples, at some point; it may not be the first adaptive response (for example, selling livestock may come first) but when other things do not work and there is no support from government, migration becomes an option (Brown, 2008a: 22).

Migration is sometimes described as a failure of adaptation rather than a form of it (Brown, 2008a: 38), giving the impression that, as people are unable to cope with the disaster, they have no option but to leave. However, the studies discussed in this report demonstrate that it can, in fact, be a carefully considered adaptation strategy enabling households to cope with extreme climatic risks (McLeman & Smit, 2004). Examples from Bangladesh or Ethiopia, where there are repeated natural disaster events, show that communities have learnt a form of adaptive behaviour to cope with disasters. Even where disasters do not lead to new outflows, previous migration may play an important role in recovery through the receipt of remittances. A survey carried out by IOM in Guatemala found that 11 per cent of persons receiving remittances reported being victims of natural disasters: "*Migrants who are already living overseas when a natural disaster strikes send more remittances. Therefore, remittances after a natural disaster can be considered vital to the household economy of families directly affected and to the country as it deals with the emergency and reconstruction phases*" (IOM, 2008b). Research on the Caribbean shows that remittances represent a significant form of post-disaster financing for affected households (Attzs, 2008: 11–12).

Experiences post-disaster

Although post-disaster experiences are not the main subject of this chapter, it is worth noting a few of the factors affecting migrants following a disaster. Migration and disaster can raise a variety of human rights issues, including access to assistance, non-discrimination, protection of women and children, trafficking, access to education, loss of documentation, participation of internally displaced persons (IDPs), voluntary return and resettlement, and property (Kalin, 2003). In addition to these risks, migrants suffer emotional trauma and find it difficult to adjust to their new surroundings. Hugo argues that involuntary migrants are more likely to find it difficult to adjust to their new destination for multiple reasons connected with the unexpectedness of the move, their own emotional state and the unfamiliarity of their new surroundings (Hugo, 2009).

Migrants in destination countries can themselves be victims of natural disasters, which may cause them to be displaced once again. The Indian Ocean tsunami and Hurricane Katrina both brought to the fore the plight of migrant workers in the face of natural disasters (Laczko & Collett, 2005). A comparison of these two disasters reveals a remarkable similarity in experience between the Burmese migrants in Thailand, post-tsunami, and the Latino migrants in the USA, post-Katrina. They both faced a loss of documentation and insecurity about migration status, fear of arrests, deportations, a lack of access to aid, insecurity about migration status and, even, in the case of the Burmese, an inability to properly grieve for and bury their dead. These experiences highlight the importance of including migrants in disaster preparedness plans (Naik, 2008).

As a final point, it is also worth reflecting on what happens to disaster-affected areas that experience mass out-migration. Researchers say that out-migration itself can leave a chronic legacy. McLeman argues that the 1930s Dust Bowl years in the USA resulted in a social decline of affected areas as young skilled families with money and social networks migrated, resulting in the abandoned communities becoming polarized between affluent property-owners, on one side, and an impoverished underclass, on the other (Brown, 2007: 23).

4. Migration into disaster-affected areas

Migration into disaster-affected areas is a little studied area but the literature does acknowledge that these movements are occurring, even if it does not probe them further. These movements can take two forms: the entry of new migrants, whether they are seeking work in reconstruction or coming in to provide assistance and support to friends and relatives, and the return of those displaced/migrated.

New migrants

Migration into disaster-struck areas can take place for various reasons: a search for work opportunities in reconstruction, a desire to assist and support family members, or simply because the affected areas continue to be prestigious and act as a draw, regardless of disaster. Migration into disaster-struck areas is a known phenomenon but it has not received much attention. Belcher and Bates, in their study of the Guatemalan earthquake and Hurricane David in the Dominican Republic in 1979, found that both disasters sped up pre-existing migration patterns and led to permanent migration, even by those who had not personally suffered loss. The disaster somehow acted as a catalyst, as opportunities were created for personal betterment. Population movements included an influx by the rural poor from areas that were not damaged by the earthquake to areas offering employment opportunities in the international programmes on housing construction and the building boom in the aftermath of disaster. The authors concluded that natural disasters directly or indirectly create a redistribution of the population; they do not alter migration patterns but speed up pre-existing trends (Belcher & Bates, 1983: 127).

This phenomenon has gained visibility in recent years with Hurricane Katrina and the Asian tsunami. In the USA, the construction industry in New Orleans quickly became a magnet for Latino immigrants who were lured by the promise of paid work and an emergency federal decree temporarily suspending immigration-enforcement sanctions and minimum-wage regulations. The media picked up on the trend

(Donato, 2006) but there were no official estimates. Unofficial estimates put new arrivals some three months after the disaster at 30,000. These arrivals included USA residents as well as new migrants from Guatemala, Honduras, Mexico and other Latin American countries (Donato, 2006). Following the tsunami, Burmese migrant workers came to Thailand's coastal areas in search of work (Naik et al., 2007: 8, 9). In both situations, there were reports of discrimination and ill-treatment of newly arrived migrant workers. Reports from both countries indicated exploitation in the labour market, poor pay or no pay, with employers reneging on agreements, inadequate health and safety standards, and hostility from locals (Donato, 2006; Naik, 2008). It is not known how long the new migrants stayed in either case. However, previous experience shows that even where new migrants come with the intention of leaving, the longer the jobs last, the more likely they are to stay permanently. In 1992, when Hurricane Andrew displaced 250,000 residents in south-eastern Florida, it triggered a construction boom that attracted large numbers of Latino immigrants who have now settled in the area. The Latino population grew by as much as 50 per cent during the 1990s (Donata, 2006).

New migrants may also be motivated for other reasons to come into disaster-affected areas. Following the tsunami in Sri Lanka, many people returned to their village of origin to help with reconstruction or to comfort and emotionally support family members that survived. These were returnees from previous migrations and their return home put further strain on the households, since the cost of sudden return and the expenses of settling into their home village increased their financial vulnerability (Gallina, 2007: 6). There were also some movements by skilled diaspora coming in to provide support (Naik et al., 2007: 8, 9). Little more is known about these types of movements, which merit further attention.

Finally, evidence from the USA shows that some areas continue to hold such appeal that the occurrence of natural disasters is not a deterrent. Florida frequently suffers from hurricanes and was battered in 2004–2005 by Hurricane Wilma – the most powerful windstorm ever measured. Florida is the top American state, in terms of net migration from other American states (i.e., the number of people moving into Florida exceeds those moving out). Migration patterns in hurricane-affected regions and Florida, as a whole, suggested no lasting longer-term impact in terms of changing the overall net flow of migrants into the state. The impacted regions continue to draw a high number of new residents. There were some movements out of immediately affected areas but not the state as a whole (despite the fact that the whole state is prone to these disaster events) (National Association of Realtors (NAR), 2006). There was a similar pattern following the Loma Prieta earthquake in California in 1989. Analyses show that short-term out-migration increased in the year after the earthquake and that in-migration decreased, with some important exceptions – for example, the county of San Francisco experienced more in-migration after the earthquake than before (Wilson, 1995: 8).

Return migration after natural disasters

The other facet of migration into disaster-affected areas is the return of those who have been displaced by the disaster. Again, this remains a much understudied area. Recent disasters and, in particular, Hurricane Katrina have stimulated some research but, even here, the focus has been on analysing the decision to evacuate in the first place rather than the return (Landry et al., 2007: 7).

Displaced persons do not always have the option of returning. ,The 1995 Montserrat volcano eruption, for example, permanently displaced half the country's inhabitants, and a response was developed by Caribbean and UK governments (Ferris, 2008: 2), which included relocation of some residents to the UK, initially for a period of two years (Montserrat Governor, 1996). But, where return is possible, it is generally accepted that those who are displaced by natural disaster are likely to return home (Raleigh et al., 2008: quoting Surkhe, 1993). A study of disasters in Central America in the 1970s found that the majority of disaster victims returned home and, moreover, that a population retention rate of 90 per cent was found in damaged and undamaged areas (Raleigh et al., 2008: 7, quoting Belcher & Bates, 1983). Research also shows that they are likely to return more rapidly than those displaced by conflicts. A study in four South Asian countries by the Calcutta Research Group found that 80 per cent of those displaced by natural disasters had been displaced one year or less, whereas 57 per cent of those displaced by armed conflict and 66 per cent of those displaced by development projects had been displaced for more than five years. But Ferris says this difference may be one of degree, as there are still Central Americans displaced from Hurricane Mitch in 1998 (both in the USA and in the region), though there is no way of tracking numbers (Ferris, 2008: 2).

Hurricane Katrina provided an opportunity for academics to focus on this subject. The research suggests that those who do not return fare worse than those who do. A study by Groen and Polivka found that evacuees who did not return to their pre-Hurricane Katrina homes fared much worse than those who returned, partly due to individual and family characteristics but primarily because they came from areas that suffered more housing damage and, thus, more disruption to their lives (Groen & Polivka, 2008: 48).

Paxson examined determinants of return to New Orleans 18 months after the hurricane. This empirical analysis shows that flood exposure was the single most important factor in determining the decision to return, especially among those who did not own their own homes or lived in the homes of relatives or friends – all groups that were less likely to return. There was no support for the idea that people did not return because they found better opportunities elsewhere. On the contrary, those who experienced flooding and did not return experienced reductions in earnings (Paxson & Rouse, 2008: 42).

Landry also looked at the decision to return and found that return migration is affected by household income, age, education level, employment, marital status and home ownership, but the results depended upon the population under consideration. The study appeared to come to a variety of conclusions. Household income increases the likelihood of returning home, for some evacuees, and households with higher income are financially better equipped to make the return trip, are more likely to own homes in areas less likely to have been flooded, and have better resources to rebuild in the event that their home has been damaged (Landry et al., 2007: 26–28). Home ownership has a strong influence on the likelihood of return. Moreover, households seem willing to return even if wages in the destination area are relatively higher than those in the home area (Landry et al., 2007: 26–28).

Elliot et al. also examined the likelihood of return and found a more nuanced picture showing that lower-income home-owners are more likely to report plans to return than higher-income home-owners, both of whom are more likely to return than renters (Elliot & Pais, 2006: 315). Class and, especially, home ownership exhibit a stronger pull, as does the ability to rebuild; on the other hand, this also creates a financial weight and obligation, especially among less affluent home-owners who are more likely to say they will return. This is consistent with findings from Hurricane Andrew in 1992 (Morrow-Jones, 1997; Elliot & Pais, 2006: 317–318). The data from these studies suggest a range of variables that might be factors in the decision to return.

Although it is impossible to make any reliable generalizations from this limited research, based on a range of variables, economic issues, once again, rise to the fore in individual decision-making. Home-owners and those with higher incomes seem more ready to return than those who are poor, who have lost everything or who never owned property. This is supported by the few studies available on returns in developing countries, which show that people's decision to stay in disaster-affected areas is influenced by economic opportunities rather than the damage done. Studies of post-disaster migrants and non-migrants in the Dominican Republic and Guatemala found that people's intention to stay in their villages was not related to the damage they experienced but, rather, to the type of work they were previously involved in. Specifically, those working in coffee plantations decided to move as their economic future looked bleak, whereas people who had invested more in their home area were less likely to move (Raleigh et al., 2008: 24, citing Belcher & Bates, 1983; Quarantelli, 1982; Morrow-Jones & Morrow-Jones, 1991; and Perch-Nielsen, 2004).

5. Conclusions and recommendations

A review of the literature yields few hard and fast facts on numbers displaced by natural disasters or migratory patterns post-disaster that can be used to predict future flows. This is partly due to the limited and ad hoc research attention this subject has received but, more importantly, because, migration post-natural disaster results in a diversity of patterns of behaviour that are influenced by myriad individual, community and national-level factors. As with all studies of human behaviour, the social complexity of human society and the idiosyncrasies in individual decision making make it impossible to pre-determine rigid patterns of movement. Nonetheless, the research does highlight certain influences on migration post-disaster that, if taken into account by policy makers, could lead to better measures for prevention and response.

The chapter firstly focused on available data on migration and natural disasters and found that no global figures were being collected systematically on this subject by any agency. There is increasingly good information on other aspects of natural disasters –for example, the fact that reports of natural disasters are increasing, that increasing numbers are being affected and that developing countries bear the greatest burden in terms of human mortality and suffering. In terms of actual numbers of those displaced, Myers's 1995 figure of 25 million 'environmental refugees' (Myers, 1995) is still quoted to this day but is in need of updating, revision and validation. Lower-end estimates put the figure at 10 million. Other than that, numbers of persons affected (which include numbers displaced) can be seen as a very top-end figure but must be used with great caution and only as a 'not more than' notation. On the basis of this, one can say that, based on data derived from EM-DAT, no more than 255 million a year are displaced by natural disasters. In fact, a recent OCHA–IDMC study of this very subject analyses EM-DAT data for 2008 and separates out 'affected' from 'displaced' and finds that some 36 million were displaced in that year alone (OCHA–IDMC, 2009).

The issue of collecting data on natural disasters is fraught with difficulties, definitional problems, politics and pragmatic obstacles. Moreover, what focus and national capacity there is on data collection tends to be deployed on cross-border refugee movements rather than internal displacement (the most likely option for natural disaster victims). Nonetheless, it also emerges that not enough is being made of what already exists. It is evident that database initiatives, such as EM-DAT and the University of Richmond's Disaster Database Project, do have some relevant information that has yet to be fully exploited or analysed in the estimation of numbers of migrants due to natural disasters. A useful pilot initiative, for instance, might involve working with these databases for a trial period to disaggregate incoming data by displacement/migration, as and when they are received. It may also be possible to do a post-facto analysis of earlier periods, depending on what types of supporting documentation are retained by these projects.

Further efforts to make a more concerted effort to collect new information on migration and displacement would clearly help to further illuminate the question of numbers. There is growing recognition of this need and the Inter-Agency Standing Committee's (IASC) informal group on Migration/Displacement and Climate Change has identified the need for significant further research and analysis regarding the scale, nature and patterns of climate-related population mobility, and regarding those who do not and cannot move (IASC, 2008).

In theory, the lack of statistics should make it difficult for decision makers to plan and respond effectively to the needs of migrant populations affected by natural disasters. However, the fact that disaster planners do not appear to seek out this type of global analysis suggests that they are managing with the information available and that, perhaps, localized knowledge and information have so far proved adequate. Alternatively, it might suggest that groups affected by displacement because of natural disaster are being overlooked. This is an area that merits closer examination.

Some practitioners, such as Hovy of the United Nations (Population Division), doubt the usefulness of collecting global information on persons displaced by natural disasters, saying that, as disasters differ significantly in nature and impact, there appears to be little added value in comparing the levels and characteristics of displacement and, in the absence of international legislation protecting the rights of those displaced in this way, there is no need for a monitoring mechanism

(Hovy, 2009). "*... it would appear that efforts should focus on assisting those who are affected by natural disasters, rather than on the fraction of victims who become internally or internationally displaced. There appears neither an operational need, nor a legal imperative, to collect detailed information on the number of persons displaced by natural disasters at global level. Efforts to improve data collection should focus on information management for particular populations at risk from and affected by natural disasters*" (Hovy, 2009).

The chapter then focused on the question of migration patterns post-natural disaster, both into and out of the affected area. Migration movements out of areas struck by disaster represent a typical response that has occurred throughout human history; thus, is it not a question of whether these movements occur but what form and shape they take. A clear finding is that migration post-disaster results in a diversity of patterns that vary according to different types of disaster, groups of people and parts of the world; the pattern may be temporary or permanent, cyclical or linear. Migration patterns depend on exposure to risk and adaptive capacity and it is important to assess migration responses in a localized context rather than making generalizations.

Migration flows were considered according to a number of spectra: voluntary–forced; temporary–permanent; internal–international; and vulnerability–resilience.

- Flight post-disaster is naturally at the forced end of the voluntary–forced continuum. Trafficking post-disaster is an extreme consequence for those with no alternatives. However, the migration experience of those who have fled natural disasters changes over time and shades into voluntary decisions on whether to return home or not. There appears to be a lack of empirical data exploring this spectrum, and a further teasing out of the distribution of forced/voluntary movements would be beneficial.

- Migration post-disaster is assumed to displace people temporarily. There are numerous examples of temporary displacement in such circumstances but it remains difficult to generalize on this point because of the complexity of setting boundaries between temporary and permanent moves (as demonstrated, for example, in the study by EACH-FOR in Mozambique) and the commonality of circular patterns. Indeed, temporary circular migration emerges as a coping strategy in places that face a recurring threat of natural disaster and examples can be found in countries as far apart as Bangladesh, Ethiopia, India, Sudan and West Africa. Furthermore, there is continuing debate about whether natural disasters themselves can ever lead to permanent migration or whether it is the deficiency of government responses that prevents people from rebuilding and returning.

- There is firmer evidence on where people go post-disaster, which shows that movements tend to be internal and, moreover, localized, to nearby safe locations. This is confirmed by examples from both developing and developed countries. The moves are not international and, in fact, research suggests that natural disasters inhibit long-distance overseas migration as this requires more planning and resources than local migration. The burden clearly falls largely on developing countries. Nonetheless, international destinations are of increasing importance and outflows from Central America to the USA, post-Hurricane Mitch, provide the largest example from recent history of cross-border migration stimulated by a natural disaster.

- It is not disasters themselves that generate risk but the state of human development that shapes vulnerability and exacerbates its effects and consequences. The difference in vulnerability between developing and developed countries is evident but there are also differences in vulnerability within countries, based on individual and community characteristics. A number of studies have looked at variables such as race, gender, class etc., to see what impact they have on migration post-disaster. Notwithstanding the problems in making comparisons between studies using different definitions and methodologies, some observations can be made.

Perhaps surprisingly, the role of poverty is not as conclusive as might be assumed. Studies do indeed confirm, as might be expected, that poorer people are more likely to migrate – especially because they are likely to live in vulnerable, densely populated areas that have suffered more damage, and they are less likely to own homes or have jobs to keep them in situ. However, there is a significant body of research to show that it is better-off groups that migrate post-disaster. These conflicting findings can perhaps be reconciled as follows. The studies showing that better-off groups migrate tend to refer to slow-onset disasters such as droughts or those that occur with regularity (e.g., floods) where people with the means at their disposal are able to plan sustainable migrations away from the affected area. These findings also confirm what is already known about migration, that it is not always the most vulnerable or the absolute poorest who migrate and that migration itself requires a modicum of resources, contacts and skills. The studies also show how poverty interplays with other factors such as race/ethnicity, thus accounting for why some groups are more likely to migrate than others. The gender distinctions post-disaster are relatively consistent: men migrate and women stay at home. Added to these factors are the individual and deeply personal reasons that motivate people to move, such as the trauma of losing a loved one, or wider macro factors that interplay with individual characteristics such as location, population density, and the economic state of the affected area.

Importantly, out-migration does not always occur. As noted above, aside from immediate flight in the face of disaster, migration may not be an option for the poorest and most vulnerable groups. Furthermore, there are factors that may encourage people to stay; studies show that effective disaster relief and recovery programmes in both the developing and developed world can serve as a brake on movements out. Populations that don't move out of areas facing repeated disaster tend to rely on a diversification of livelihoods to help them cope with disaster; migration is one option and is therefore a means of adapting rather than a failure to adapt.

Although human rights are not the main focus of this paper, it is worth noting that migrants post-natural disaster often face violations or loss of their human rights (e.g., lack of access to assistance, loss of documentation, trafficking, lack of access to education, lack of protection of women and children etc.) and, as forced migrants, find adjustment in destination areas more difficult than those who have moved voluntarily in a planned way. Furthermore, migrants who experience natural disasters in destination areas tend to be a forgotten group, facing some unique challenges.

Migration into disaster-affected areas has been even less studied than out-migration. However, it is known that new migrants are drawn to affected areas by the promise of jobs in reconstruction. This was observed during the disasters in Central America in the late 1970s and witnessed again in the USA post-Katrina and in tsunami-affected parts of Thailand. These new migrants face exploitation in the workplace and problems relating to health, safety and migration status. Some very prestigious and desirable locations, such as certain parts of the USA, continue to be a draw, despite the fact that they are vulnerable to natural disasters. There is little literature on the return of victims of natural disasters. Indications are that migrants from natural disasters tend to return but more empirical evidence is needed on this point. Hurricane Katrina provided an opportunity for academics to focus on this aspect. The ensuing studies revealed that those who do not return fare worse and that, once again, economic factors, such as resources, jobs and home ownership, are key incentives to return and rebuild.

Overall, the literature indicates an immense diversity in migration patterns shaped by individual, community and macro-level factors. What is evident is that economic factors play a critical role in both exacerbating the risk of natural disaster and impeding an effective response to it. Developing countries are worst affected, and poor, marginal communities in both the developing and developed world suffer more and worse damage, are more likely to face the upheaval of displacement and are less likely to return – all largely due to economic factors, such as lack of income, jobs and homes. In areas that face repeated crises, communities have adopted innovative methods of coping with disaster and used migration as a tool to help see them through a crisis. The most critical lesson to be learned from this literature review is that investment in economic development can help mitigate risks, prevent the upheaval of displacement and help facilitate returns.

Predictions of migration flows caused by environmental factors are impossible to make on the basis of this literature. Myers's projection of 200 million 'environmental refugees' by 2050 (Myers, 1995) has been dismissed as apocalyptic and based on no more than anecdotal evidence and intuitive judgement (Castles, 2004; Hugo, 2009: 47, quoting Lonergan & Swain, 1999). Making accurate predictions is complicated by the lack of baseline data, as shown by this report (due to problems of causality, capacity etc.), as well as unknown factors regarding future population growth and the evolution of climate change, including the scale of future emissions (Brown, 2008a: 10). Although the impact of climate change on future movement is not known, it will obviously not be possible or desirable to seek to prevent all migration flows. Populations from small island states facing submergence, for example, will require resettlement and other areas, too, may become unsuitable for human habitation. However, where limiting out-migration is an appropriate policy option, it is clear that investment in economic development is key to minimizing flows and to restoring areas to their pre-disaster status. The following are some recommendations and suggestions for furthering the work on migration and natural disasters:

- More efforts could be made to collect data on migration and natural disasters but these should build on existing initiatives. A partnership with databases such as EM-DAT should be pursued and could involve:

  - extrapolation and analysis of existing statistics to see what additional lessons can be learned about natural disasters and migration (these data can be further disaggregated by sex, age, etc.);
  - review of original documentation collected by databases collecting information on natural disasters and related issues to see how they can inform both numbers and patterns of movement;
  - efforts to collect better statistics on migration and natural disasters. This would involve agreeing definitions and methodologies and working with organizations managing relevant databases to add 'migration' as a research criterion for their collections. This work could be carried out on a trial basis to see whether it is indeed feasible to collect reliable data on this topic or if, as some would suggest, this is not possible at all. At the time of finalizing this paper, OCHA–IDMC had embarked on just such an exercise, looking at data on natural disasters and migration (OCHA–IDMC, 2009). Such efforts need to ensure that both internal and cross-border movements are tracked and disaggregated.

There are numerous recommendations on how data collection can be improved (BESR et al., 2007; Reed et al., 1998). They are beyond the scope of this chapter but they should be borne in mind in any further initiatives undertaken.

- There are still gaps in knowledge about patterns of movement that merit further attention. Although all aspects of migration require further research, particular areas where knowledge would be useful to policy makers include whether post-disaster migration is temporary or permanent, what factors influence return, and what factors build resilience. There are several ways in which patterns of movement post-migration can be studied further:

- review and analyse original documentation collected by natural disaster databases to see what patterns of movement emerge;
- carry out real-time studies during natural disaster crises to track migration outwards and inwards in more detail;
- carry out a longitudinal study over a finite period (e.g., a year) to track migration responses to all natural disasters within a certain timeframe (inevitably relying on the proactive collection of secondary source material of the type collected by the existing databases);
- carry out empirical studies on a sample of natural disasters covering different disaster types and regions but using the same methodology and variables in order to produce comparable results.

• The research also highlights the need to work more closely with disaster planners and policy makers:

  - It is necessary to carry out an analysis of how well humanitarian organizations are able to respond to the needs of those migrating post-natural disaster. This aspect was beyond the scope of this particular report and would involve consultations and partnerships with key organizations responsible for humanitarian delivery in such situations, in order to map out what is being done. This is crucial to understanding what the response is, what the gaps are, what agencies feel they need in terms of data and information, and what migrants or vulnerable groups themselves feel they need in such situations. Research following Hurricane Katrina and the Indian Ocean tsunami suggests that migrants are falling through the cracks, in some respects.
  - Another central message, which may be of particular interest to policy makers in developed countries concerned about potential influxes of migrants due to climate change, is that economic development to reduce vulnerability clearly emerges as the best way of mitigating risks, minimizing damage and speeding recovery and return.

6. References

Adow, M.
2008 "Pastoralists in Kenya", *Forced Migration Review*, 31:34, Oxford. Available at: http://www.fmreview.org/climatechange.htm

Afifi, T.
2009 *Egypt case study report*, Environmental Change and Forced Migration Scenarios (EACH-FOR). Available at: http://www.each-for.eu/documents/CSR_Egypt_090130.pdf

Afifi, T.
2009 *Niger case study report*, Environmental Change and Forced Migration Scenarios (EACH-FOR). Available at: http://www.each-for.eu/documents/CSR_Niger_090130.pdf

Afolayan, A.A. & Adelekan, I.O.
1998 "The role of climate variations on migration and human health in Africa", *The Environmental*, 18: 213–218.

Alscher, S. et al.
2009 *Mexico case study report*, Environmental Change and Forced Migration Scenarios (EACH-FOR). Available at: http://www.each-for.eu/documents/CSR_Mexico_090126.pdf

Alvarez Gila, O. et al.
2009 *Ecuador case study report*, Environmental Change and Forced Migration Scenarios (EACH-FOR). Available at: http://www.each-for.eu/documents/CSR_Ecuador_090126.pdf

Attzs, M.
2008 *Natural Disasters and Remittances: Exploring the Linkages between Poverty, Gender and Disaster Vulnerability in Caribbean SIDS*, Research Paper No. 2008/6, United Nations University, World Institute for Development Economics Research (UNU-WIDER), Helsinki. Available at: http://www.wider.unu.edu/publications/working-papers/research-papers/2008/en_GB/rp2008-61/

Baker, J.
2008 "Hotspots – predictions and action", *Forced Migration Review*, 31:44–45, Oxford. Available at: http://www.fmreview.org/climatechange.htm

Basher, R.
2008 "Disasters and what to do about them", *Forced Migration Review*, 31:35–36, Oxford. Available at: http://www.fmreview.org/climatechange.htm

Bates, D.
2002 "Environmental Refugees? Classifying Human Migrations Caused by Environmental Change", *Population and Environment*, 23(5), Springer, The Netherlands.

Battistella, G. & Asis, M.M.B.
2003 *Unauthorized Migration in Southeast Asia*, Scalabrini Migration Center, Quezon City, Philippines.

Belcher, J.C. & Bates, F.L.
1983 "Aftermath of natural disasters: Coping through residential mobility", *Disasters*, 7(2):118–128.

BESR (Board on Earth Sciences and Resources) et al.
2007 *Tools and Methods for Estimating Populations at Risk from Natural Disasters and Complex Humanitarian Crises*, National Academies Press, Washington, DC. Available at: http://www.nap.edu/catalog.php?record_id=11895#orgs

Black, R.
2001 *Environmental Refugees*: Myth or Reality? UNHCR Working Papers 34:1–19, UNHCR, Geneva.

Black, R.
2008 *Demographics and climate change: future trends and their policy implications for migration, Department for International Development (DFID), London, UK*. Available at: http://www.unitarny.org/mm/file/Migration/2008/DfID%20report_summary%2013%20Jan.pdf

Bleibaum, F.
2009 *Senegal case study report*, Environmental Change and Forced Migration Scenarios (EACH-FOR). Available at: http://www.each-for.eu/documents/CSR_Senegal_090126.pdf

Brown, O.
2007 *Climate change and forced migration: Observations, projections and implications*, UNDP, New York. Available at: http://www.iisd.org/pdf/2008/climate_forced_migration.pdf

Brown, O.
2008a *Migration and Climate Change*, International Organization for Migration, Geneva. Available at: http://www.iom.int/jahia/webdav/site/myjahiasite/shared/shared/mainsite/published_docs/serial_publications/MRS-31_EN_correct.pdf see

Brown, O.
2008b "The numbers game", *Forced Migration Review*, 31:8–9, Oxford. Available at: http://www.fmreview.org/climatechange.htm

Castles, S.
2002 *Environmental Change and Forced Migration: Making Sense of the Debate*, UNHCR Working Papers 70 :1–14, UNHCR, Geneva.

Castles, S.
2003 "The International Politics of Forced Migration", *Development*, 46(3): 11–,20(10), Palgrave Macmillan.

Castles, S.
2004 "Confronting the Realities of Forced Migration", *Migration Information Source*. Available at:
http://www.migrationinformation.org/Feature/display.cfm?ID=222

Chinedu, U.O.
2008 "Internal displacement in Nigeria", *Forced Migration Review*, 31: 37, Oxford. Available at:
http://www.fmreview.org/climatechange.htm

Davy, M.
2006 "The Central American Foreign Born in the United States", *Migration Information Source*. Available at: http://www.migrationinformation.org/USFocus/display.cfm?ID=385

De Jong, G. F. & Fawcett, J.
1981 "Multidisciplinary frameworks and models of migration decision making" in *Migration Decision Making Multidisciplinary Approaches to Microlevel Studies in Developed and Developing Countries*, De Jong, G.F & Gardner, R.W. (Eds), Pergamon Press, New York: 13–58.

Delaney, P.L. & Sharder, E.
2000 *Gender and Post-Disaster Reconstruction: The Case of Hurricane Mitch in Honduras and Nicaragua*, World Bank, Washington, DC. Available at: http://www.genderandenvironment.org/admin/admin_biblioteca/documentos/272_3_40_40_reviewdraft.doc

Deshingkar, P.
2005 *Maximizing the benefits of internal migration for development*, background paper prepared for the Regional Conference on Migration and Development in Asia, Lanzhou, China, 14–16 March.

de Vries, D.
2006 *Internal Migration after Hurricane Floyd Flooding in Eastern North Carolina*, University of North Carolina. Available at: http://www.ncsu.edu/wrri/conference/2006ac/posters/DeVries.pdf

Donato, K. & Hakimzadeh, S.
2006 "The Changing Face of the Gulf Coast: Immigration to Louisiana, Mississippi and Alabama", *Migration Information Source.* Available at: http://www.migrationinformation.org/Feature/display.cfm?ID=368

Dun, O.
2008 *Migration and displacement triggered by flooding events in the Mekong Delta, Vietnam*, United Nations University Institute for Environment and Human Security (UNU-EHS), Bonn. Available at: http://www.ehs.unu.edu/file.php?id=432

Dun, O. & Gemenne, F.
2008 "Defining 'environmental migration", *Forced Migration Review*, 31:10–11, Oxford. Available at: http://www.fmreview.org/climatechange.htm

Dun, O. & Stal, M.
2008 *The linkages and dynamics between flooding and forced relocation: Fieldwork Experiences from Vietnam and Mozambique*, Geophysical Research Abstracts, Vol. 10, United Nations University Institute for Environment and Human Security (UNU-EHS), Bonn.

Dun, O.
2009 *Vietnam case study report*, Environmental Change and Forced Migration Scenarios (EACH-FOR). Available at: http://www.each-for.eu/documents/CSR_Vietnam_090212.pdf

Elliot, J.R. & Pais, J.
2006 "Race, class and Hurricane Katrina: social differences in human responses to disaster", *Social Science Research*, 35(2) :295–321.

El-Hinnawi, E.
1985 *Environmental Refugees*, United Nations Environment Programme (UNEP), Nairobi.

EM-DAT (International Emergencies Disaster Database)
2009 Available at: http://www.emdat.be/index.html

Enarson, E.
2000 *Gender and Natural Disasters*, InFocus Programme on Crisis Response and Reconstruction, Working Paper No. 1, ILO, Geneva. Available at: http://www.unisdr.org/eng/library/Literature/7566.pdf

Enarson, E.
2001 "*We Want Work*": Rural Women in the Gujarat Drought and Earthquake, Natural Hazards Research and Applications Information Center, University of Colorado. Available at: http://www.colorado.edu/hazards/research/qr/qr135/qr135.html

Ezra, M.
2001 "Rural out-migration in the drought prone areas of Ethiopia: a multilevel analysis", *International Migration Review*, 35(2): 749–771.

Fermin, A.
2009 *Spain case study report*, Environmental Change and Forced Migration Scenarios (EACH-FOR). Available at: http://www.each-for.eu/documents/CSR_Spain_090130.pdf

Ferris, E.
2007 *Making sense of climate change, natural disasters and displacement: a work in progress (lecture)*, Brookings Institution, Washington, DC. Available at: http://www.reliefweb.int/rw/lib.nsf/db900sid/PANA-7AHDLV/$file/brookingsbern-dec2007.pdf?openelement

Ferris. E.
2008 *Displacement, natural disasters and human rights*, Brookings Institution, Washington, DC. Available at: http://www.brookings.edu/speeches/2008/1017_natural_disasters_ferris.aspx?p=1

Findley, S.E.
1994 "Does drought increase migration? A study of migration from rural Mail during the 1983–1985 drought", *International Migration Review*, 28(3):539–53.

Gainesville Sun
2004 "After hurricanes, Sunshine State loses sheen", 14 September, p.3B.

Gallina, A.
2007 *Migration and Development Linkage in Sri Lanka: A Post-Tsunami and Civil Conflict Approach*, Federico Caffe Centre Research Report No. 2006/131694–1. Available at: http://papers.ssrn.com/sol3/papers.cfm?abstract_id=1285999

Gemenne, F. & Reuchlin, P.
2008 "Central Asia case study', *Forced Migration Review*, 31:15. Available at: http://www.fmreview.org/climatechange.htm

Groen, J.A. & Polivka, A.E.
2008 "The effect of Katrina on the labor market outcomes of evacuees", *American Economic Review*, 98(2): 43–48, Nashville, Tennessee.

Grote, U. et al.
2006 *Migration due to the tsunami in Sri Lanka – analysing vulnerability and migration at the household level*, United Nations University Institute for Environment and Human Security (UNU-EHS)/The Center for Development Research (ZEF), Bonn. Available at: http://ageconsearch.umn.edu/handle/7117

Guha-Sapir, D. et al.
2004 *Thirty Years of Natural Disasters 1974–2003: The Numbers, Centre for Research on the Epidemiology of Disasters, Presses Universitaires de Louvain, Brussels.* Available at: http://www.emdat.be/Documents/Publications/publication_2004_emdat.pdf

Halliday, T.J.
2007 *Migration, risk and the intra-household allocation of labour in El Salvador*, University of Hawaii, Manoa. Available at: http://www.iza.org/conference_files/worldb2007/halliday_t3381.pdf

Hansen, B.
2008 *Three cases of migration and conflict: Hurricane Katrina, Bangladesh and The Dust Bowl*, Ecowar – notes on links between conflict and natural resources. Available at: http://ecowar.blogspot.com/2008/08/three-cases-of-migration-and-conflict.html

Haque, C.E. & Zaman, M.Q.
1989 "Coping with Riverbank Erosion Hazard and Displacement in Bangladesh: Survival Strategies and Adjustments", *Disasters*, 13(4): 300–314. Available at: http://www3.interscience.wiley.com/journal/119441201/abstract

Hedman, E.
2005 "Tsunami: learning from the humanitarian response", *Forced Migration Review*, Special Issue, July 2005, Oxford University. Available at: http://repository.forcedmigration.org/show_metadata.jsp?pid=fmo:4495

Hovy, B.
2009 Private communication with author in February 2009.

Hugo, G.
1996 "Environmental concerns and international migration," *International Migration Review*, 30(1): 105–131.

Hugo, G.
2009 *Migration, development and environment*, International Organization for Migration (IOM), Geneva.

Hunter, L.
2005 "Migration and Environmental Hazards", *Population and Environment*, 26(4), Springer, The Netherlands.

IASC
2008 *Climate change, migration and displacement: who will be affected*, working paper submitted by the informal group on Migration/Displacement and Climate Change of the IASC, Geneva.

IFRC
2001–
2006 *World Disasters Report*, International Federation of Red Cross and Red Crescent Societies (IFRC), Kumarian Press, Bloomfield, CT. Available at: http://www.ifrc.org/publicat/

IOM
2004 *Glossary on Migration*, International Organization for Migration (IOM), Geneva. Available at: http://www.iom.int/jahia/webdav/site/myjahiasite/shared/shared/mainsite/published_docs/serial_publications/Glossary_eng.pdf

IOM
2008a Speech by IOM Director General, Brunson McKinley, at the Institute for Public Policy Research Conference on Climate Change and Forced Migration. Available at: http://www.iom.int/jahia/Jahia/cache/offonce/pid/1336?entryId=17096

2008b *Seventh Annual Survey on Migrant Remittances to Guatemala*. Available at: http://www.iom.int/jahia/Jahia/pbnAM/cache/offonce?entryId=19554

IOM
2009 *Commonly used migration terms. Available at:* http://www.iom.int/jahia/Jahia/about-migration/migration-management-foundations/terminology/commonly-used-migration-terms

Irianni, M. et al.
2009 *Argentina case study report*, Environmental Change and Forced Migration Scenarios (EACH-FOR). Available at: http://www.each-for.eu/documents/CSR_Argentina_090126.pdf

IRIN
2009 *Global: Nowhere to run from nature*. Available at: http://www.irinnews.org/Report.aspx?ReportId=78387

Jacobsen, K. et al.
2008 *Internal Displacement to Urban Areas: the Tufts-IDMC Profiling Study, Case: Santa Marta, Colombia*. Available at: http://www.reliefweb.int/rw/RWFiles2008.nsf/FilesByRWDocUnidFilename/AMMF-7J7GYL-full_report.pdf/$File/full_report.pdf

Jayaweera, S. et al.
2002 *Returnee Migrant Women in Two Locations in Sri Lanka*, CENWOR, Colombo.

Jeffery, S.
1982 "The creation of vulnerability to natural disaster: Case studies from the Dominican Republic", *Disasters*, 6(1): 38–43, Wiley/Overseas Development Institute.

Johnson, R. et al.
2008 *Impacts of the 2005 Gulf Coast Hurricanes on Domestic Migration*, U.S. Census Bureau, Washington, DC. Available at: http://paa2008.princeton.edu/download.aspx?submissionId=80690

Kadirbeyoglu, Z.
2009 *Turkey case study report*, Environmental change and Forced Migration Scenarios (EACH-FOR). Available at: http://www.each-for.eu/documents/CSR_Turkey_090126.pdf

Kalin, W.
2003 *Natural disasters and IDPs' rights*. Available at: http://www.migrationforcee.org/textOnlyContent/FMR/Tsunami/03.doc

Kalin, W. & Dale, C.H.
2008 "Disaster risk mitigation – why human rights matter', *Forced Migration Review*, 31:38–39, Oxford. Available at: http://www.fmreview.org/climatechange.htm

Kayastha, S.L. & Yadava, R.P.
1985 "Flood induced population migration in India: a case study of Ghaghara Zone", *Population redistribution and development in South Asia* (edited by Leszek, A. et al.), The Netherlands: 79-88.

Keane, D.
2004 "Environmental causes and consequences of migration: A search for the meaning of 'environmental refugees'", *Georgetown International Environmental Law Review.* Available at: http://findarticles.com/p/articles/mi_qa3970/is_200401/ai_n9353848/pg_1

Kelman, I.
2008 "Island evacuation", *Forced Migration Review*, 31: 20, Oxford. Available at: http://www.fmreview.org/climatechange.htm

Koerber, K.
2006 *Migration Patterns and Mover Characteristics from the 2005 ACS Gulf Coast Area Special Products*, U.S. Census Bureau, Washington, DC. Available at: http://www.census.gov/Press-Release/www/emergencies/gulf_migration.pdf

Kugler, A.
2006 *Effects of Low-Skilled Immigration on U.S. Natives: Evidence from Hurricane Mitch*, University of Houston. Available at: http://www.uh.edu/~adkugler/Mitch_Kugler_4.pdf

Laczko, F. & Collett, E.
2005 "Assessing the tsunami's effects on migration", *Migration Information Source*, Migration Policy Institute, Washington, DC. Available at: http://www.migrationinformation.org/Feature/display.cfm?ID=299

Landry, C. et al.
2007 *Going Home: Evacuation-Migration Decisions of Hurricane Katrina Survivors*, Appalachian State University. Available at: http://econ.appstate.edu/RePEc/pdf/wp0703.pdf

Leitner, M.
2007 *The Impacts of Hurricanes Katrina and Rita on Public Safety in Louisiana – a spatial and temporal analysis of reported crimes.* Available at: http://www.ojp.usdoj.gov/nij/events/maps/pittsburgh2007/papers/Leitner2.pdf

Lonergan, S.
1998 "The role of environmental degradation in population displacement", *Environmental Change and Security Project Report*, Issue 4 (Spring 1998).

Louisiana Recovery Authority
2006 *2006 Louisiana Health and Population Survey*. Available at: http://popest.org/popestla2006/files/MigrationReport_FINv4.pdf

Mahler, S.J. & Ugrina, D.
2006 "Central America: Crossroads of the Americas", *Migration Information Source*. Available at: http://www.migrationinformation.org/Feature/display.cfm?ID=386

Mayell, H.
2002 *Ancient Chile Migration Mystery Tied to Drought*, National Geographic News.

McLeman, R. & Smit, B.
2004 *Commentary No. 86: Climate Change, Migration and Security*, Canadian Security Intelligence Service. Available at: http://www.csis-scrs.gc.ca/pblctns/cmmntr/cm86-eng.asp

McLeman, R.
2006 "Global warmings huddled masses", *The Ottawa Citizen*, 23 November 2006.

Meze-Hausken, E.
2000 "Migration caused by climate change: how vulnerable are people inn dryland areas?" *Mitigation and Adaptation Strategies for Global Change*, 5(4), Springer, The Netherlands.

Migration Information Source
2005 *Record Numbers Displaced by Natural Disasters.* Available at: http://www.migrationinformation.org/Feature/display.cfm?ID=359

Montserrat Office of the Governor
1996 *Volcano Alert: Voluntary Evacuation to the United Kingdom.* Available at: http://www.geo.mtu.edu/volcanoes/west.indies/soufriere/govt/monmedia/1996/gis_042396.html

Morrissey, J.
2008 "Rural-urban migration in Ethiopia", *Forced Migration Review*, Issue 31:28–29, Oxford. Available at: http://www.fmreview.org/climatechange.htm

Morrow-Jones, H. & Morrow-Jones, C.
1991 "Mobility due to natural disaster: Theoretical considerations and preliminary analyses", *Disasters*, 15(2): 126–132.

Murray, B.
2009 *Mass migration – a worldwide phenomenon* (summary of a presentation by D. Yang), Analysis Online. Available at: http://www.analysisonline.org/site/aoarticle_display.asp?sec_id=140002434&news_id=140001396&issue_id=1

Myers, C. et al.
2008 Social vulnerability and migration in the wake of disaster: the case of Hurricanes Katrina and Rita, *Population and Environment*, Vol. 29, Number 6/July 2008, Springer, The Netherlands.

Myers, N.
1995 *Environmental exodus: an emergent crisis in the global arena*: 14–17, Climate Institute, Washington, DC. Available at: http://www.climate.org/PDF/Environmental%20Exodus.pdf

Naik, A. et al.
2007 *Migration, development and natural disasters: Insights from the Indian Ocean tsunami*, IOM Migration Research Series, IOM, Geneva. Available at: http://www.iom.int/jahia/webdav/site/myjahiasite/shared/shared/mainsite/published_docs/serial_publications/MRS30.pdf

Naik, A.
2008 "A tale of two disasters: the plight of migrants caught up in Hurricane Katrina and the Asian Tsunami", *Journal of Humanitarian Assistance*, Feinstein Centre. Available at: http://jha.ac/2008/11/04/a-tale-of-two-disasters-the-plight-of-migrants-caught-up-in-hurricane-katrina-and-the-asian-tsunami/

NAR
2006 *The Impact of Hurricanes on Housing and Economic Activity: A Case Study for Florida*, National Association of Realtors (NAR). Available at: http://www.realtor.org/research/research/hurricanepaper

Nasritdinov, E. et al.
2009 *Kyrgyzstan case study report*, Environmental Change and Forced Migration Scenarios (EACH-FOR). Available at: http://www.each-for.eu/documents/CSR_Kyrgyzstan_090204.pdf

Naude, W.
2008 *Conflict, disasters and no jobs: reasons for international migration from Sub-Saharan Africa*, UNU-WIDER, Volume 2008/85.

NDMC (National Drought Mitigation Center)
2006 *Understanding Your Risk and Impacts: Impacts of Drought*. Available at: http://drought.unl.edu/risk/impacts.htm

Oberoi, P.
2005 "Humanitarian response towards Myanmarese migrant workers", *Forced Migration Review*, 23: 40–1.

OCHA-IDMC
2009 *Monitoring disaster displacement in the context of climate change,* OCHA-IDMC, Geneva.

ODI/UNDP
2005 *ODI/UNDP Cash Learning Project Workshop in Aceh, Indonesia: Workshop Report*, Overseas Development Institute (ODI)/ Humanitarian Policy Group (HPG), London. Available at: http://www.odi.org.uk/HPG/meetings/Cash_learning_workshop.pdf

Oliver- Smith A.
1992 *The Martyred City, Death and Rebirth in the Andes*. Waveland Press, Prospect Heights, Illinois.

Oliver-Smith, A.
2006 *Disasters and Forced Migration in the 21st Century*, Social Science Research Council (SSRC), New York. Available at: http://understandingkatrina.ssrc.org/Oliver-Smith/

Paul, B.K.
2005 "Evidence against disaster-induced migration: the 2004 tornado in north-central Bangladesh", *Disasters*, 29(4): 370–385. Available at: http://www3.interscience.wiley.com/journal/118655339/abstract

Paxson, C. & Rouse, C.E.
2008 "Returning to New Orleans after Hurricane Katrina", *American Economic Review*, 98(2):43–48, Nashville, Tennessee.

Petersen, W.
1958 "A general typology of migration", *American Sociological Review*, 23: 256–266.

Petersen, W.
1975 *Population*, MacMillan Publishing, New York.

Poncelet, A.
2009 *Bangladesh case study report*, Environmental Change and Forced Migration Scenarios (EACH-FOR). Available at: http://www.each-for.eu/documents/CSR_Bangladesh_090126.pdf

Raleigh, C. et al.
2008 *Assessing the impact of climate change on migration and conflict*, World Bank, Washington, DC. Available at: http://siteresources.worldbank.org/EXTSOCIALDEVELOPMENT/Resources/SDCCWorkingPaper_MigrationandConflict.pdf

Ratha, D. & Shaw, W.
2007 "Causes of south-south migration and its socioeconomic effects", *Migration Information Source*. Available at: http://www.migrationinformation.org/Feature/display.cfm?ID=647

Rayhan, I. & Grote, U.
2007 "Coping with floods: Does rural-urban migration play any role for survival in rural Bangladesh?" *Journal of Identity and Migration Studies*, 1(2): 82–98. Available at: http://www.e-migration.ro/jims/Vol1_no2_2007/JIMS_vol1_no2_2007_RAYHAN_GROTE.pdf

Reed, H. et al.
1998 *The Demography of Forced Migration: summary of a workshop*, The National Academies Press, Washington, DC. Available at: http://www.nap.edu/catalog.php?record_id=6187

Richmond, A.
1993 *The Environment and Refugees: Theoretical and Policy Issues*, revised version of a paper presented at the meetings of the International Union for the Scientific Study of Population, Montreal.

Rosenberg, M.
2006 *Louisiana Post-Hurricane Demographic Changes*. Available at: http://geography.about.com/od/obtainpopulationdata/a/postkatrina_2.htm

Saldaña-Zorrilla, S.O. & Sandberg, K.
2009 "Spatial econometric model of natural disaster impacts on human migration in vulnerable regions of Mexico", *Disasters*, ODI, London, Wiley InterScience. Available at: http://www3.interscience.wiley.com/journal/121636960/abstract

Smith, C. et al.
2008 "Predictive modelling", *Forced Migration Review*, 31:59, Oxford. Available at: http://www.fmreview.org/climatechange.htm

Smith, S. & McCarty, C.
1996 "Demographic effects of natural disasters: a case study of Hurricane Andrew", *Demography*, 33(2): 265–275. Available at: http://www.crid.or.cr/digitalizacion/pdf/eng/doc12120/doc12120.pdf

Speare, A. Jr.,
1974 "Residential satisfaction as an intervening variable in residential mobility", *Demography*, 11(2): 173–188.

Stal, M.
2009 *Mozambique case study report*, Environmental Change and Forced Migration Scenarios (EACH-FOR). Available at: http://www.each-for.eu/documents/CSR_Mozambique_090126.pdf

Stavroppoulou, M.
2008 "Drowned in definitions?" *Forced Migration Review*, 31: 11–12, Oxford. Available at: http://www.fmreview.org/climatechange.htm

Stover, E. & Chusri, D.
2005 *After the Tsunami – Human Rights of Vulnerable Populations*: 75–92, Human Rights Centre, University of California/Berkeley, East-West Center. Available at: http://www.reliefweb.int/rw/rwb.nsf/db900SID/VBOL-6JEE3A?OpenDocument

Task Force on Climate Change, Vulnerable Communities and Adaptation
2003 *Livelihoods and climate change: Combining disaster risk reduction, natural resource management and climate change adaptation in a new approach to the reduction of vulnerability and poverty*, The International Institute for Sustainable

Development, Winnipeg.

UNDP
2008 *Human Development Report 2007–2008, Fighting climate change: Human solidarity in a divided world*, United Nations Development Programme (UNDP), Palgrave Macmillan, New York. Available at: http://hdr.undp.org/en/media/HDR_20072008_EN_Complete.pdf

UNHCR
1995 *Sexual violence against refugees: guidelines on prevention and response*, United Nations High Commissioner for Refugees (UNHCR), Geneva.

UNHCR
2008 *Climate change, natural disasters and human displacement:* a UNHCR perspective, UNHCR, Geneva.

United Nations
1998 *Recommendations on Statistics of International Migration, Revision 1*, United Nations Department of Economic and Social Affairs, Statistics Division, New York. Available at: http://unstats.un.org/unsd/publication/SeriesM/SeriesM_58rev1E.pdf

University of Bielefeld
2009 *Hispaniola Island case study report*, Environmental Change and Forced Migration Scenarios (EACH-FOR). Available at: http://www.each-for.eu/documents/CSR_Hispaniola_090126.pdf

UNU-EHS
2008 *Environmental migrants: conference aims to build consensus on their definition, support and protection*, United Nations University Institute for Environment and Human Security, Bonn.

UNU
2008 *EACH-FOR Update*, United Nations University. Available at: http://www.ehs.unu.edu/file.php?id=486

van der Geest, K.
2009 *Ghana case study report*, Environmental Change and Forced migration scenarios (EACH-FOR). Available at: http://www.each-for.eu/documents/CSR_Ghana_090126.pdf

Wahlström, M.
2007 "The Humanitarian Impact of Climate Change," *UN Chronicle, Online Edition*. Available at: www.un.org/Pubs/chronicle/2007/issue2/0207p30.htm

Warner, K. et al.
2008 "Field observations and empirical research", *Forced Migration Review,* 31:13–15, Oxford. Available at: http://www.fmreview.org/climatechange.htm

Warner, K. & Laczko, F.
2008 "A global research agenda", Forced Migration Review, 31:60, Oxford. Available at: http://www.fmreview.org/climatechange.htm

Wilson, P.
1995 *The effect of the Loma Prieta earthquake on California migration,* Demographic Research Unit, State of California. Available at: http://www.dof.ca.gov/HTML/DEMOGRAP/ReportsPapers/documents/DISASTER.PDF

Wolpert, J.
1966 "Migration as an adjustment to environmental stress", *Journal of Social Issues*, 22(4): 92–102.

World Bank
2001 *Hurricane Mitch: The Gender Effects of Coping with Crisis*, The World Bank, Washington, DC.

Migration and slow-onset disasters: desertification and drought

Michelle Leighton[1]

[1] Director Human Rights Programs, University of San Francisco School of Law; E-mail: mleighton@usfca.edu

1. Introduction

The Intergovernmental Panel on Climate Change ("IPCC") predicts that global warming will lead to major shifts in weather patterns, ocean currents, and possibly ecosystems. In addition to higher temperatures and rising sea levels, scientists forecast that rainfall will become more variable, drought more prevalent and prolonged. This will exacerbate soil erosion and desertification around the world. In some geographic regions, these events will combine with a higher incidence of rapid-onset disasters in the wet season, causing more violent and destructive storm surges, floods and hurricanes.

The changes in climate now anticipated (and which some presume is already occurring) will disrupt and perhaps permanently alter how and where food is grown. A significant number of countries could lose one-third to one-half of their capacity for agricultural production over the coming decades. Countries in equatorial Africa may lose as much as 60 per cent.

The rural poor dependent on agriculture for their subsistence and employment are due to suffer most. The loss of crops, livestock and demand for farm labour due to longer dry seasons and water scarcity may force people to migrate in search of better livelihood or more durable human security. There is now a growing understanding that those in rural communities struggling with persistent drought and desertification use migration as a coping strategy.

How many of the world's 191 million international migrants have left their communities because of climate-related disasters, or how many will do so in the future is unknown for certain. Some speculate in the tens of millions, others in the hundreds of millions. Estimates are questionable without deeper research. However, anticipating that climate change will lead to larger numbers of displaced populations than would otherwise occur, some governments are seeking support for their adaptation strategies to address migration (see Martin Chapter 7 in this volume).

As decision-makers begin to formulate policies within the climate context, it will be important to consider what research data and tools exist to explain the relationship between climate variability and migration, to predict patterns of human mobility or to identify "hot spots" of the most vulnerable communities. The goal of this chapter is to consider succinctly the available information on drought and desertification-related migration to gain a better understanding of how predicted impacts of global warming may affect migration in the future. The chapter considers some of the scientific forecasts for drought and agricultural impacts, followed by an analysis of the field studies and literature. It will then identify the most significant data gaps and highlight the key policy challenges in addressing migration in the context of climate adaptation strategies.

The chapter will conclude with observations and suggestions for a new research agenda within the international community to address the most significant concerns. Several aspects, such as considering new migration management frameworks and investment in targeted interdisciplinary research, require immediate action.

2. The changing climate and human vulnerability

At the beginning of 2009, scientists made an alarming pronouncement: that humankind's alteration of the climate may be crossing an "ecological threshold," altering the ecosystems upon which life on earth depends, and this may be irremediable.[2] In 2008, new scientific models suggested that more rapid changes in climate were occurring and the world could experience temperature increases of 3–6°C over the next eight or nine decades (Clark et al., 2008). Similarly, examining the impact of temperatures on ocean currents and the rates in thawing of the Greenland and West Antarctic ice sheets, scientists have suggested that sea level could rise by 1.3 m in this same period, a rate that occurred only during the last ice age. (Clark & Weaver, 2008; see also Grinsted et al., 2009).

The Stern Review has identified that poorer countries will fare worst from these environmental changes (2006). Their low level of economic development means that they are more vulnerable and less able to adapt to climate disasters. From a biophysical perspective, they are geographically disadvantaged since most are located in lower latitudes where predicted temperature increases, longer dry seasons, and water scarcity will greatly restrict their capacity to grow food and raise livestock.

For many, agricultural production could drop significantly in the next several decades. The African continent is likely to suffer most harshly. The United Nations Development Programme estimates that up to 90 million hectares of drylands in sub-Saharan Africa could experience drought (UNDP, 2009: 18). With all but 4 per cent of Africa's croplands being rainfed (Heltberg & Siegel, 2008: 5), drought can have devastating

[2] In a report examining new data on climate change and ecosystem impacts released in 2009, the US Climate Change Science Program cautioned that "more ecosystems may be getting pushed toward response thresholds simultaneously, and ... little is known regarding where the tipping points are." (Schlesinger et al., 1990; Reynolds & Stafford Smith, 2002, cited in, US Climate Science Program/USGS, 2009: 74–75).

effects on most of the continent's rural people who subsist off the land and have little or no access to stored water supply for irrigation.

Overall, African countries are expected to lose a great deal of their agricultural capacity. Analyzing a series of climate-related agricultural models, Cline moderately estimates that by 2080, production losses in countries within the equatorial belt could rise to 60 per cent (Cline, 2008). Losses may reach 56 per cent in Sudan, 52 per cent in Senegal, 39 per cent in Morocco, 36 per cent in Algeria and Mali, 31 per cent in Ethiopia, 19 per cent in Nigeria and 47 per cent in other countries in the Southern African region (ibid). Losses are also expected to be significant in parts of Asia and Latin America: 43-44 per cent in northern India, 41 per cent in Iraq, 36 per cent in Pakistan, and 35 per cent in Malaysia, and Mexico, and 43 per cent on average for some countries in South America (ibid).

By 2080, international agencies estimate that food shortages could lead to the malnutrition of 600 million people (UNDP, 2009: 18), The humanitarian crises that could result from food scarcity would be further compounded by the lack of clean water, spread of disease, and potential conflicts. The Stern Review estimates that nearly two billion people could be affected (2006).

In light of the predicted environmental, economic and social impacts, populations are likely to be displaced or to migrate in order to survive. How temporary or permanent this displacement or migration will be and where populations will go is uncertain. As discussed in the following section, mining the studies on drought-related migration can provide some initial insights.

3. Drought, desertification and migration[3]

Scholars, social scientists, and international bodies have debated how best to distinguish from the general category of economic migrants those environmental migrants where the decision to migrate was partially influenced by a change in the physical environment or natural resource base. Less clear is what percentage of the total migrant pool is comprised of environmental migrants, in today's debate, climate-related migrants. There have been numerous attempts to better define and characterize these migrants as a specific subset of all other economic migrants or refugees in need of special protection (see Zetter Chapter 8 in this volume).

The effort to distinguish these migration patterns is a challenge because it involves the simultaneous consideration of space and time, dynamic human attitudes and the national and international character of the movement. These can change over time. Studies must consider whether migration is seasonal (for work), temporary or permanent, related to individual, family or household decisions, and whether it is undertaken for primarily economic reasons or others, such as family reunification (see Kniveton et al. Chapter 2, Bilsborrow Chapter 3 and Warner et al. Chapter 4 in this volume).

A number of studies, however, particularly related to drought have yielded some important data on the links between climate variability and migration. What is relatively unknown is how and where populations will move as climate variability *increases* in frequency and volatility. This raises a number of questions. Is there a direct relationship between the most critical climate-affected areas and increased migration, such that migration will necessarily be more significant in places where food and water security are most threatened? Will the

[3] The case studies are based in large part on the author's survey in, 'Desertification and Migration,' published in Governing Global Desertification, eds., Johnson, P.M., Mayrand, K., and Paquin, M. (Ashgate Press, London, U.K.).

increase in migration translate into more internal, local migration, or to more people crossing borders? Are there 'hot spots' that warrant greater attention?

To better understand the state of knowledge in these areas, this section examines the findings of field research and case studies related to drought and desertification in rural agricultural communities. The case studies available for review span the African and North and South American continents and a few have been completed in Central Asia and Europe. Following a summary of key findings is a discussion of the gaps in data and methodology that, if filled, could notably enhance understanding of climate-related migration in the future.

Findings of research and investigation

Desertification has been a major problem for dryland environments for centuries (Herrmann and Hutchinson, 2006). It involves land degradation in arid, semi-arid and dry sub-humid areas, otherwise defined by international agreements as "desertification" resulting from climate and human activities which reduces soil fertility and the ability of vegetation to thrive (ibid; see also UNCCD, 1994). Contributing factors of human activity include the overexploitation of lands and water, and deforestation for agricultural land expansion causing soil degradation (Drigo, 1999). Low rainfall or extreme rainfall variability and high temperatures can create drought. "On the ground, droughts manifest themselves in vegetation stress and ultimately loss of green vegetation cover, decreases in stream flow, and the dying out and cracking of soil surfaces" (Herrmann and Hutchinson, 2006: 16). Drier soils are more susceptible to wind and water erosion. These biophysical changes result in less soil fertility, less food production and less plant life for grazing livestock. Thinner herds are more susceptible to disease and, subsequently, less profitable.

While drylands are routinely subject to moisture deficits, including droughts, and thus susceptible to desertification processes, (ibid), the concern today is that the intensity, incidence and severity of drought and desertification are accelerating. This is leading to major environmental and socio-economic problems, particularly in parts of Africa where more than 70 per cent of Africa's rural poor live in and depend on dryland agriculture for food and income. The concern is that "[t]he incidence in poverty in sub-Saharan Africa is increasing faster than the population" (IPCC, 2006). At the time of writing, the drought affecting countries in east Africa, particularly Kenya, Ethiopia, Uganda and Somalia, is moving through its fifth year, "driving more than 23 million east Africans in seven countries towards severe hunger and destitution..." (Reuters, "East Africa Drought in Fifth Year, Millions Hungary", Sept. 29, 2009).

The resources of humanitarian groups are becoming more strained. Oxfam's East Africa Director has noted that , "[t]his is the worst humanitarian crisis Oxfam has seen in east Africa for over ten years," citing that failed and unpredictable rains are becoming more common in the region, and "that broader climate change meant wet seasons were becoming shorter. Droughts have increased from once a decade to every two or three years" (Ibid, citing Paul Smith Lomas).

As drought and desertification threaten rural household income sources and food security, (either directly by affecting land assets or indirectly by contributing to the decline of agricultural employment), many rural agricultural families are forced to diversify their income streams to survive. This can be accomplished by having one or more family members migrate (Bilsborrow, 1992; see also Bilsborrow Chapter 3 in this volume). Generally, income decline and poverty (though not absolute impoverishment) are key determinants of migration from rural areas (Leighton, 2006; de Janvry 1997). Those facing the loss of livelihood are more likely to move than those who are relatively well off. The increased likelihood or incidence of poverty can influence whether a family considers migration as a coping strategy.

The migration involved tends to be seasonal or temporary migration, meaning that migrants will be returning to their communities of origin at the end of the agricultural employment season or, at least, on a regular basis. In some cases, they return to tend to their household agricultural activity, using the migration income to support their own investment.

Better documented are instances of drought-related migration from rural to urban destinations, or to international destinations where seasonal labourers may cross a nearby border of a neighbouring state for employment. This type of internal or closer-proximity migration related to drought is better understood than longer-distance (overseas) migration related to drought. International migration requires more planning, extended social networks, and is generally more costly than local migration (Perch-Nielson, 2004). Several studies discussed below suggest that if there are sufficient networks, opportunities and motivation, migrants will undertake a more distant international move as a longer-term or even permanent survival strategy (Skeldon, 2008).

The resources needed for any migration, whether local or international, are often considerable, given that most rural migrants are unskilled. Unless faced with immediate safety issues, it is generally not the poorest of the poor who can migrate from rural areas but those who can find the means to migrate (Massey, 1993; Bilsborrow, 1992). Moreover, a migrant's income or poverty relative to others in the same community can be more important than his or her absolute deprivation (de Janvry, 1997).

Studies also show that gender can be a determinant of local migration. In 2010, the number of international migrants is expected to reach 214 million, of which women will account for 49 per cent (UN DESA, 2009). While women comprise nearly half of all international migrants, it is estimated that men may account for the greater share of seasonal migrants (Knabe & Nkoyok, 2006).

Migration "within" borders

Seasonal movements or circular migration, leading migrants home at the end of the growing season, in response to droughts appear to be fairly common. This type of migration has been documented in Burking Faso (Henry et al., 2004), Ethiopia (Ezra, 2001), Mali (Findley, 1994) and Senegal (Seck, 1996) among other studies in sub-Saharan countries. Others report that drought is a factor that has combined with high population growth to increase the incidence of poverty such that droughts in the Sahel that lasted years or decades, e.g., the 1968–1973 and 1982–1984 droughts, led to the use of migration as a systemic coping strategy (including in Burkina Faso, Chad, Djibouti, Ethiopia, Eritrea, Mali, Mauritania, Niger, Nigeria, Senegal, Somalia and Sudan) (see Tomadong-Helin & Helin, 1991; Sahel Club, 1984).

In a recent study, Henry, Schoumaker and Beauchemin found that in analyzing certain migration data in Burkina Faso, the inclination to migrate from one rural area to another is three times higher for men living in drought-afflicted areas (poor agro-climatic areas) than for those living in areas with higher rainfall averages (Henry et al., 2004: 25). The length of drought also appeared to play a role: the data suggested that if the drought had occurred in the prior three years, men in the drier areas had a 60 per cent higher chance of migrating (ibid).

The impetus to migrate before and during drought events is relayed in case studies of Ghana. Data gleaned from recent interviews with migrants suggest that both land failure and environmental degradation in the communities of origin, as well as the promise of better land in the south, prompted people to migrate (Kees van der Geest, 2008; Black et al., 2008) (where there is now more distress migration in drought years).

In Senegal, migration is a consistent coping response to long-term drought and desertification (Seck, 1996). Since the 1960s, more frequent drought and soil erosion have diminished crop yields, leading to a lack of farm employment and large-scale emigration to Dakar and other urban centres. By the early 1990s, it was reported that 90 per cent of the Tambacounda region's men between 30 and 60 years old had migrated at least once in their lifetime (ibid). Since the 1980s, other communities have engaged in migration as a coping strategy in times of drought (see Knerr, 2004).

Short-term, seasonal migration in response to drought is also documented in other regions of the world (Argentina (Adamo, 2003); India (Rogaly, 2002); Kazakhstan (Glazovsky & Shestakov, 1994; Shestakov & Streletsky, 1998); Niger (Afifi, 2008); Turkey (Zeynep Kadirbeyoglu, 2008)). Massey found that in Nepal, for example, the impairment of natural resources led to migration to nearby communities (Massey, 2007). He reported a much weaker correlation with cross-border international migration, with migrants being members of lower castes and non-Hindus (ibid).

Migration across borders

A number of studies suggest that rural communities affected by drought and desertification may respond by engaging in internal or international migration, particularly where it is easy to cross neighbouring borders. The case studies while fewer in number, reveal that international migration has become more important for rural communities though overall it remains a much smaller portion of population mobility. In this context, much of the documented international migration has been the movement of people to neighbouring states within the Sahel region, and from Mexico to the United States.

During the 1968-1973 Sahelian drought, an estimated one million people migrated from Burkina Faso, at least temporarily, to other countries of the Sahel (Sahel Club, 1984, cited in Tamondong-Helin & Helin, 1991). Though it is believed that the majority of the movement was within the Sahel region, it is uncertain how many of these migrants may have returned to their communities or settled permanently in other countries. Findley reports that Malians also moved to other areas in the Sahel roughly during this same period in response to drought (Findley, 1994: 539-542).

In Senegal's Tambacounda region, drought-related migration is reported to have begun as a more local or seasonal form of migration, discussed above, but eventually flowed to other African states and to Europe, in particular France where more extensive migrant networks exist (Seck, 1996). Although many migrants eventually returned home, the absence of men for extended periods made it difficult to rehabilitate degraded lands and increased the economic burden on the remaining women and children. Remittances became critically important to the community for survival, contributing 75 per cent of family incomes in 1993 and helping to finance schools, post offices and social service centres (Seck, 1996).

A few studies document declines in longer-distance international migration during drought for particular communities, while reporting that more local and even proximate cross-border migration increases. Findley's study, for example, reports that local and Sahelian country migration increased while longer distance international migration from Mali to France actually declined during the most significant drought years (Findley, 1994). Possibly, the financial capital needed to undertake such travel serves as a barrier after years of drought that reduces farm income. It has been postulated that community members wait for improved economic conditions before migrating overseas (see findings of Henry et al., 2004: 26; Kniveton et al., 2008); Kees van der Geest, 2008), choosing to migrate locally instead.

Studies of desertification affecting communities in Asian countries such as Bangladesh, Kazakhstan, India, Iran, Syria and Uzbekistan were similar to those reported in the African context above; playing a role in both internal and international migration (India: Maloney, 1991; Kazakhstan: Glazovsky & Shestakov 1994; Shestakov & Streletsky 1998; Syria: Escher, 1994; Goria, 1998). The precise role of drought in driving migration is uncertain, as most studies report a variety of interrelated factors. For example, Escher reports that a combination of drought, low agricultural production and population growth led to migration among the Syrian Druze (1994). Shestakov and Streletsky report that in Kazakhstan, migration was driven by lower household income after the pollution of water resources and erosion of the Aral Sea which caused desertification of pastures and farm lands (1998: 68).

In Latin America, case studies in Mexico document the prominent influence of drought and desertification on migration flows from rural to rural and rural to urban areas both inside the country and to the United States. Over two-thirds of the country's lands are affected by desertification (Leighton, 2006; and 1997). Recurrent drought, coupled with poor land management practices, contributes to soil erosion that reduces household income. Many rural families find it necessary to undertake migration in order to cope with diminished incomes. In some cases, people may migrate directly from their rural town to a destination in the United States, particularly where they have a family member or strong social network already established (Leighton, 1997; Munshi, 2003). Some of the states undergoing rapid desertification are also those with rapidly accelerating rates of migration, particularly Oaxaca and Tamaulipas (Leighton, 1997).In other contexts, the ability to obtain land for agricultural activity may determine migration. This is illustrated by Bilsborrow's study of migration in the Ecuadorian highlands (see Bilsborrow Chapter 3 in this volume).

In sum, research in the field conducted since 1990 has suggested that weather patterns, or climate variability that affects human survival, can influence rural-to-rural and rural-to-urban migration domestically. Studies have also documented that migrants may also cross nearby borders in response to recurrent drought, as particularly demonstrated in countries of the Sahel where borders are more porous. Though much fewer in number, there are studies that report the influence of drought on longer distance or overseas migration. A few anomalies exist as well that indicate that the severity of drought may not always instigate longer-distance migration. These findings may suggest that financing such longer distance migration may be less feasible after years of low agricultural returns so that more temporary or seasonal migration becomes a more feasible choice. This suggests that the role of "climate" in overseas migration is very context-specific. Finally, networks and financial means play a key role in all forms of migration but may be more important in determining international or overseas migration.

Gaps in Data and Methodology

When analyzing the various case studies above, it is clear that gaps in data collection, analysis and methodology make it difficult to generalize on the extent to which environmental factors are the primary considerations in the decision to migrate. These gaps also make it difficult to build a more accurate picture of potential future climate-related migration.

The complex interactions between environmental, agricultural, social and economic factors means that further studies are required that transcend a single discipline or approach. Therefore, interdisciplinary research and field studies are needed which can able to capture both local level and national patterns (Heltberg & Siegel, 2008). Taking on such research implies high costs and time requirements and further constraints due to gaps in existing methodologies. One challenge lies in capturing the decision-making process of migrants. Some household surveys are unable to do this as migrants or their family members, when interviewed, may not report all of the factors influencing or leading to their decision to migrate, thereby potentially skewing the answers obtained (Bilsborrow, 1992).

Better methods to document and measure the environment–migration nexus are needed (see Kniveton et al. Chapter 2, Bilsborrow Chapter 3, Warner et al. Chapter 4 in this volume). It will be important to use interdisciplinary models, though barriers to the availability of, and access to, environmental time series data and socio-economic data related to migration exist. Environmental time series data (e.g., land use change and biodiversity loss comparable over time), is often not collected at the community level, and at scales comparable to municipal-level demographic and socio-economic data. Available socio-economic data may not include information about migration in key communities.

Challenges also exist with regard to using census data. This does not usually include information on the decision-making processes of migrants or their flows into more than one destination. Nevertheless, if data can be collected and made available, there are models developed to integrate and correlate biophysical and socio-economic information which could be tested in environmentally induced migration contexts (Leighton, 2002). This would better equip decision-makers with knowledge to improve community adaptation strategies in anticipation of future climate crises.

To conclude, case studies from various regions of the world have illustrated that drought and desertification have an impact on the movement of people. Such movements are most often internal, cross-border and, to a lesser extent, international migration beyond the nearest border. Establishing trends or making generalizations are difficult as the case studies are context-specific. However, existing studies serve as a baseline for better understanding the linkages between migration decision- making processes and their relation to drought and land degradation. In the future, a combination of improved national-level data collection and interdisciplinary research would greatly enhance our understanding of these migratory trends.

4. Other policy challenges: migration and climate adaptation

Beyond the need to close research and data gaps, policy makers seeking to construct appropriate climate- migration related migration strategies, including in adaptation programmes, face additional challenges. These are two-fold. The first relates to whether migration is viewed as a positive element of development or as a form development failure. In practicality, the answers may be similar to those posed by other experts as to whether migration is a failure of adaptation or a form of adaptation (see Naik Chapter 5, Martin Chapter 7 in this volume). Without seeking to directly address that collateral issue, this section highlights the debate surrounding the extent to which environment-related migration can be beneficial for development. Understanding the key elements of the debate may help decision-makers better manage migration issues in the context of adaptation.

The second challenge may be to identify the most critically affected communities in terms of climate-related migration, in order to determine where to concentrate more immediate research and investigation if resources are limited. This section highlights one potential frame for identifying future "hotspots."

Remittances: a role in future adaptation?

As global migration has grown, so have remittance flows, and so has their potential impact on development. It has been suggested that migration has the affect of reducing poverty in the country of origin because the remittances sent home by migrants to their families can substantially improve household income and assets (Skeldon, 2008). It has been well documented that the amount of remittances transferred globally consistently exceeds foreign direct investment (FDI) and overseas development assistance, and are often more stable and resilient than the latter (ibid.;World Bank & Economic Commission for Africa, 2006). Remittances in the Latin America and Caribbean region, for example, have outpaced foreign direct investment and overseas development assistance flows combined since 2002, totaling US$66.5 billion for 2007 (Vargas-Lundias, 2007: 14).

In 2007, the amount of global remittances reached US$317 billion, ten times the amount calculated in 1990 (World Bank, 2008). For some African countries, in particular sub-Saharan Africa, the share of remittances as a percentage of GDP is also steadily increasing, rising from US$4.9 billion in 2000 to US$8.1 billion in 2006 (Economic Commission for Africa, 2006: 40–42, 75). Remittances for all countries on the continent now annually exceed US$14 billion (ibid). In fact, these official figures reported probably under-represent the actual total value, as there are presumably large amounts of informal transfers taking place that are not captured in the official numbers.

Impacts at the community and household level

There are varied considerations when examining the impacts of remittances on development. For the most part, remittances are personal, cash transfers from a migrant worker or immigrant to an individual or household in the country of origin. They can also be funds invested, deposited or donated by the migrant to the community of origin. These funds can play a central role in maximizing the benefits of migration for migrants and their families as well as for countries of origin and destination, because they are perhaps, as illustrated above, the largest economic benefit of migration for origin countries.

Remittances flow mainly to poor and marginalized families. In many cases, remittances make up a large percentage of total household income, acting as a substitute for earned income lost due to a variety of reasons including because of natural disasters, crop failure, etc. (IOM, 2007). Even in cases were remittances are not a regular part of monthly income, they often act as a safety net in the case of an emergency. For example, remittances were especially important during the Asian tsunami disaster of 2004 in terms of supplementing family income and aiding in reconstruction (Laczko & Collet, 2005). In Mexico, one study revealed that each additional family member migrating to the USA who sent home remittances increased the receiving family's income by 10 per cent (Vargas-Lundias, 2007: 32, citing Yunex-Naude, 2001: 3).

Remittances can also stimulate agricultural investments in communities that lack agricultural financial or insurance markets as well as be used for reinvestment in rural livelihoods. If they are invested in purchasing yield-increasing inputs and in shifting production to labour-intensive, high-value crops, this may increase the demand for farm labour in the community of origin thus reducing the future incentive for migration (Vargas-Lundias, 2007). Knerr's review of earlier studies in sub-Saharan

Africa noted that remittances from seasonal migration were used to support the agricultural activities of communities of origin (2004).

At the same time, it may be easy to overstate the redirection of remittances toward broader social and economic development. Migration, for example, can result in the loss of human resources in sending communities, particularly in the case of international migration, as it may prevent the return of family members to help with the local harvest (Vargas-Lundias, 2007: 30–31). It has also been argued that remittances may contribute to inequality in rural communities between those with relatives migrating and those without, and as between rural and urban areas (Skeldon, 2008). It is postulated that this could generate more migration among the relatively less well off which may or may not produce similar benefits (See Liu Yang, 2004; Black et al., 2005).

Research in one Moroccan community demonstrates how dynamic this is as an issue. In the case study, the migration of men away from traditional villages, along with the decline of nomadism, resulted in the neglect of lands and hydraulic systems in the community. This led to sand accumulation in once-active irrigation furrows. Families left behind struggled with diminished productivity and income (de Haas, 1998: 12). Yet, as families received and then began to invest remittances in water-harvesting technologies, farming in the area was revitalized (de Haas, 2003: 260). Over time, however, it has become clear that, the benefits were only enjoyed by families whose household participated in that reinvestment.

The sending and receiving of remittances is also a gendered process. Women headed households due to the migration of men may on the one hand be empowered or on the other may be further disadvantaged if discrimination exists which hinders their access to agricultural inputs (such as financing and credit) that could improve productivity. The level of impact with regard to the latter depends on several factors including geographic distance and the ability of the migrant to return periodically or seasonally.

In sum, countries of origin may tend to view migration as a beneficial strategy for both development and climate adaptation if migration is already having a substantial beneficial impact, especially if climate disasters are expected to adversely affect agriculture and employment. The caution lies in whether communities have freedom of choice to migrate or are forced to separate from their families and community involuntarily, and in avoiding additional hardship to those already most vulnerable to climate disasters. These are critical challenges for those promoting long-term durable adaptation solutions, and for receiving country governments seeking to manage migration.

Identifying Future Trends: Targeting "hotspots"

While there remain some key gaps in data available, existing information does help to signal which countries or communities may be potential "hot spots" that at some preliminary level warrant further research on climate-related migration as a priority. This section provides a few observations in this context.

Some countries are due to be affected by climatic events more severely than others and each country's unique pattern of internal and international migration may, in turn, be disrupted. Outside of any consideration of climate change, migration will increase as the global population increases. The current rate of international migration is about 3 per cent of world population. At this rate, international migration could reach 263 million by 2050.[4] While rates for each country are reported, what portion of expected migration will be comprised of climate-motivated migrants is uncertain. For purposes of prioritizing research to identify which countries may experience the greatest challenges associated with climate-related migration, those countries with existing high rates of migration and where migration internationally is due to increase are countries that warranting an initial evaluation.

In this context, "hot spots" for international migration purposes should include the leading net emigration countries in which a number of significant slow-onset climate events (droughts, desertification) are expected to occur in roughly the same time period as the number of international migrants is expected to rise. Should these variables converge, the humanitarian crisis may become significant. For example, using this rough template, there are at least a dozen countries that are predicted to suffer severe agricultural and food production declines from prolonged droughts (where production may drop by over 30% and up to 60%) and that are expected to have significantly

[4] Calculated based upon data provided by UN DESA, Migrant Stocks at 2300. (UN DESA, 2008).

higher international migration rates, irrespective of global warming, in the next few decades, 40 per cent to 150 per cent increases over the current number of international migrants. These countries include Ecuador, Ethiopia, Equatorial Africa, India, Iraq, Mali, Mexico, Morocco, Pakistan, Peru, Syria, Senegal, Sudan, Venezuela, Zambia and Zimbabwe. In prioritizing future research, these countries would be key.

A similar matrix could be constructed for the set of low-income countries that are threatened by both slow- and rapid-onset climate disasters (from droughts to floods to hurricanes to landslides), and that are also experiencing high rates of internal and/or international migration. This would generate a somewhat different list of countries,[5] but a number of countries make both lists, including Ecuador, India, Iraq, Mali and Senegal, as well as other countries in Equatorial Africa. Various matrices could be constructed to consider future scenarios. At present, few researchers concerned with climate change hot spots have built models that include migration factors or population mobility trends (see Arnold et al., 2006).

Building a research agenda that will include methods for anticipating how much of the increase in migration flows will be related to slow-onset climate disasters and where these additional flows will occur will help governments seeking to effectively manage such flows. Adaptation programmes are likely to have a higher chance of success if they consider how to give people a choice to remain in their communities or to migrate.

[5] Depending on the combination of disasters selected, these countries likely include Afghanistan, Bangladesh, most of Central America, Cambodia, Colombia, Ecuador, Equatorial Africa, India, Iraq, Kazakhstan, Mali, Mozambique, Myanmar, Niger, Nigeria, Peru, the Philippines, Senegal, Thailand and Viet Nam. They may be viewed as key hot spots because they already face enormous social and economic challenges: a growing population that is food-insecure, unemployed and expected to increase its levels of migration in the future.

5. Conclusion

Existing research suggests that droughts, floods and other disasters, as they impact livelihoods and contribute to poverty, can play a significant role in migration, mixed within a more complex set of variables including finance and social networks between sending and receiving communities. Migration may be one of several adaptation strategies employed by households in response to droughts and other slow-onset disasters. The slow-onset nature of drought may provide household's with an opportunity to consider their options and thus be considered more of a voluntary undertaking. However, in some instances it may be the only viable solution for a household's survival. Seasonal migration versus more permanent migration tends to be the predominant coping strategy.

Migrants who leave their rural agricultural lands (due to slow-onset impacts, such as drought) usually move to rural areas or to nearby urban centers as discussed in the cases studies above. In some cases, they move across borders to neighboring countries, particularly where agricultural lands, water availability and jobs are more abundant. This follows more general migration patterns. For example, of the 17 million African international migrants worldwide, 63 per cent move within the African continent. Other movements across borders to neighbouring states occur among countries in Eastern Europe, Central and South Asia, parts of Latin America and between the USA and Mexico. However, longer-distance international migration incurs greater financial costs and planning and is therefore less frequent.

Climate events in the future may changes these patterns. For example, the decline of agricultural industries will likely affect the economies of the countries listed in Section III above and, to some extent, their neighbours. If agricultural employment declines over an entire region migrants may be forced to move to more distant destinations to find employment. The unsustainability of agricultural lands and increasing drought in the African continent may already be contributing to a growing trend in this regard. Both regular and irregular migration is increasing to the western and northern African coasts, and from there into European jurisdictions.

Higher levels of international migration due to environmental change pose challenges for developing and developed countries alike. International migration from developing countries is now equally divided between movement to other developing countries and to developed countries. Developing countries will encounter more pressure to handle higher levels of both immigration and emigration flows if rural people are forced to move longer distances in search of livelihood. The more frequent incidence and longer duration of droughts, the greater the challenges for sending and receiving communities, particularly where local infrastructure, employment and social services are already stretched. Understanding how and where environmental change will tip the balance toward additional migration for vulnerable communities is therefore of great importance.

Recommendations for future policy -oriented research

An evaluation of the research to date suggests that there are wide gaps in the information available to policy makers. These gaps exist in content (how and when environmental changes become a primary driver of migration, and which communities are most vulnerable), scale and methodology (breadth of studies, and methods for interdisciplinary analysis), and frameworks for appropriate migration management strategies.

As a relatively new topic among social scientists, little research capital has been invested in broad-scale environment-migration studies. Those available are typically community-level studies and largely, though not always, anecdotal in nature. While these help to clarify context-specific socio-economic behaviors, their conclusions cannot readily be scaled-up to assist policy at the national level or used to create systems for early warning. The lack of statistically relevant data at the national or regional levels constrains the design of policies that could build resilience and promote adaptation among vulnerable communities.

Investing in the development of both short and long-term research, data collection, and monitoring projects could help to close these gaps. Three key areas warrant further attention.

1. Analysis of policy options for managing environment-related migration: The information presented in this chapter suggests that while precise numbers are not known, some portion of the migration occurring internally and internationally is resulting from slow-onset climate-events such as drought, flooding and desertification. While basic research and data collection is

needed, as recommended below, a further critical need is a cogent analysis of options for countries to manage environment-related migration before future crises emerge. Anticipating that migration will continue to pose serious challenges, countries have already begun to discuss migration as an adaptation planning strategy within their National Adaptation Programmes for Action (NAPAs). As yet, however, there is little analysis on what standards, policies or programmes are most appropriate for managing this category of internal or international migration flows, particularly among destination countries. Research that undertakes a robust exploration of the potential frameworks on the ways and means for better managing environment-related migration flows within countries, and as between sending and receiving countries, is critically needed. While global level evaluation of best practices is clearly warranted, research on a regional, rather than the global, scale may be of more immediate value to affected countries.

2. Content of research and identification of tipping points: Large data gaps exist in the literature and case studies as this chapter documents. Further research and field-work is needed on how and when biophysical variables such as drought, desertification, storms, and floods become a primary driver of migration, and on the impacts of that migration for sending and receiving areas. How and why families engage in migration is often the result of a complex set of variables, including the viability of farm employment, environmental stresses, economic policies, family and community networks between sending and destination area, and availability of financial resources to invest in migration. Relatively little is understood about the environmental "tipping points" that can lead to migration decisions. Case studies suggest, for example, that while prolonged drought can instigate international migration in some cases, in others it influences shorter-distance, seasonal migration. These anomalies can exist among communities in the same country. In-depth and comparative case studies are needed to better explain these relationships.

3. Scale of research, methodologies and identification of hot spots: Collecting and integrating analysis of national level data on environmental change related to migration could further help policy makers better identify communities at risk and hot-spots that need critical attention. However, national census and other statistics gathering efforts related to migration do not include environmental information that could help serve as identifiers in migration. In order to consider these relationships

it is necessary to evaluate different environmental and socio-economic data sets.

This poses a number of challenges, not just in the collection of data but in methodology for integrating and analyzing the data:

i. Researchers can differ in their selection of the indicators they believe best represent the environment-migration relationship. Among the biophysical variables the level of rainfall, vegetation loss, deforestation and or land-use change are all relevant but results may differ depending on which variable is chosen for analysis. Among socio-economic indicators, demographic data, levels of employment, agricultural land development, potable water distribution, education, access to social services, and community infrastructure are among a myriad of possible indicators that can influence migration.

ii. Even after indicators are harmonized, a constraint is the availability and accessibility of data, particularly in developing countries. Not all data needed may have been collected consistently in the communities sought to be studied. Jurisdictions may have historically differed in their data collection, or simply not collected data at all for certain indicators.

iii. There remains the challenge of comparing and integrating the various environmental, social and economic data sets because these are often collected by different ministries at different points in time and at different scales (e.g., municipal versus statewide levels).

Promoting interagency and interdisciplinary data collection and data sharing could strengthen the capability of governments to observe and analyse migration patterns. Allowing researchers better access to official data could also enhance study results. One opportunity to improve the level of information available in this area most immediately would be to collect environmental data via the national census process. In this way, data would more easily be comparable with migration and other demographic data. It would also allow for comparability of information in sending and receiving communities.

Finally, a larger investment in the development of research methodologies for this type of interdisciplinary research that can help to support the content-based research recommended above. These methods can be shared among governments, researchers and affected communities to generate long-lasting benefits for appropriate policy reform.

6. References

Adamo, S.
2003 *Vulnerable people in fragile lands: migration and desertification in the drylands of Argentina – the case of the department of Jáchal,* Ph.D. thesis, University of Texas, Austin.

Afifi, T.
2008 *Niger case study report*, UNU/EHS, Each-For Project. Bonn, Germany.

Arnold, M. et al. (Eds.)
2006 *Natural disaster hot spots case studies*, World Bank, Washington, DC.

Bilsborrow, R.E.
1992 "Rural Poverty, Migration, and Environment in Developing Countries: Three Case Studies", *Country Economics Department Paper No. 1017*, World Bank, Washington DC.

Black, R., Natali, C. & Skinner, J.
2005 "Migration and inequality", background paper prepared for World Development Report 2006: *Equity and Development*. Available at: http://siteresources.worldbank.org/INTWDR2006/Resources/477383-1118673432908/Migration_and_Inequality.pdf.

Brown, O.
2008 "Migration and climate change", *International Organization for Migration Research Series No. 31*, IOM. Geneva. Available at: http://publications.iom.int/bookstore/index.php?main_page=product_info&products_id=96.

Castles, S.
2002 *Environmental change and forced migration: making sense of the debate*, Working paper No. 70. UNHCR, Refugee Study Centre, Oxford University, Oxford.

Clark, P.U. Weaver, A.J. Co-Lead Authors
2008 *Abrupt climate change: a report by the US Climate Change Science Program & the Subcommittee on Global Change Research*, Brook, E., Cook, E.R., Delworth, T.L., Steffen, K. (Chapter Lead Authors), U.S. Geological Survey, Reston, VA.

Cline, W.
2007 *Global Warming and Agriculture: Impact Estimates by Country*, Peter G. Peterson Institute for International Economics. Washington, DC.

Coleman, D., & Rowthorn, R.
2004 "The economic effects of immigration into the United Kingdom", *Population and Development Review* 30(4): 579-624.

de Haas, H.
1998 "Socio-economic Transformations and Oasis Agriculture in Southern Morocco", Blaikie and Han (Eds.), *Looking at Maps in the Dark: Directions for Geographical Research in Land Management and Sustainable Development in Rural and Urban Environments of the Third World*, Royal Dutch Geographical Society and Faculty of Environmental Sciences, University of Amsterdam, Utrecht and Amsterdam.

de Haas, H.
2003 *Migration and Development in Southern Morocco: The Disparate Socio-economic Impacts of Out-Migration on the Todgha Oasis Valley*, Ph.D. thesis. University of Amsterdam.

de Haas, H.
2006 *Trans-Saharan Migration to North Africa and the EU: Historical Roots and Current Trends*, Oxford, Nov, MPI. Available at: www.migrationinformation.org/Feature/display.cfm?id=484.

de Janvry, A., Sadoulet, E., Davis, B., Seidel, A. & Winters, P.
1997 "Determinants of Mexico-U.S. migration: the role of household assets and environmental factors", CUDARE Working Paper No. 853. University of California at Berkley, Leighton, M., *Environmental degradation and migration: The US-Mexico case study*, US Congressional Commission on Immigration Reform, Washington, DC.

Dia, I.
1992 "Les migrations comme stratégie des unités de production rurale: Une étude de cas au Sénégal", Blokland, A. & van der Staay, F. (Eds.), *Sustainable Development in Semi-Arid Sub-Saharan Africa*, Ministry of Foreign Affairs, Netherlands, The Hague.

Drigo, R. & Marcoux, A.
1999 *Population dynamics and the assessment of land use changes and deforestation*, FAO, Rome. Available at: www.fao.org/sd/wpdirect/wpan0030.htm.

El-Hinnawi, E.
1985 *Environmental refugees*, Nairobi, United Nations Environment Programme.

Escher, A.
1994 *Migrant Network: An Answer to Contain Desertification; A Case Study of Southern Syria (Gabal al-Arab),* Paper presented at the International Symposium on Desertification and Migrations, 9–11 February. Almeria, Spain.

Ezra, M.
2001 *Ecological Degradation, Rural Poverty, and Migration in Ethiopia: A Contextual Analysis*, Policy Research Division Working Paper No. 149. Population Council, New York. Available at: http://www.popcouncil.org/pdfs/wp/149.pdf.

Findley, S.
1994 "Does drought increase migration? a study of migration from Rural Mali during the 1983–1985 drought", *International Migration Review* 28(3): 539–553.

Glazovsky, N. & Shestakov, A.
1994 *Environmental migration caused by desertification in Central Asia and Russia*, International Symposium on Desertification and Migrations, 9–11 February. Almeria, Spain. Available at: www.geographytsu.freehomepage.com/CIS%20env%20migration.htm.

Goria, A.
1998 *Desertification and migration in the Mediterranean: an analytical framework*, Fondazione Eni Enrico Mattei, Milan.

Grinsted, A., Moore, J. C. & Jevrejeva, S.
2009 "Reconstructing sea level from paleo and projected temperatures 200 to 2100 ad.", *Journal Climate Dynamics*, Jan. 6, 2009.

Haug, S.
2008 "Migration Networks and Migration Decision Making", *Journal of Ethnic and Migration Studies* 34(4): 585–605.

Heltberg, R., Jorgensen, S., & Siegel, P.B.
2008 *Climate change: challenges for social protection in Africa*, World Bank. Available at: http://papers.ssrn.com/sol3/papers.cfm?abstract_id=1174774.

Henry, S., Schoumaker, B., & Beauchemin, C.
2004 "The impact of rainfall on the first out-migration: a multi-level event-history analysis in Burkino Faso", *Population and Environment* 25(5): 423–460.

Herrmann, S. & Hutchinson, C.
2006 "Links between land degradation, drought, and desertification", Johnson, P.M., Mayrand, K., & Paquin, M. (Eds.), *Governing Global Desertification*, Ashgate Press, London.

Hugo, G.
1995 "Environmental Concerns and International Migration", *International Migration Review* 30(1): 105–142.

Intergovernmental Panel on Climate Change
2007 *Climate change 2007: climate change impacts, adaptation, vulnerability*, Working Group II Contribution to the Intergovernmental Panel on Climate Change Fourth Assessment Report Summary for Policymakers, IPCC WGII Fourth Assessment Report.

International Organization for Migration
2007 *IOM and Remittances: Factsheet*, Geneva.

Kees van der Geest
2008 *Migration and natural resources scarcity in Ghana*, UNU/EHS, Each-For Project, Bonn, Germany.

Knabe, F. & Nkoyok, J.
2006 "Overcoming barriers: promoting women's local knowledge", *KM4D Journal* 2(1): 8-23.

Knerr, B.
2004 "Desertification and human migration", Werner, D.(Ed.), *Biological Resources and Migration*, Springer, Berlin, pp. 317–338.

Kniveton, D., Schmidt-Verkerk, K., Smith, C., & Black, R.
2008 *Climate change and migration: improving methodologies to estimate flows*, International Organization for Migration, Geneva. Available at: www.iom.int/jahia/webdav/site/myjahiasite/shared/shared/mainsite/published_docs/serial_publications/MRS-33.pdf.

Laczko, F. & Collett, E.
2005 "Assessing the tsunami's effects on migration", *Migration Information Source*, Migration Policy Institute, Washington, DC.

Leighton, M.
2006 "Desertification and migration", Johnson, P.M., Mayrand, K., & Paquin, M.: *Governing Global Desertification*, Ashgate Press, London, UK.

Leighton, M.
2002 "An indicator model for dryland ecosystems in Latin America", *Final Terminal Report*, GEF Project No. GF/1040-00-10. Global Environment Facility, Washington, DC.

Leighton, M.
1998 *The U.S.-Mexico case study, environmental change and security project report*, Woodrow Wilson Center, Washington, DC.

Leighton, M.
1997 *Environmental degradation and migration: The U.S.-Mexico case study*, US Congressional Commission on Immigration Reform, Washington, DC.

Maloney, C.
1991 "Environmental Displacement of Population in India", *Field Staff Reports* No. 14. UFSI Inc. and National Heritage Institute. San Francisco.

Martin, P. & Zucheher, G.
2008 "Managing migration, the global challenge", *Pop Bulletin* 3(1-3).

Massey, D.S., Axinn, W. & Ghimire, D.
2007 "Environmental change and out-migration: evidence from Nepal", *Popultion Studies Center Research Report* No. 07-615. January 2007. University of Michigan, Ann Arbor.

Massey, D.S., Singer, A.
1998 "The social processes of undocumented border crossing among Mexican migrants", *The International Migration Review*, Fall 1998. 32(3): 561.

Massey, D.S. et al.
1993 "Theories of international migration: a review and appraisal", *Population and Development Review*, Sep. 1993. 19(3):.431-466. Available at: www.jstor.org/stable/2938462.

Munshi, K.
2003 "Networks in the modern economy: Mexican migrants in the U.S. labour market", *Quarterly Journal of Economics* 118(2): 549-599.

Papademetriou, D.
2005 *The global struggle with illegal migration: no end in sight*, Migration Policy Institute. Available at: www.migrationinformation.org/Feature/display.cfm?id=336.

Perch-Nielson, S.
2004 *Understanding the effect of climate change on human migration: the contribution of mathematical and conceptual models*, Diploma Thesis, Department of Environmental Sciences, Swiss Federal Institute of Technology.

Reuters
2009 "East Africa Drought in Fifth Year, Millions Hungary", Sept. 29, 2009.

Rogaly, B., Coppard, D., Ratique, A., Rana, K., Sengupta, A. & Biswas, J.
2002 "Seasonal migration and welfare/illfare in Eastern India: a social analysis", *Journal of Development Studies* 38.5: 89–114.

Sahel Club
1984 *Environmental change in the West African Sahel* (Transformation de l'environnement dans le Sahel ouest africain), Club de Sahel, Paris.

Seck, E.S.
1996 *Désertification: effets, lutte et convention. Environnement et Développement du Tiers Monde,* (ENDA-TM), Dakar.

Selvaraju, R., Subbiah, A.R. Baas, S. & Juergens, I.
2006 *Livelihood adaptation to climate variability and change in drought-prone areas of Bangladesh. Developing institutions and options*, United Nations, FAO Rural Development Division, Rome.

Shestakov, A. & Streletsky, V.
1998 *Mapping of risk areas of environmentally induced migration in the Commonwealth of Independent States*, United Nations High Commissioner for Refugees, International Organization for Migration, and Refugee Policy Group, Geneva.

Skeldon, R.
2005 "Migration and mobility: the critical population issue of our time", *Asia-Pacific Population Journal*, 20(3).

Skeldon, R.
2008 "International Migration as a Tool in Development Policy: A Passing Phase?", *Population and Development Review*, 34(1): 1-18.

Stern, N.
2006 *Review on the economics of climate change*, H.M. Treasury, UK, October, Available at: www.sternreview.org.uk.

Tamondong-Helin, S. & Helin, W.
1991 "Migration and the environment: interrelationships in Sub-Saharan Africa", *Field Staff Reports* No. 22. Environmental Change in the West African Sahel, p. 1, 86. Natural Heritage Institute and National Academy Press. Washington DC, 1991.

United Nations Convention to Combat Desertification
1994 *Text of the United Nations Convention to Combat Desertification*. Available at: www.unccd.int/convention/history/INCDresolution.php.

United Nations Convention to Combat Desertification
2005a *Reports submitted by Africa*. Available at: www.unccd.int/cop/reports/africa/africa.php.

United Nations Convention to Combat Desertification
2005b *Combating desertification in Africa*. Available at: www.unccd.int/publicinfo/factsheets/showFS.php?number=11.

United Nations Department of Economic and Social Affairs, Population Division
2006 *The world population in 2300*. Available at: www.un.org/esa/population/publications/longrange2/WorldPop2300final.pdf.

United Nations Department of Economic and Social Affairs, Population Division
2009 *Trends in global migration stock: the 2008 revision*.
2007 *Tends in global migration stock: the 2005 revision*. Available at: www.un.org/esa/population/publications/migration/UN_Migrant_Stock_Documentation_2005.pdf.

United Nations Department of Economic and Social Affairs, Division for Sustainable Development
1992 *Agenda 21, ("U.N. Agenda 21")*. Available at: www.un.org/esa/sustdev/documents/agenda21/index.htm.

United Nations Development Program
2009 "Fighting climate change: human solidarity in a divided world", *Human Development Report 2007/2008*, UN Development Program.

United Nations Economic Commission for Africa
2006 *International migration and development: implications for Africa*, ECA, Addis Ababa.

United Nations Framework Convention on Climate Change
2007 *Climate Change Impacts, Vulnerability and Adaptation in Developing Countries*. Available at: http://unfccc.int/resource/docs/publications/impacts.pdf.

US Climate Change Science Program and the Subcommittee on Global Change Research
2009 *Final report, synthesis and assessment product 4.2.*, Lead agency: U.S. Geological Survey, Contributing agencies National Oceanic and Atmospheric Administration, National Science Foundation, U.S. Department of Agriculture, Forest Service, U.S. Department of Energy, Environmental Protection Agency, US Climate Change Science Program/USGS. 2009.

van Moppes, D.
2006 "The African Migration Movement: Routes to Europe", *Working papers, Migration and Development Series*, Report No. 5, Research Group Migration and Development, Radboud University, Nijmegen, the Netherlands.

Vargas-Lundias, R., Lanly, M., Viallareal, M., & Osorio, M.
2007 *International migration, remittances and rural development*, International Fund for Agricultural Development and Food and Agriculture Organization. Available at: www.ifad.org/pub/remittances/migration.pdf.

World Bank
2008 *Migration and Remittances Fact Book*. Available at: http://go.worldbank.org/QGUCPJTOR0.

World Bank
2008 *World Development Report, Agriculture for development*, International Bank for Reconstruction and Development, 2007, Washington, DC.

Yang, L.
2004 *Unequal provinces but equal families? An analysis of inequality and migration in Thailand*, Essays on the Determinants and Consequences of Internal Migration.

Zeynep Kadirbeyoglu
2008 *Turkey case study report*, UNU/EHS, Each-For Project.

Managing environmentally induced migration[1]

Susan F. Martin[2]

[1] My thanks go to Kathryn Plummer, for her very valuable research assistance.

[2] Donald G. Herzberg Associate Professor of International Migration, School of Foreign Service Georgetown University; E-mail: martinsf@georgetown.edu

1. Introduction

"Human migration, forced or otherwise, will undoubtedly be one of the most significant consequences of environmental degradation and climate change in decades to come ... Current knowledge about the social consequences we should expect from these processes is still quite sparse. Targeted research and assessment are of course essential to achieve a better understanding of the issue but we cannot afford to wait. It is critical that we start immediately to translate existing knowledge into humanitarian policies and practices."

Achim Steiner
UN Under-Secretary General and Executive Director
United Nations Environment Programme (UNEP)
in Forced Migration Review No. 31, October 2008

In making this call for a combined research and action approach, Under-Secretary Steiner summarized the dilemma facing international organizations, national governments and non-governmental organizations concerned about developing policies to manage the current and potential intersections between migration and climate change. It is only in the recent past that serious attention has been paid to developing the knowledge base about the inter-relationship between environmental pressures (including climate change) and migration, needed to develop informed policy and programmes. Yet there is a pressing need to implement strategies that will help ensure that migration that occurs as a result of climate change does not pose humanitarian, economic and security challenges to the affected countries.

Experts generally agree that the environment is but one of the many reasons that people migrate, sometimes operating on its own but more often through other mechanisms – particularly loss of livelihoods affected by environmental disruption. Since the 1980s, when the term 'environmental refugees' was coined, experts within the environmental and migration fields have differed in their characterization of the phenomenon. Brown (2008) identifies two groups among those concerned with the interconnections: alarmists and sceptics. The alarmists see the environment as a principal cause of population movements, emphasizing the forced nature of the migration (thus, using the term 'refugee'), and often projecting that hundreds of millions of persons will be affected, often without differentiating between those who will move short distances to safer ground and those who may move thousands of miles to new countries. The sceptics, by contrast, raise questions about the models used to generate estimates of those who would be forced to migrate and emphasize that pull factors in destination locations are often more important than push factors at home in determining whether, where and in what volume people will migrate. Perhaps not surprisingly, some environmentalists have been particularly alarmist, often using the threat of mass migration as a reason that immediate action should be taken to address climate change and other environmental problems. Migration experts, concerned about a potential backlash against migrants and misuse of such terms as 'refugee', which is carefully defined in international law, have tended to join the camp of the sceptics.

In more recent years, scholars and activists from the two communities have come together to develop the knowledge base needed to determine the causal mechanisms at work, as well as the potential numbers of people who might be affected. The potential role of climate change in affecting movements of people has been a catalyst for establishing a more scientific basis for estimating numbers. Four paths, in particular, whereby climate change may affect migration have been identified: 1) intensification of natural disasters, such as hurricanes and cyclones that destroy housing and livelihoods and require people to relocate for shorter or longer periods; 2) increased warming and drought that affects agricultural production, diminishing people's livelihoods and access to clean water; 3) rising sea levels that render coastal areas uninhabitable; and 4) competition over natural resources that may lead to conflict, which, in turn, precipitates displacement. For examples of this literature, see Raleigh et al., 2008; Renaud et al., 2007; Brown, 2008; Hugo, 2008; Kniveton et al., 2008.

Recognizing such complexity, and the broader context in which the environment affects population movements, the International Organization for Migration (IOM) offered the following definition of environmental migrants:

"Environmental migrants are persons or groups of persons who, for compelling reasons of sudden or progressive change in the environment that adversely affects their lives or living conditions, are obliged to leave their habitual homes, or choose to do so, either temporarily or permanently, and who move either within their country or abroad."

(IOM, 2007)

Policy makers have been slow to identify potential responses to environmentally induced migration that take these more complex models into account. The recent literature on environmentally induced movements emphasizes that migration can have positive as well as negative consequences – a factor that affects how policies are formulated. The negative impacts stem particularly from emergency mass movements that are generally related to intensified natural disasters and to competition for resources that may result in low- or high-intensity conflict. These movements most closely resemble refugee movements and often require large-scale humanitarian assistance. Negative impacts may also occur if large numbers of people spontaneously relocate from rural to urban areas that are not ready to absorb them, in terms of housing, infrastructure, jobs and services. Most of these movements are likely to be internal, and many will likely challenge the resources of already impoverished countries.

The more positive impacts occur when migration is a voluntary coping strategy that allows people time to weigh alternatives and to use migration as a way of reducing household risk. Morton et al. (2008) describe this situation as follows: "Leaving environmentally degraded and agriculturally unsustainable regions can be seen as a legitimate coping strategy for affected populations. In addition, migration could potentially help slow the process of environmental degradation and allow those who remain in affected communities to adjust their livelihood strategies by changing their agricultural practices or, for instance, shifting to non-agricultural activities."

This chapter reviews frameworks for managing environmentally induced migration in countries likely to experience large-scale internal migration and international emigration resulting from environmental hazards, with particular focus on climate change. It also examines policies adopted by potential destination countries. It is divided into five major sections. The first section presents the stages of environmentally induced migration, in order to establish the context for managing such movements of people. The stages include prevention, mitigation and adaptation to environmental hazards, migration (planned, spontaneous and emergency), return or settlement in another location, and integration into the home or new location.

The second section examines policies adopted by at-risk low-income developing countries, generally through promulgation of national adaptation programmes of action, to avert large-scale displacements resulting from climate change and other environmental hazards. This section further examines national climate change strategies of middle-income countries, with particular focus on three countries with significant levels of internal and international migration that has been attributed, at least in part, to environmental factors – China, India and Mexico. The third section discusses plans for relocation of at-risk populations in the context of other planned resettlement programmes, precipitated by development projects and natural disasters. The fourth section focuses on the policies of potential destination countries related to admission of persons affected by environmental hazards and natural disasters. The chapter concludes with recommendations as to what can be done to improve policy frameworks for managing environmentally induced migration and what can be done to improve the information base needed to formulate more effective management policies.

2. Life cycle for managing environmentally induced migration

Different policies and responses are needed at each stage of environmentally induced migration. The first stage is pre-migration, when actions to prevent, mitigate and help individuals adapt to environmental hazards takes place. It is outside of the scope of this chapter to explore the steps being taken by localities, nations and the international community to reverse current environmental problems and to avert future environmental shocks that may arise out of climate change. It is clear that prevention of the underlying causes of environmentally induced migration is the most critical requirement in managing the issues covered in this chapter, but it will require considerable political will, time and resources to take the steps that are needed to protect the environment.

Adaptation and disaster risk reduction deal more specifically with migration. Adaptation refers to "adjustment in natural or human systems in response to actual or expected climatic stimuli or their effects, which moderates harm or exploits beneficial opportunities." (Intergovernmental Panel on Climate Change (IPCC), 2007: 869). Similarly, disaster risk reduction involves "systematic efforts to analyse and manage the causal factors of disasters, including through reduced exposure to hazards, lessened vulnerability of people and property, wise management of land and the environment, and improved preparedness for adverse events" – according to the United Nations International Strategy for Disaster Reduction (UNISDR, 2009). Identifying vulnerabilities is essential in each case since the "characteristics and circumstances of a community, system or asset ... make it susceptible to the damaging effects of a hazard" (ibid). As will be discussed in greater detail below, adaptation and disaster risk reduction can involve steps to reduce the need for individuals to migrate to get out of harm's way, or it can involve migration as an adaptation/risk reduction strategy that allows a community or household to cope with changes and, perhaps, reduce risk for others.

Migration is the second stage of the life cycle. Migration can be planned or spontaneous, involving individuals and households or entire communities. It can be internal, with people moving shorter or longer distances to find new homes and livelihoods within their own countries, or it can be international, with people seeking to relocate to other countries. It can proceed as an orderly movement of people from one location to another, or it can occur under emergency circumstances. It can be temporary, with most migrants expecting to return home when conditions permit, or it can be permanent, with most migrants unable or unwilling to return. Each of these forms of migration requires significantly different approaches and policy frameworks. Depending on the specific situation, the environmental migrants may resemble labour migrants, seeking better livelihood opportunities in a new location, or they may resemble refugees and internally displaced persons who have fled situations beyond their individual control.

The third stage of the life cycle involves return or settlement in another location. The decision as to whether return is possible involves a range of variables, including the extent to which the environmental causes – either direct or through other channels – is likely to persist. Policies in the receiving communities and countries, depending on whether the migration is internal or international, will also affect the likelihood of return or settlement in the new location. In addition to immigration policies, the policies affecting return and settlement include land use and property rights, social welfare, housing, employment and other frameworks that determine whether individuals, households and communities are able to find decent living conditions and pursue adequate livelihoods.

The final stage of the life cycle involves (re)integration into the home or new location. The policy frameworks outlined above will be key determinants of integration, influencing the access of displaced populations to housing, livelihoods, safety and security. Integration is also affected by plans and programmes to mitigate future dislocations from environmental hazards, coming full circle on the life cycle to a focus on prevention, adaptation and risk reduction.

3. Strategies in developing countries to manage environmental migration

To identify strategies adopted by developing countries to manage environmentally induced migration, we turn to a review of the principal planning documents that governments have promulgated to address the challenges posed by climate change, natural disasters and, more generally, poverty reduction. This section is divided into two parts. The first focuses on least developed countries, many of which will be most adversely affected by environmental hazards arising from climate change. The second focuses on middle-income developing countries, with particular emphasis on China, India and Mexico.

Least developed countries

The National Adaptation Programmes of Action (NAPAs) are the principal frameworks adopted by low-income developing countries to manage environmentally induced migration. According to the United Nations Framework Convention on Climate Change (UNFCCC), NAPAs "provide a process for Least Developed Countries (LDCs) to identify priority activities that respond to their urgent and immediate needs to adapt to climate change – those for which further delay would increase vulnerability and/or costs at a later stage".[3]

As of October 2008, 38 countries had submitted plans.[4] In preparing the NAPAs, countries are to prepare syntheses of available information, undertake a participatory assessment of vulnerability, identify key adaptation measures and criteria for prioritizing activities, and select a prioritized short list of activities. The NAPAs have serious limitations as a mechanism for identifying the full range of adaptation needs and plans. The United Nations Development Programme (UNDP) assessed the NAPAs in its 2007/2008 *Human Development Report*:

[3] http://unfccc.int/adaptation/items/4159.php

[4] All NAPAs referenced herein can be accessed at: http://unfccc.int/national_reports/napa/items/2719.php

"Many of these plans include useful analytical work, providing important insights on priorities. However, they suffer from two basic shortcomings. First, they provide a very limited response to the adaptation challenge, focusing primarily on 'climate-proofing' through small-scale projects: the average country financing proposal generated in the plans amounts to US$24 million. Second, the NAPAs have, in most countries, been developed outside the institutional framework for national planning on poverty reduction. The upshot is a project-based response that fails to integrate adaptation planning into the development of wider policies for overcoming vulnerability and marginalization."

(UNDP, 2007:4).

In relation to this last point, the *Human Development Report* notes that the NAPAs have generally not been integrated into the Poverty Reduction Strategies that these same developing countries prepare as part of their national development planning.[5]

The NAPAs nevertheless remain one of the few planning instruments for least developed countries that are facing the prospect of large-scale dislocations due to climate change. A review of the 38 NAPAs indicates awareness in many countries that climate change may well affect migration patterns. Repeatedly, countries claim that loss of habitat and livelihoods could precipitate large-scale migration, particularly from coastal areas that may be affected by rising sea levels and from areas susceptible to increased drought, flooding or other environmental hazards that will affect agriculture. A number of the NAPAs cite examples of migration already occurring in response to environmental events:

- Bangladesh notes that the high depth of standing water is preventing crop cultivation during Kharif season, affecting jobs and livelihoods and leaving limited food sources, leading to migration to cities for jobs and livelihoods.
- Cambodia states that farmers depend on subsistence rain-fed rice farming, which is vulnerable to floods and droughts.

[5] For a discussion of the integration of migration issues into Poverty Reduction Strategy Papers, see the Working Paper prepared for the Roundtable on Policy and Institutional Coherence of the Global Forum on Migration and Development in Manila. Available at: http://government.gfmd2008.org/index2.php?option=com_docman&task=doc_view&gid=25&Itemid=45

Increased crop losses have led to increased food shortages and poor health, serving as a catalyst for rural-urban migration and cross-border migration.

- Cape Verde notes the thousands of its residents who have emigrated because of devastating famines resulting from the interplay of environmental and population pressures. Its NAPA also refers to frequent torrential rains that have provoked large losses of infrastructure and agricultural production – and, at times, displacement of families or loss of human lives.
- Eritrea notes that individual coping strategies include extensive seasonal movement, particularly for casual labour in urban areas, and movements to cooler uplands and/or raised grounds.
- Ethiopia states that recurrent drought events in the past have resulted in huge loss of life and property as well as migration of people. Ethiopia also notes that "traditional and contemporary coping mechanisms [used in response] to climate variability and extremes in Ethiopia include changes in cropping and planting practices, reduction of consumption levels, collection of wild foods, use of inter-household transfers and loans, increased petty commodity production, *temporary and permanent migration in search of employment* (emphasis added), grain storage, sale of assets such as livestock and agricultural tools, mortgaging of land, credit from merchants and money lenders, use of early warning system, food appeal/aid, etc.
- Gambia states that unpredictable rainy seasons and dry spells result in lower crop yield, reduced availability of forest products, and poor animal pasture, which leads to decreased rural household incomes and serves as a catalyst for rural-urban migration.
- Guinea-Bissau notes increased pressure on the uplands as the longer dry seasons, particularly in countryside regions (eastern part of the country), are causing displacement of whole villages. Populations have to abandon rice fields due to salt-water invasion. Many farmers are seeking new lands and transforming them into rice fields. Others, from the southern littoral, are migrating to the north or Guinea. Migratory movements are also happening in the east, northwest and some locations in the south of the country.
- Haiti cites the migration of large numbers of people from rural areas to Port au Prince, due to a combination of poverty, population growth and environmental problems.
- Mali refers to the migration from north to south within the country and towards coastal countries and the west as a

spontaneous adaptation strategy for dealing with drought, but acknowledges that the internal migration was stressing the already fragile ecosystem.

- Mauritania has experienced a massive rural exodus of livestock herders, many of whom have given up their nomadic lifestyle due to loss of livestock as a result of decreased rainfall.
- Sudan considers drought to be a cause of internal displacement, with some herders being forced to move southwards in search of grazing land.
- The United Republic of Tanzania cites erosion and rising sea levels, leading to loss of settlements in coastal areas, with a potential adaptation activity being the relocation of these vulnerable communities to other areas.
- Uganda considers drought and soil erosion to cause rural–urban migration. Also, displacement occurs due to floods and the subsequent impacts on clean water and sanitation resulting in the spread of disease.

A number of NAPAs link climate change to the intensification of natural disasters that displace large numbers, often in emergency circumstances. Mozambique cites flooding as a cause of mass displacement and indicates that the numbers of displaced persons from these events have been used as a rationale for prioritizing projects. Tuvalu notes that coastal areas and human settlements are exposed to coastal current force and prone to natural tragedies such as strong force winds from storms, cyclones and tidal surges due to climate change. Bangladesh's NAPA notes that climate change will intensify a range of natural disasters (cyclones, floods) that cause displacement.

The majority of NAPAs see the adaptation strategies they describe as ways to reduce migration pressures and allow people to remain in their original settlements. The strategies generally seek to adapt agricultural practices, management of pastoral lands, infrastructure such as dykes and coastal barriers, fishing patterns and other strategies to reduce pressures on fragile ecosystems, thereby allowing populations to remain in place. Bangladesh, for example, seeks to combat salinization, arguing that it will help reduce migration to cities for jobs and other livelihood possibilities, and help halt the "social consequences of mass-scale migration to cities". Guinea Bissau proposes a project – Protection of Salt-Water Rice against High-Tide Invasion – to stem migration. Central African Republic designated a project entitled Management of Native Lands for Rehabilitation of Pastoral Spaces as a way of reducing nomadic practices that are shifting towards more permanent settlement. Mali proposed enhancing durable production of fish and diversifying activities of fishing communities to reduce migration pressures.

Other NAPA approaches focus on early warning and emergency preparedness to reduce displacement due to natural disasters associated with climate change. Tuvalu proposes a project – Strengthening Community Disaster Preparedness and Response Potential – that includes a post-disaster resettlement and rescue plan. Mozambique proposes to establish an early warning system that will help identify risky and vulnerable areas and resettle/relocate the affected populations from flood- and cyclone-prone areas. Bangladesh's NAPA reflects policies also promulgated in its 2005 Poverty Reduction Strategy Paper, which presented the need for a Comprehensive Disaster Management Programme, with the following objectives: professionalizing the disaster management system; mainstreaming risk management programming; strengthening community institutional mechanisms; expanding risk reduction programming across a broader range of hazards; and strengthening emergency response systems (Bangladesh, 2005).

In some cases, the NAPA identifies migration as an adaptation strategy in itself. This perspective appears in two contexts. First, some countries see migration as a way to reduce population pressures in places with fragile ecosystems. Second, countries recognize that resettlement of some populations may be inevitable, given the likely trends, and should be accomplished with planning. In the first category, the NAPAs often provide very little information about the ways in which resettlement of population may reduce further environmental problems. Gambia, for example, refers to resettlement of people as an adaptation strategy to address limited water resources and to rehabilitate mangrove areas, but there is no further discussion of the issues.

More prevalent is the second type of adaptation strategy involving migration – resettlement to mitigate the harm accompanying climate change, particularly flooding and rising sea levels. Sao Tome and Principe, for example, proposes an infrastructure project entitled Displacement of Local Communities, arguing that torrential rains, floods, and rising sea levels put fishermen and farmers at risk, interrupt their livelihoods, and force them to migrate. The NAPA cites the government's intention to construct new homes, noting that displacement of the communities of Malanza, Santa Catarina and Sundy will be necessary in the context of climate change. Coastal populations at risk from floods and landslides will be relocated to protected areas, and the communities will be compensated for the harmful effects of climate change.

Samoa's NAPA also notes that relocation of families is a current adaptation strategy in the village community sector. Potential adaptation activities in the NAPA include assistance for relocation of communities inland. A plan entitled Implement Coastal Infrastructure Management Plans for Highly Vulnerable Districts Project envisions incremental relocation of community and government assets outside coastal hazard zones.

Similarly, Solomon Islands presents projects focused on relocating at-risk populations. One project, entitled Human Settlement, recognizes that island communities' main adaptation option is relocation. The project will enhance communities' capacity to manage the impacts of climate change and sea-level rise and plan for adaptation.

The first priority for adaptation in the Maldives NAPA is implementation of the Safer Island Strategy, which would resettle communities from the smaller, more vulnerable islands into larger, better protected ones, elevate areas of the islands and protect costal zones. The NAPA notes that, "given that the average height of Maldivian islands is 1.5m above MSL [mean sea level], sea-level rise would cause regular tidal inundations in most islands, even at the medium prediction. The high prediction could cause inundations recurrently in almost all islands." Complicating the situation, the "scarcity of land in the Maldives, the smallness of the islands and extreme low elevation makes retreat inland or to higher grounds impossible." The NAPA also notes that population density in certain islands is contributing to the environmental degradation. Therefore, relocation of the population from the more fragile, vulnerable islands to safer ones may be the only solution.

The NAPA states that the Maldives will undertake detailed hazard and vulnerability assessments for five of the proposed safer islands and will develop a hazard-mitigation and vulnerability reduction action plan. Specific reference is made to developing the human resource and institutional capacity at atoll and island levels to manage coastal zones. No reference is made, however, to developing plans for the resettlement of the population to be moved to the safer islands.

Subsequent to the publication of the NAPA, the Maldives has gone further in identifying resettlement as a potential adaptation strategy. President Mohamed Nasheed announced at the end of 2008 that the Maldives was establishing a sovereign wealth fund that could be used to purchase a new island for the country's population. According to Nasheed, "this trust fund will act as a national insurance policy to help pay for a new homeland, should future generations have to evacuate a country disappearing under the waves."[6] Hoping that the funds would never be used for this purpose, Nasheed used the announcement as a call for renewed action to reduce greenhouse gas emissions.

Anote Tong, president of Kiribati, has also made it clear that the population of his island might be forced to relocate en masse. His focus has been on identifying immigration possibilities for Kiribati nationals in nearby countries, particularly Australia and New Zealand. In a recent trip to New Zealand, he suggested that the best educated Kiribatis should emigrate first, in an orderly fashion, and then establish communities that others could join as the situation requires.

Middle-income countries

While the NAPAs are valuable in assessing low-income countries' understanding of the potential ramifications of climate change for migration, they do not provide information on the ways in which middle-income countries see the connections between climate change and migration. This section of the report examines plans (or lack thereof) to address climate change and environmental hazards in three of the major source countries of international migration – China, India and Mexico. This analysis focuses on the National Action Plans on Climate Change. In contrast to the NAPAs, these plans provide little or no perspective on migration, either as a consequence of climate change or as part of an adaptation strategy.

China's National Climate Change Programme, promulgated in 2007, discusses migration in only one context – the "proper relocation" of migrants in the context of hydro-power projects. Hydro-power construction is, itself, seen as a mitigation strategy. The plan provides no indices of what "proper relocation" entails.

India's National Action Plan on Climate Change also has little information on migration, focusing on eight national missions, mostly related to mitigation, not adaptation. Only the mission on strategic knowledge on climate change references migration, citing the need for more research on such socio-economic impacts of climate change

[6] Science News, 28 February 2009. Available at: http://www.sciencenews.org/view/feature/id/40789/title/First_wave

as migration. The technical paper accompanying the national action plan also refers to displacement that may accompany hydro-power construction, noting that resettlement of populations as a result of the construction of dams "has to be attended to with care" (Government of India, 2008: 40). India's five-year development plan illuminates the concerns that may have motivated the inclusion of resettlement as an issue: "Our practices regarding rehabilitation of those displaced from their land because of development projects, conflicts or calamities are very deficient. These have caused many people to feel vulnerable and there is anger because of forced exclusion and marginalisation" (Government of India, 2006). The plan goes on to outline steps needed to redress the situation: "To give displaced people, especially women, their due rights, it is necessary to frame a transparent set of policy rules that address compensation, proper resettlement, and rehabilitation and also gives project affected persons a permanent stake in project benefits. Moreover, these rules need to be given a legal format in terms of the rights of the displaced" (Government of India, 2006).

Mexico's National Strategy on Climate Change (Mexico, 2007) makes even less reference to migration, using the term only in the context of species migration and biodiversity, not human migration. The National Development Plan of Mexico (2007–2012) does discuss migration, but it is fully in the context of the plan for effective democracy and responsible foreign policy. The plan recognizes that people migrate, particularly to the United States of America (USA), in search of better economic opportunities, outlining ways to improve the rights of the migrants and to create new opportunities at home. The section of the plan that focuses on sustainable development does not address the potential for environmentally induced migration.

Since there has been an abundance of literature on the environment and migration in all three countries since the 1990s (see, for example, Leighton Schwartz and Notini, 1994; Kelin, 1998; Raju & Maloney, 1992), the absence of any serious attention to the phenomenon of environmentally induced migration in the strategies on climate change or development planning is all the more surprising. Whereas the least developed countries pay some attention to migration in their climate change and poverty reduction strategies, the three middle-income countries that represent some of the highest levels of international migration (at least in absolute numbers) and remittance flows have not integrated these issues into their planning for sustainable development.

Managing planned resettlement

While planned resettlement in the context of climate change is a relatively new idea, it has a long history in the development field, although there are few examples of international, rather than internal, relocation. The experience of planned resettlement programmes raises many questions about the effectiveness of such initiatives in managing environmentally induced migration. As early as the nineteenth century, transmigration programmes in Indonesia sought to move people from islands with high population density to those with more ample land and natural resources (Robinson, 2003). After independence, the Government of Indonesia accelerated these programmes, moving thousands of settlers from the islands of Bali and Java to the outer islands of Kalimantan, Papua New Guinea, Sulawesi and Sumatra. These programmes were highly controversial, with the indigenous populations of the outer islands accusing the central government of trying to extend its authority through its population redistribution policies. The movements had environmental consequences, often leading to destruction of rainforests and other environmental hazards, particularly in areas with less fertile farming opportunities than existed in Java. In some instances, the transmigration programmes led to violent clashes between the original residents and the new settlers, even leading to secessionist movements and civil conflicts.

In 1985, in the midst of massive food shortages, the Government of Ethiopia announced its intention to resettle 1.5 million people from drought-affected areas to more fertile regions of the country. Within a year, 600,000 had been moved. The country also embarked on a villagization movement that involved movement of peasants into larger settlements, presumably to ease distribution of services and support collectivization of agriculture. Both programmes were criticized for the way in which the relocations occurred as well as the impacts on the affected populations (Robinson, 2003). The government used heavy-handed mechanisms that often involved significant violations of the human rights of those forced to relocate. Critics contended that the government was motivated principally by political concerns, to eliminate opposition groups engaged in insurgency campaigns. The programmes generated large-scale flight and likely exacerbated the famine that killed thousands during the 1980s (Martin, 1991).

The most comparable experiences are the programmes that resettle persons displaced by dams, reservoirs, urban renewal, mining and other development programmes. In these cases, the homes and/ or livelihoods of people are destroyed when areas are flooded or otherwise rendered uninhabitable. Under the worst-case scenarios, when the long-term needs of the relocated are not taken into account, the displaced are at serious risk of "becoming poorer than before displacement, more vulnerable economically, and disintegrated socially" (quoted in Robinson, 2003). Cernea cites eight interrelated risk factors associated with resettlement from development projects: landlessness, joblessness, homelessness, marginalization, food insecurity, increased morbidity and mortality, loss of access to common property, and social disintegration.

In response to such findings, the World Bank and the regional development banks have promulgated guidelines for measuring the adequacy of resettlement plans. These guidelines are pertinent to the management of resettlement in the environmental context. The World Bank recommends that baseline surveys precede resettlement, identifying two types of surveys: a census of all affected persons and assets, and a survey of the socio-economic conditions of the affected persons. The baseline surveys are important in developing the resettlement plans and measuring the impact that resettlement ultimately has on the socio-economic status of the affected persons.

When requested to fund projects involving involuntary resettlement, the World Bank requires a Resettlement Action Plan, which consists of several basic features: a statement of policy principles; a list or matrix indicating eligibility for compensation and other entitlements or forms of assistance; a review of the extent and scope of resettlement, based upon a census/survey of those affected by the project; an implementation plan establishing responsibility for delivery of all forms of assistance and evaluating the organizational capacity of involved agencies; a resettlement timetable coordinated with the project timetable, ensuring (among other things) that compensation and relocation are completed before initiation of civil works; and discussion of opportunities for those affected to participate in the design and implementation of resettlement programmes, including grievance procedures.

Consultation with the affected populations – those who are resettled and the communities they join – is an essential part of managing resettlement. The Inter-American Development Bank (IADB, 1999) describes the benefits of an effective participatory process:

"Participation can facilitate the provision of information and helps ensure that the resettlement plan reflects the needs and aspirations of those affected. It promotes greater transparency and encourages the community to take a more active role in economic development and in the operation and maintenance of local infrastructure. Effective consultation is also essential to avoid the creation of undue expectations and speculation."

The planning process should pay particular attention to the restoration of livelihoods in the new location, provision of housing, security of persons, and other needs related to effective integration.

These resettlement guidelines are consistent with the *Guiding Principles on Internal Displacement* (Office of the High Commissioner of Human Rights (UNHCHR), 1998), which are based on international human rights and humanitarian law. The Guiding Principles were promulgated largely to address problems arising from conflict-induced displacement, but internally displaced persons include others "who have been forced or obliged to flee or to leave their homes or places of habitual residence." The Guiding Principles affirm that all persons have the "right to be protected against being arbitrarily displaced from his or her home or place of habitual residence." In the case of development-induced displacement, arbitrary displacement includes situations in which individuals are forced to flee for reasons that "are not justified by compelling and overriding public interests". In the case of natural disasters, such displacement is arbitrary, "unless the safety and health of those affected requires their evacuation." In the worst-case examples discussed in the NAPAs, it is likely that planned relocation programmes would meet these standards, but the Guiding Principles also state that "the authorities concerned shall ensure that all feasible alternatives are explored in order to avoid displacement altogether. Where no alternatives exist, all measures shall be taken to minimize displacement and its adverse effects."

The Guiding Principles also reiterate the need for consultation with the affected parties, emphasizing that the free and informed consent of those to be displaced shall be sought. The authorities responsible for displacing persons are encouraged to involve those affected, particularly women, in the planning and management of their relocation. In particular, care should be taken to ensure that "proper accommodation is provided to the displaced persons, that such displacements are effected in satisfactory conditions of safety, nutrition, health and hygiene, and that members of the same family are not separated."

Guidance provided to state authorities regarding displacement due to natural disasters is particularly relevant to the issues covered in this chapter. *Human Rights and Natural Disasters: Operational Guidelines and Field Manual on Human Rights Protection in Situations of Natural Disaster* (Brookings-Bern Project, 2008), issued by the UN Emergency Relief Coordinator and the Secretary General's Special Representative on Internally Displaced Persons, defines the conditions for the return of displaced persons:

"...the return of persons displaced by the disaster to their homes and places of origin should only be prohibited if these homes or places of origin are in zones where there are real dangers to the life or physical integrity and health of the affected persons. Restrictions should only last as long as such dangers exist and only be implemented if other, less intrusive, measures of protection are not available or possible."

Conversely, people should not be required to return to areas in which their safety may be compromised: "Persons affected by the natural disaster should not, under any circumstances, be forced to return to or resettle in any place where their life, safety, liberty and/or health would be at further risk" (ibid).

Most of the NAPAs reviewed for this chapter do not set out a process of consultation with the affected populations. An exception is the NAPA prepared by the Solomon Islands. The islands' Human Settlement project envisions that the communities themselves will be deeply involved in adaptation assessments. Key vulnerabilities and adaptation options, strategies and measures will be identified. The NAPA recognizes that the consultation process will help determine the effectiveness of any relocation strategy:

> The biggest risk is that land owners and resource owners may not agree to the terms and conditions of relocation and also may claim compensation to the amounts that could be prohibitive for the government. It is therefore imperative to engage the relocating people and the resource owners at the very early stage of planning. Such engagement and continuous dialogue will ensure the long term sustainability of this programme (Government of Solomon Islands, 2008: 87).

The NAPA also recognizes the important role of the government, noting that relocation of the most vulnerable populations will necessarily become the responsibility of the government because of problems associated with land resources, tenure and management systems. The aim of the project is to develop and implement plans, including promulgation of specific legislation and legal frameworks to guide the process of relocation. It has little detail, however, on how the actual relocation would be accomplished, particularly if the consultative processes yield opposition from the affected populations.

Destination country policies

With only a few exceptions, mostly related to island countries at risk of rising sea levels, the relocation strategies identified in NAPAs assume that people will move internally in search of safer alternatives. Rural–urban migration is the principal focus of the NAPAs. International migration of environmentally induced migrants has received considerably less attention and will be an unlikely solution for most persons affected by climate change. Whether hampered by financial resources, distance, lack of networks in destination countries, or other factors, many would-be migrants will not have the ability to migrate internationally.

That is not to say that international migration will be absent; many of the countries that will experience loss of livelihoods and habitat related to climate change, and many that will likely suffer from intensified natural disasters, are already countries of emigration with well established patterns of labour migration. It is therefore important to look at the policies in potential countries of destination to address what may be increased pressures due to an influx of migrants from other countries.

The immigration policies of most destination countries are not conducive to receiving large numbers of environmental migrants, unless they enter through already existing admission categories. Typically, destination countries admit persons to fill job openings or to reunite with family members. Employment-based admissions are usually based upon the labour market needs of the receiving country, not the situation of the home country. Family admissions are usually restricted to persons with immediate relatives (spouses, children, parents and, sometimes, siblings) in the destination country.

Humanitarian admissions are generally limited to refugees and asylum seekers – that is, those who fit the definition in the *UN Convention*

Relating to the Status of Refugees (UNHCR, 1951) persons with a well-founded fear of persecution on the basis of race, religion, nationality, membership in a particular social group or political opinion. Most environmental migrants will be unlikely to meet the legal definition of a refugee, as they will be forced to flee because of loss of livelihood or habitat and not because of persecutory policies.

Some countries have established special policies that permit individuals whose countries have experienced natural disasters or other severe upheavals to remain at least temporarily without fear of deportation. The USA, for example, enacted legislation in 1990 to provide temporary protected status to persons "in the United States who are temporarily unable to safely return to their home country because of ongoing armed conflict, an environmental disaster, or other extraordinary and temporary conditions". Environmental disaster may include "an earthquake, flood, drought, epidemic, or other environmental disaster in the state resulting in a substantial, but temporary, disruption of living conditions in the area affected." In the case of environmental disasters, as compared to conflict, the country of origin must request designation of 'temporary protected status' (TPS) for its nationals.

Importantly, TPS only applies to persons already in the USA at the time of the designation. It is not meant to be a mechanism for responding to an unfolding crisis in which people seek admission from outside of the country. It also only pertains to temporary situations. If the environmental disaster has permanent consequences, a TPS designation is not available, even for those already in the USA, or it may be lifted. When the volcano erupted in Montserrat in 1997, TPS was granted to its citizens and was extended six times. In 2005, however, it was ended because "it is likely that the eruptions will continue for decades, [and] the situation that led to Montserrat's designation can no longer be considered 'temporary' as required by Congress when it enacted the TPS statute."

Another significant factor is that the designation is discretionary, to be granted by the Secretary of Homeland Security. Countries or parts of countries are designated, allowing nationals only of those countries to apply. Currently, the designation is in effect for citizens of El Salvador, Honduras and Nicaragua. TPS was originally triggered following the earthquakes in El Salvador and Hurricane Mitch in Honduras and Nicaragua. It has been extended until 9 September 2010 (for El Salvador) and 5 July 2010 (for Honduras and Nicaragua). Notably, TPS was not triggered for the hurricanes that destroyed large parts of Haiti.

Given the temporary nature of the grant and its application only to those already in the country, TPS has only limited utility in addressing environmentally induced migration.

At the European Union level, the "Temporary Protection Directive establishes temporary protection during 'mass influxes' of certain displaced persons. The term 'mass influx' refers to situations where masses of people are suddenly displaced and where it is not feasible to treat applicants on an individual basis. It was decided that 'mass influx' was to be defined on a case-by-case basis by a qualified majority of the Council" (quoted in Kolmannskog, 2009).

Finland and Sweden have included environmental migrants within their immigration policies. Sweden includes within its asylum system persons who do not qualify for refugee status but have need for protection. Such a person in need of protection "has left his native country and does not wish to return there because he has a fear of the death penalty or torture, is in need of protection as a result of war or other serious conflicts in the country, is unable to return to his native country because of an environmental disaster." The decision is made on an individual, not group, basis. Although many of those granted this status are presumed to be in temporary need of protection, the Swedish rules foresee that some persons may be in need of permanent solutions. Similarly, according to the Finnish Aliens Act, "aliens residing in the country are issued with a residence permit on the basis of a need for protection if [...] they cannot return because of an armed conflict or environmental disaster" (quoted in Kolmannskog, 2009).

A number of countries provide exceptions to removal on an ad hoc basis for persons whose countries of origin have experienced significant disruption because of natural disasters. After the 2004 tsunami, for example, Canada, Switzerland and the United Kingdom of Great Britain and Northern Ireland suspended deportations of those from such countries as India, Indonesia, Maldives, Seychelles, Somalia, Sri Lanka and Thailand.

To date, there are no examples of legislation or policies that address migration of persons from gradual climate changes that may destroy habitats or livelihoods in the future. For the most part, movements resulting from slow-onset climate change and other environmental hazards that limit economic opportunities are treated in the same manner as other economically motivated migration. Persons moving outside of existing labour and family migration categories are considered to be irregular migrants. In the absence of a strong humanitarian basis for exempting them from removal proceedings (which is unlikely in the slow-onset scenario), these migrants would be subject to the regular systems in place for mandatory return to their home countries. As their immediate reasons for migrating would be similar to those of other irregular migrants – that is, lack of economic opportunities at home and better economic opportunities abroad – there would be little reason for destination countries to manage these movements outside of their existing immigration rules.

In only a few cases has there been any serious discussion of new immigration policy frameworks for those displaced by climate change; even in this context, however, the focus has been on disaster-related, not slow-onset, movements. The Green Party in Australia launched an initiative in 2007 to establish a 'climate refugee visa' in immigration law.

The visa would be available to persons who had been displaced as a result of a "climate change-induced environmental disaster", which, in turn, was defined as:

"...a disaster that results from both incremental and rapid ecological and climatic change and disruption, that includes sea level rise, coastal erosion, desertification, collapsing ecosystems, fresh water contamination, more frequent occurrence of extreme weather events such as cyclones, tornadoes, flooding and drought and that means inhabitants are unable to lead safe or sustainable lives in their immediate environment."[7]

A determination that a disaster exists would have to be made personally by the Minister of Immigration and Citizenship, using the following criteria: (a) the geographical scope of the disaster; (b) adaptation options and long-term sustainability; (c) the capability of the country and neighbouring countries to absorb displaced persons; and (d) international efforts to assist.

The bill was defeated in 2007 but members of the Green Party intend to reintroduce it or a similar bill. The governing party has indicated that it sees international displacement of environmental migrants as a

[7] A bill for an act to recognize refugees of climate change-induced environmental disasters, and for related purposes.

last resort. When asked if Australia intended to resettle those likely to be affected by rising sea levels in the Pacific, Immigration Department Deputy Secretary Peter Hughes responded:

"I think the general view that has emerged about climate change displacement is that, first and foremost, the activities of governments ought to be aimed at mitigation of the climate change factors that might displace people, adaptation within countries where that is possible (and internal relocation could be part of that adaptation process) and, lastly, as a last resort, if needed, international resettlement as a response."

(quoted in AdelaideNow, 26 October 2008).

New Zealand, under similar pressures regarding the potential need for resettlement of Pacific Islanders affected by rising sea levels, has not established a specific category of admissions either. The government has introduced a Pacific Access Category (PAC), whereby 75 people from Kiribati, 75 from Tuvalu, and 250 from Tonga may immigrate to New Zealand each year. The programme is based on employment, however, not environmental factors. The immigrants must be between 18 and 45 years old, have an offer of employment in New Zealand, have English skills, meet minimum income requirement, undergo a health check, and have no history of illegal entrance. The programme is not intended to provide access to those who may be most vulnerable to climate change-induced displacement – the elderly or the infirm, for example.

Conclusions and recommendations

Discussion of mechanisms for managing environmental migration is in its infancy. With increasing understanding of the various ways that environmental change affects migration patterns, and vice versa, governments are beginning to think about how to manage the implications of these interconnections. The National Adaptation Programmes of Action often discuss the ways in which migration has been used as a coping strategy when environmental factors impinge on people's livelihoods and security. Many of them also reflect concerns that climate change-induced environmental hazards will intensify such migration. Increased rural–urban migration is seen as problematic, particularly when urban centres are unable to absorb large numbers of internal migrants who have lost their means of livelihood. The NAPAs often propose land use policies and programmes that would have the effect of stabilizing populations in areas that might experience large-scale out-migration in the absence of such measures.

Fewer NAPAs focus on migration as an explicit adaptation process in its own right, either to help preserve fragile ecosystems by reducing population pressures or to protect populations affected by natural disasters or rising sea levels. Where resettlement is referenced, there is little detail as to how it will be accomplished. The lessons of previous planned resettlement programmes do not appear to have been integrated into planning for what are often seen as inevitable relocations. As this chapter suggests, involuntary resettlement can be fraught with perils for both the migrants and the receiving communities, necessitating a process that involves far more consultation and planning than described in the NAPAs.

In general, countries expect to manage environmental migration internally. The expectation is that most people will move within the borders of their own countries, with the exception of small island countries that may experience such significant rising sea levels that international migration will be inevitable. Few potential destination countries have explicit policies to manage such flows of people, unless they migrate through the normal immigration policies that give preference to family reunification and employment-based admissions. With the exception of some discussions in Australia and New Zealand regarding admissions from the Pacific small islands developing States, no destination countries have considered establishing special labour admissions programmes for persons affected by loss of livelihood as a result of slow-onset climate change or other environmental hazards. While potential destination countries have asylum and/or resettlement systems to manage the admission of persons who cannot return home because of a well-founded fear of persecution, none have systems in place to manage the admission of persons who cannot remain or return home because of environmental threats. At best, destination countries have policies to defer deportation of those coming from countries experiencing natural disasters, but these are generally post-disaster and ad hoc in their implementation. In sum, no major destination country has a proactive policy designed to resettle persons adversely affected by environmental hazards.

Given the current gaps, more attention needs to be placed on identifying and testing new frameworks for managing potential movements. Attention needs to be given to both sides of the environment and migration nexus: 1) identifying adaptation strategies that enable people to remain where they currently live and work; and

2) identifying resettlement strategies that protect people's lives and livelihoods when they are unable to remain. Since internal migration is the most likely outcome for those affected by climate change and other environmental hazards, highest priority should be given to policies and programmes aimed at managing these issues within the most affected countries.

Nevertheless, some international migration may well be needed, particularly for the citizens of island nations, necessitating identification of appropriate admissions policies in potential destination countries. Highest priority should be given to determining mechanisms for the admission of those who cannot be relocated within their home countries, either because of widespread habitat destruction (again, as in the case of certain island states) or because relocation would pose security risks that could provoke violence or even conflict. Some attention should also be given to the slower-onset climate change scenarios in which loss of livelihoods generates emigration pressures. In the absence of legal opportunities to immigrate, at least some portion of those who lose their livelihoods as a result of climate change and other environmental hazards will likely become irregular migrants. The challenge in these cases is determining whether these individuals should be given consideration over others who migrate in search of better opportunities. It is likely that many destination countries will answer this question in the negative With the exception of their refugee and asylum policies, countries tend to frame their admissions policies around their own national interests, prioritizing the admission of persons who will contribute to economic growth, meet labour shortages or have close family ties in the destination country. While exceptions may be made for environmentally induced migrants whose situation most resembles that of refugees, there is less likelihood that governments will make an exception for those whose situation resembles economic migrants'.

In moving towards more coherent frameworks, the lessons of the past will be useful. More systematic examination of previous planned resettlement programmes – in the context of transmigration, villagization and development projects – would help ensure that climate change-induced resettlement programmes do not fall victim to the same problems identified in these initiatives. Identification of best-case examples of resettlement – that is, programmes that respected the rights of the resettled and resulted in an improved economic and social situation – is as important as identification of pitfalls experienced in programmes that failed. Guidelines promulgated to protect those who are involuntarily resettled from development projects or who are displaced from natural disasters should be examined systematically to determine their applicability to the resettlement programmes proposed in the NAPAs. Technical assistance and training to the ministries that may be responsible for resettlement is essential to ensure that all alternatives are exhausted before people are required to relocate, affected populations are involved in the planning, and all steps are taken to ensure appropriate preparations and implementation.

Although most migration is likely to be internal, the potential scale of movements will necessitate international action in support of the affected countries. Putting the issues discussed in this chapter onto the agenda of international forums on both environment and migration will be a key ingredient in the development of more effective strategies to manage movements. In the immediate term, it would be useful to generate discussion of environmental migration at the Global Forum on Migration and Development, scheduled for Athens in November 2009 and Mexico in 2010, and the United Nations Climate Change Conference in Copenhagen in December 2009 and beyond. International cooperation in mitigating harmful migration while planning for movements that will be an essential component of adaptation strategies will help ensure the protection of those who will be most affected by environmental change.

4. References

AdelaideNow
2008 "Tuvalu refugees 'last resort'", 26 October 2008. Available at: http://www.news.com.au/adelaidenow/story/0,22606,24553193-2682,00.html

Bangladesh, Government of
2005 *Unlocking the Potential: National Strategy for Accelerated Poverty Reduction*. Available at http://www.imf.org/external/pubs/ft/scr/2005/cr05410.pdf

Brookings-Bern Project
2008 *Human Rights and Natural Disasters: Operational Guidelines and Field Manual on Human Rights Protection in Situations of Natural Disaster.* Available at: http://www.unhcr.org/refworld/docid/49a2b8f72.html

Brown, O.
2008 *Migration and Climate Change*, Migration Research Series No. 31, International Organization for Migration (IOM), Geneva.

Hugo, G.
2008 *Migration, Development and Environment*, Migration Research Series No. 35, IOM, Geneva.

IADB
1999 *Involuntary Resettlement in IDB Projects: Principles and Guidelines*, Inter-American Development Bank (IADB), Washington, DC.

India, Government of
2006 *Towards Faster and More Inclusive Growth: An Approach to the 11th Five Year Plan (2007–2012)*. Available at: http://planningcommission.nic.in/plans/planrel/app11_16jan.pdf

India, Government of
2008 *National Action Plan on Climate Change*. Available at: http://pmindia.nic.in/Pg01-52.pdf

IOM
2007 *Discussion Note: Migration and the Environment*, MC/INF/288, International Organization for Migration (IOM). Available at: http://www.iom.int/jahia/webdav/site/myjahiasite/shared/shared/mainsite/microsites/IDM/workshops/evolving_global_economy_2728112007/MC_INF_288_EN.pdf

IPCC
2007 *Climate Change 2007: Impacts, Adaptation and Vulnerability*, Cambridge University Press, Cambridge. Available at: http://www.ipcc.ch/publications_and_data/publications_ipcc_fourth_assessment_report_wg2_report_impacts_adaptation_and_vulnerability.htm

Kelin, W.
1998 "Environmental Migration and Sustainable Development: A Case Study of Southwest China", *Chinese Geographical Science* 8(3): 271–280.

Kniveton, D. et al.
2008 Climate Change and Migration: Improving Methodologies to Estimate Flows, Migration Research Series No. 33, IOM, Geneva.

Kolmannskog, V.
2009 "*Climate change-related displacement and the European response*", paper presented at SID Vijverberg Session on Climate Change and Migration, The Hague, 20 January 2009.

Leighton Schwartz, M. & Notini, J.
1994 *Desertification and Migration: Mexico and the United States*, Research Paper, US Commission on Immigration Reform, Washington, DC. Available at: http://www.utexas.edu/lbj/uscir/respapers/dam-f94.pdf

Martin, S.F.
1991 "Development and politically generated migration," in Sidney Weintraub and Sergio Diaz-Briquets, *Determinants of Emigration from Mexico, Central America, and the Caribbean*, Westview Press, Boulder.

Mexico, Government of
2007 *El Plan Nacional de Desarrollo 2007—2012 [National Development Plan 2007–2012]*. Available at: http://pnd.calderon.presidencia.gob.mx/

Morton, A. et al.
2008 "Human security policy challenges", in Climate Change and Displacement, *Forced Migration Review* No.31. Available at: http://www.fmreview.org/FMRpdfs/FMR31/05-07.pdf

Raju, K.V. & Maloney, C.
1992 *Environmental Refugees in India*, Paper presented at the IOM/RPG Seminar on Migration and the Environment, Nyon, Switzerland.

Raleigh, C. et al.
2008 *Assessing the impact of climate change on migration and conflict*, The World Bank, Washington, DC.

Renaud, F. et al.
2007 *Control, Adapt or Flee: How to Face Environmental Migration?* UN University, Bonn.

Robinson, W.C.
2003 Risks and Rights: The Causes, Consequences, and Challenges of Development-Induced Displacement, Brookings Institution, Washington, DC.

Solomon Islands, Government of
2008 *Solomon Islands: National Adaptation Programmes of Action*. Available at: http://unfccc.int/resource/docs/napa/slb01.pdf

Steiner, A.
2008 Foreword to "Climate Change and Displacement", *Forced Migration Review* No. 31. Available at: http://www.fmreview.org/climatechange.htm

UNDP
2007 *Fighting Climate Change: Human Solidarity in a Divided World*, United Nations Development Programme (UNDP), New York.

UNHCHR

1998 *Guiding Principles on Internal Displacement*, United Nations Office of the High Commissioner for Human Rights, Geneva. Availableat:http://www.unhchr.ch/html/menu2/7/b/principles.htm

UNHCR

1951 *United Nations Convention Relating to the Status of Refugees*. Available at: http://www.unhcr.org/cgi-bin/texis/vtx/search?page=search&docid=3b66c2aa10&query=Convention%20Relating%20to%20the%20Status%20of%20Refugees

UNISDR

2009 *Terminology on Disaster Risk Reduction*, United Nations International Strategy for Disaster Reduction (UNIDSR).Available at: http://www.unisdr.org/eng/terminology/terminology-2009-eng.html

The role of legal and normative frameworks for the protection of environmentally displaced people[1]

Professor Roger Zetter[2]

[1] I thank Dr. Camillo Boano (University College London) and James Morrissey (DPhil research student, University of Oxford) for background papers for this report.

[2] Director, Refugee Studies Centre, University of Oxford; E-mail: roger.zetter@qeh.ox.ac.uk; www.rsc.ox.ac.uk

1. Introduction

The current and potential environmental impacts of climate change, together with policies for mitigation and adaptation, raise a variety of human and civil rights challenges and concerns (ICHRP, 2008:1). However, until recently, mainstream debate and policy making on the social impacts of climate change have paid little attention to these rights-based implications.[3]

Among the most significant of these emerging rights-based concerns are those pertaining to people who may be induced, or compelled, to migrate in response to changing environmental conditions. As far back as 1990, the Intergovernmental Panel on Climate Change (IPCC) predicted that "the gravest effects of climate change may be those on human migration." [4] Of equal concern are the rights of the much larger number of people who will be affected by climate change but who do not, or cannot, migrate in response. Yet, almost 20 years later, in 2008, an Inter-Agency Standing Committee (IASC) Working Paper noted, "Neither the UN Framework Convention on Climate Change, nor its Kyoto Protocol, includes any provisions concerning specific assistance or protection for those who will be directly affected by the effects of climate change."[5] This chapter considers these protection needs for people displaced by the environmental impacts of climate change. While there is no agreed definition for such displaced groups, the following description will be used in this chapter:

[3] As the International Council on Human Rights notes (2008:2), this situation is changing. The United Nations Human Rights Council passed a resolution UN Doc. A/HRC/7/L.21/Rev.1 (26.03.08) calling on the Office of the High Commissioner for Human Rights to undertake a "detailed study of the relationship between climate change and human rights".

[4] First IPCC Assessment Report of 1990, Impacts Assessment of Climate Change – Report of Working Group II. Available at: http://www.ipcc.ch/ipccreports/assessments-reports.htm

[5] *Climate Change, Migration and Displacement: Who will be affected?* Working paper submitted by the informal group on Migration/Displacement and Climate Change of the IASC, 31 October 2008.

"People who are displaced from or who feel obliged to leave their usual place of residence, because their lives, livelihoods and welfare have been placed at serious risk as a result of adverse environmental, ecological or climatic processes and events."

Gorlick, 2007

Regardless of whether the resulting displacement is permanent or temporary, this definition has the merit, among those that are widely used, of specifically identifying climatic variables, rather than referring generically to environmental change.

Specifically, the chapter explores the very significant gap that exists in the research and policy-making literature on environmental displacement: this is the role that legal and normative frameworks of protection might play in mitigating the impacts of displacement on people who are compelled to move due to environmental pressures, or in assisting strategies of adaptation and resilience for those at risk of displacement. Rights-based concerns are driven by the need to protect lives and livelihoods threatened by climate-induced environmental changes but, as this chapter notes, these concerns encompass a range of socio-economic, political and cultural rights.

The chapter considers these rights-based issues within a framework of protection and human security. It focuses on developing countries where the migratory impacts of climate change are likely to be most severe and the resources for mitigation, adaptation and protection most constrained.

The chapter also aims to assist development, humanitarian and migration policy makers and practitioners working in governments, intergovernmental and non-governmental agencies, as well as civil society organizations, by proposing the enhancement of the significant body of frameworks of norms, legal instruments and policy responses that exist at national, regional and international levels for the protection of displaced people such as refugees and internationally displaced persons (IDPs). These frameworks, available for other migrant groups, offer a solution to the protection gap with respect to migrants displaced by environmental change.

Focused on developing countries, where the phenomenon of environmentally induced migration is likely to be significant, this chapter addresses three questions:

i. Does a case exist for developing the capacity of national, regional and/or international legal apparatus (based on norms of protection, human rights, and human security) to support the needs of environmental migrants?

ii. To what extent can existing legal and normative frameworks afford effective protection to environmental migrants and what is the scope for enhancing these frameworks?

iii. To what extent can legal and normative frameworks, in general, support the capacity of local and regional governance and civil society structures to implement adaptation and resilience strategies in order to avert potential negative effects of environmental degradation and climate change?

These questions are explored respectively in the following three sections. The chapter then concludes with a review of research needs.

2. The case for developing the capacity of rights-based norms and instruments of protection to support the needs of environmental migrants

This section of the chapter focuses on the viability of developing doctrines and protocols to support the protection needs of people compelled to migrate because of the environmental impacts of climate change. It explores conceptual issues and the nature of rights that may be relevant to this category of migrants.

The 1948 *Universal Declaration of Human Rights* provided a comprehensive framework to promote and protect human and civil rights. But, since that time, intergovernmental organizations and national governments have found it increasingly necessary to extend and reinforce this framework for specific groups or categories of people. Thus the protection of displaced people, particularly where migration appears to be forced rather than voluntary, is well established both as a concept and through norms and legal instruments in domestic and international law.

Numerous international human rights instruments and norms conventions and covenants deal with forced displacement due to persecution, conflict and disasters – notably the 1951 *Geneva Convention Relating to the Status of Refugees* and the 1967 *Protocol* and, more recently, the 1998 *Guiding Principles on Internal Displacement*. The Guiding Principles extended, through non-binding norms, similar provisions for the protection of internally displaced people to those existing for the much smaller number of refugees covered by the refugee Convention and Protocol.

The provisions of the 1951 Convention/1967 Protocol and the 1998 Guiding Principles are elaborated in regional and, especially, national instruments where the main protection responsibilities principally lie. But rights protection for refugees and IDPs is becoming increasingly disputed and fragile. For these reasons, the Responsibility to Protect (R2P) agenda of the International Commission on Intervention and State Sovereignty (ICISS) is a newly emerging phase of the protection policy discourse. It seeks to address a major constraint in the current practice of protection in situations of persecution and conflict: the reluctance or inability of some states to discharge their protection responsibilities.

Alongside refugees and disaster victims, the rights of those subject to development-induced displacement – for example, as a result of large-scale infrastructure, dams, urban renewal – are also increasingly the subject of protection through resettlement policies and instruments. Although not specified in international conventions or covenants (see also Martin Chapter 7 in this volume), normative rights are embedded in the DIDR (Development-induced Displacement and Resettlement) and the IRR (Impoverishment Risks and Recovery) models pioneered by the World Bank over the last two decades.

The rights of indigenous and mobile peoples to be protected from involuntary or forced displacement and its impact on livelihoods and cultural identity are rising up the agenda. The Dana Declaration on Mobile Peoples and Conservation (Dana, Jordan, 2002), leading to the *UN General Assembly Resolution in 2007*[6], exemplifies how protection norms have been extended to another specific category of migrant.

These brief examples of protection regimes for mitigating the impacts of displacement for different categories of involuntary migrants yield two conclusions. First, rights-based protection in response to forced displacement is an accepted, long-standing and expanding principle embedded in international and states' responsibilities. Second, despite the reactive nature of protection 'machinery', it is important to emphasize the role of human agency, even in contexts of forced migration – a point that will be revisited at several stages.

It is against this backdrop that the call for states, and the international community, to articulate and address the protection of peoples' rights, in relation to environmentally induced displacement, has become a pressing issue. Key issues include the global scale of environmental change (and thus the scale of potential impacts that migration might produce), and the fact that human agency is unarguably at the centre of environmental change and, therefore, the migratory consequences.

Thus, the question of what forms of protection for environmentally displaced people currently exist and, more significantly, should be

[6] The United Nations General Assembly Resolution *Declaration on the Rights of Indigenous Peoples*, UN Doc. GA/10612 (13.09.07).

developed as these migratory processes increase, has recently been posed by a number of intergovernmental agencies and NGOs: IASC, IOM, EC, NRC, UNHCR and the Hague Debates.[7] In this respect, it is surprising that the *UN Framework Convention on Climate Change* has not, as yet, given significant attention to the protection needs of those who will migrate because of environment-related changes to their livelihoods and surroundings.

With these considerations in mind, this part of the chapter will consider the conceptual challenges in developing rights-based protection for people displaced by environmental change. This objective will, however, involve confronting at least five conceptual challenges that will be discussed below and will be recurring themes throughout this chapter.

Environmental change and causality

With the evidence that climate change can generate highly vulnerable, mobile populations and the argument that climate change places a special burden on society, regarding the protection of affected people, the focus has been on identifying migrants that will be displaced as a result. Thus, developing viable protection solutions to their potential insecurity requires identifying links between climate change, subsequent changes to environmental conditions and migration.

However, the conceptual and empirical challenges involved in identifying such a cause-and-effect 'deterministic' link are profound. Conceptually, it is extremely difficult to separate climate change impacts from other environmental impacts that cause migration, but where

[7] - Inter-Agency Standing Committee (IASC) Principals Meeting, *Humanitarian Action and Climate Change*, Geneva, April 2008 PR/0804/2645/7.
- IASC, *Climate Change, Migration and Displacement: Who will be affected?* Working paper submitted by the informal group on Migration/Displacement and Climate Change of the IASC, 31 October 2008. Also, IASC Principals meeting, 30 April 2009, briefing and discussion on *Strengthening the Humanitarian Response to Forced Displacement and Migration resulting from Climate Change.*
- International Organization for Migration (IOM), *Migration and Climate Change*, Geneva, March 2008.
- European Council *Climate Change and International Security – Paper from the High Representative and the European Commission to the European Council,* Brussels, 14.03.2008, S113/08.
- Norwegian Refugee Council (NRC), *Future Flood of Refugees: A Comment on Climate Change, Conflict and Forced Migration*, Oslo, April 2008.
- "When home gets too hot", The Hague Debates, 22 May 2008. Available at: http://blogs.rnw.nl/thehaguedebates/debate-1-when-home-gets-too-hot/
- UNHCR "Climate Change Natural Disasters and Human Displacement A UNHCR Perspective", Antonio Gutteres, UN High Commissioner for Refugees, Geneva, 23 October 2008.

climate change is not a factor. Disaggregating environmental change from the socio-political and economic processes and contexts in which such change occurs further complicates this conceptual challenge. Moreover, for the term 'climate change' to mean anything, it has to be defined as a change in average climatic conditions (described by a number of climatic proxies) over a specified period.

Thus, while there is general agreement that climate change impacts should be distinguished from more general processes of environmental change, there appears to be an appreciation of the interrelatedness of climate change, general environmental change and socio-contextual factors driving the decision to migrate and thus the desire to invoke some form of institutionalized protection for people (or groups) experiencing damaging environmental change. However, the search for an overarching definition often neglects climate change. For example, Myers (2005) chooses not to focus on the role of climate change in his definition of environmentally displaced people as those "who can no longer gain a secure livelihood in their homelands because of drought, soil erosion, desertification, deforestation and other environmental problems, together with associated problems of population pressures and profound poverty" (Myers, 2005). Lopez (2007) is similarly broad, defining an environmentally displaced person simply as: "an individual forced to leave his or her home due to environmental reasons" (Lopez, 2007). Likewise, IOM, although making the important distinction between the impact of sudden-onset environmental change and the impact of slow-onset change, does not specifically identify climate-induced change in its definition of an environmental migrant: "persons or [a] group of persons who, for compelling reasons of sudden or progressive changes in the environment that adversely affect their lives or living conditions, are obliged to leave their habitual homes, or choose to do so, either temporarily or permanently, and who move either within their country or abroad" (IOM, 2007).

These arguments and contested definitions are not to deny the significance of environmental changes in people's decisions to migrate. Nor, in extreme circumstances, can we ignore directly instrumental effects, such as rising sea levels, which will compel people to leave their habitual environments. Rather, we should recognize that both the roots and the impacts of changing environmental conditions must be set within a wider context of social, economic and political factors that induce people to migrate.

Accordingly, empirical evidence of the relationship between climate change, environmental degradation and migration points towards complicated processes and interactions. It appears that not only do socio-political and economic forces generate environmental change and vulnerability to the impacts of such change, but they also, through their differential generation of vulnerability, mediate the potential for a migratory response (Belcher & Bates, 1983). In addition, an important feature of the relationship between environmental change and migration is that decisions to migrate are not usually the result of the direct impacts of environmental conditions on economic livelihoods, but because of subsequent indirect impacts and broader social forcings (Hampshire, Randall, 1999).

Increasingly, therefore, current academic discourse and policy debate – and, indeed, historic evidence of so-called environmentally induced migration – privileges multi-causal explanations of migration over the mono-causal impact of changing environmental conditions (see Kniveton et al. Chapter 2 in this volume; Castles, 2002; Lonergan, 1998; Lee, 2001; Boano et al., 2008).

The addition of climate-induced environmental change to the 'migration nexus' does not diminish the claims for rights protection. However, it poses new, more complex and challenging circumstances in which to apply measures for rights-based protection and, more debateable, the need for new protection instruments. Given the difficulty of determining environmental causality, national governments and the international organizations may be better advised to build on existing norms and instruments for protecting migrants, shaping them to incorporate the emerging rights-based challenges posed by environmental change. In this way, the issue of agreeing on specific causal relationships and definitions becomes less problematic but without diminishing the principal concern of protecting rights.

Environmental change and forced migration

The second, and related, challenge is the extent to which environmental change can be considered to 'force' migration. The importance of this contention lies in the fact that protection norms and instruments for many other migrant categories (such as refugees, IDPs and trafficked people) are predicated on notions of force and involuntariness. The apparatus of protection in these cases is specifically designed to tackle the loss or abuse of rights brought about by forcible displacement, albeit difficult to define in practice.

Some researchers and policy makers have similarly defined the problems that cause environmental migration, in spite of the evidence of the interrelatedness of economic, social and political factors noted earlier. It could still be argued that cases exist where environmental change will be so dramatic and so all-encompassing of livelihoods that, regardless of livelihood strategy or socially constructed differences in wealth, most or all inhabitants of an impacted area will be forced to migrate. Such people would be easily identifiable because of the direct link between severe environmental change and migration generated by the scale of the impact. In the specific contexts such as extreme hydro-meteorological hazard events, or armed conflict over shrinking natural resources, the notion of 'forced' migration is plausible. The most significant case is sea-level rise. Island states are particularly compelling because they will be submerged regardless of the socio-economic or environmental conditions that might have occurred beforehand. The case of island state inundation is also striking because the forced migrations that will ensue will necessarily result in cross-border movement, making the title 'refugee' inappropriate.

However, it is largely slow-onset environmental degradation and incremental change that is taking place. It is much more difficult to ascribe the depletion of environmental resources as a factor 'compelling' or 'forcing' displacement. This is not to suggest that environmental change cannot encourage, or even compel, migration. Rather, it is to recognize that identifying who is 'forced', *uniquely*, by deteriorating environmental conditions, as opposed to a *combination* of decisive factors including environmental change, is a substantial challenge in pursuing the stated goal of protecting human rights and security in a future altered physical climate.

Caution is therefore needed in developing a protection regime for environmental migrants that replicates established normative and legal instruments whose aims are to protect, specifically, those whose displacement is 'forced' by readily recognizable and generally accepted causes such as conflict, persecution and disasters. As before, given the difficulty of singling out environmental change as uniquely precipitating, let alone 'forcing', migration, we can conclude that a protection regime should draw on these existing concepts and norms but embed them in a pluralistic framework of national and international conventions, protocols and norms to protect human rights.

Terminology and labels – extending the 1951 Geneva Convention and 1967 Protocol

Drawing again on notions of 'forced migration' to prepare protection instruments for people displaced because of environmental change, the populist use of the term 'environmental refugees' has given impetus to an argument for extending the *1951 Geneva Convention on the Status of Refugees* and the *1967 Protocol* to include this new 'category' (NRC, 2008; Christian Aid, 2007; Republic of Maldives, 2006[8]). The Convention-based refugee regime rests on a definition that is both a normative category and social designation. It is built on the concept of protecting people forced to flee a well founded fear of (state-sponsored) persecution, although it has been progressively extended to include generalized violence – notably by regional conventions such as the 1969 OAU Convention and the 1984 Cartagena Declaration.

Extending the definition to include so-called 'environmental refugees' is deeply problematic. It is erroneous to consider environmental change as a persecutory agent in the Convention sense, much less a state-sponsored process. Moreover, only in extreme cases, where competition for depleting resources might lead to conflict, would people be forced to flee (Rueveny 2007). Thus the term 'refugee' should not be used to describe those who are displaced, either in part or entirely, by environmental factors (Renaud et al., 2007; Keane, 2004).

Renegotiating the Convention to incorporate 'environmental refugees' would, inevitably, introduce greater complexity and confusion into status-determination procedures. Moreover, in the current political climate, distorting the definition in this way would risk reducing still further states' responsibility for, and standards of, protection and assistance for refugees (Castles, 2002: 10; Kibreab, 1997: 21; Lopez, 2007: 367; McGregor, 1993: 162; Suhrke, 1994: 492).

Furthermore, except in border regions where traditional patterns of migration often ignore national boundaries, the majority of people displaced by the environmental impacts of climate change are unlikely to cross international borders – the defining characteristic of a refugee, in international law. They will remain in their own countries, moving to

[8] First Meeting on Protocol on Environmental Refugees: Recognition of Environmental Refugees in the 1951 Convention and 1967 Protocol Relating to the Status of Refugees organized by the Republic of Maldives, August 2006.

urban or rural areas where environmental resource depletion is less intense. Again, it is critical to avoid referring to them as refugees. Given that the majority will remain internally displaced, they will thus fall within the rubric of national norms and legal instruments to protect their human rights. In these circumstances, the case for extending or adapting the 1998 *Guiding Principles on Internal Displacement* is much more compelling.

Even if it were possible to establish legal causation, the Geneva Convention poses an additional hurdle for those displaced by climate change. Persecution is on account of an individual's race, religion, nationality, political opinion, or membership of a particular social group. But migration precipitated by climate change is likely to be indiscriminate – at least with respect to the five conditions of the Geneva Convention. Thus, McAdam and Saul (2008) argue that it is difficult to establish connection by an immutable characteristic. On the other hand, it is possible to argue that characteristics such as social class may increase vulnerability.

Finally, the concept of refugees is often predicated on the contention that those who are forcibly displaced will 'go home' – the ideal solution among three possible 'durable' solutions to refugee displacement – resettlement or third country settlement being the other two. Although people displaced by rapid-onset disasters precipitated by climate change, such as floods and hurricanes, may well return home, those who are displaced by slow-onset and permanent environmental change will not return home and the term refugee will, once again, be misleading.

For all these reasons, the protection of rights takes on a significantly different meaning than it does in the case of refugees.

Humanitarian and restorative justice – the locus of responsibility

Arguments linking climate change to migration have encouraged, and been encouraged by, a popular opinion that climate change places a special moral obligation on society to protect those people made vulnerable by such change.

Furthermore, awareness of the inevitability, rather than the potential impacts, of climate change has tended to shift the focus away from protecting the environment per se, and has accentuated advocacy for the protection of those impacted by environmental change, such as migrants. Incontrovertible evidence for the anthropogenic causes of climate change (Oreskes, 2004), and the lack of international agreement about emission cuts, has further emphasized the responsibility of national governments and international organizations to find ways of protecting those displaced as a result of substantial changes to their local environments.

This profound shift in focus from the environment to people, in the conceptual and policy-making discourse, raises another concern: where should the locus of such protection lie and how should it be discharged?

The spotlight has fallen on tracing out the moral imperatives for affording such people protection, and on generating the tools by which protection might be implemented (Conisbee & Simms, 2003; Bell, 2004; Lopez, 2007; See, 2007). Arguments focusing on theories of justice have principally highlighted either *humanitarian* motivations for protecting groups who will be affected by climate change or, equally compellingly, *restorative* justice. Buttressing arguments for *restorative* justice, other authors cite *security* concerns in addressing the protection challenges resulting from the link between migration and climate change (Campbell, 2008).

The popular conception of the difference between these two arguments – humanitarian and restorative – has significant implications for the locus of responsibility for protection and, by extension, the capacity to deliver protection.

Notions of 'forced' displacement, associated with the 'irresistible' impact of environmental change and images of vulnerability and impoverishment, have leant considerable weight to the *humanitarian* domain of the migration protection discourse, although, as we have seen, some of the premises on which this migration is based are increasingly challenged. Nevertheless, a number of the empirical studies set in the context of the environment-migration nexus have found that migrants (particularly those for whom the decision to migrate was taken less voluntarily) tended to occupy both highly marginal positions in their home and receiving societies and were vulnerable to a variety of social and physical threats (Adger, 2000, 2006; Findley, 1994; de Bruijn & van Dijk, 2003). Manifestations of climate change in depleting environmental conditions draw attention to social, economic and civil rights related to matters of poor governance, population pressure, livelihood vulnerability, poverty and 'failed development'.

These phenomena buttress the humanitarian imperative for protection machinery. In this humanitarian formulation, rights protection is primarily the duty of states, rather than a global obligation. This mirrors the current locus of refugee and IDP protection, although this duty may be supported by external stakeholders, especially given that countries most susceptible to migration induced by climate change are least responsible for the causes of this process.

Accordingly, arguments for *restorative* justice approach the challenge differently. They propose that countries that have historically caused climate change and, consequently, find themselves far less vulnerable to its impacts, have some responsibility to protect those individuals and countries that did not cause such change but will experience its most severe impacts, such as population displacement (Conisbee & Simms, 2003; Bell, 2004; ICHRP, 2008). From this perspective, the strength of the climate change argument lies in a common conception that specific moral burdens rest on global society – global burdens that do not readily appear to exist for the other, more localized, categories of migrants, such as refugees and IDPs.

However, part of the problem with this contention lies in the fact that it may not be possible, even theoretically, to fully protect or compensate people who have been forcibly displaced. Consequently, the moral impetus has shifted to the need to introduce mitigation measures that may prevent the likelihood of displacement from occurring in the first place, such as reducing carbon emissions (Bell, 2004).

This line of reasoning, which shifts the focus of debate from protection to mitigation and compensatory remedies (such as carbon trading), is reinforced by those invoking *security* concerns. The scale of potential migration is thought to be so large that countries less affected by climate change will find it impossible to secure their borders so as to prevent the entry of migrants fleeing such change. In addition, even migrations that occur within developing countries represent a potential source of local conflicts and wider threats to global security (Baechler, 1999; Reuveny, 2007). Again, the most appropriate course of action is considered to be the reduction of emissions in developed countries, in order to reduce the potential for large-scale displacement and the resulting threats to future security (Reuveny, 2008; Myers, 2005).

Discharging restorative justice is, therefore, an essential component in developing protection norms and frameworks on grounds of both responsibility and self-interest. However, while it may reduce the propensity to migrate, it is unlikely to prevent it. For the foreseeable future, the irreversibility of climate change – the legacy effect – suggests that substantial levels of migration will prevail.

Thus, even where restorative justice is invoked, there is still a compelling case for ensuring that remote measures, such as the reduction of carbon emissions, do not detract from the production of rights-based protection measures that directly address the needs of actual and potential migrants.

In the case of restorative justice, there appears to be an obligation and a duty to aid and assist, whereas humanitarianism is undertaken voluntarily and as a virtuous act. We are thus confronted with discharging both humanitarian *and* restorative responsibilities for protecting the rights of the displaced, although the motivation is founded on contrasting principles of justice. Here, the argument now shifts again to considering the locus of responsibility to act and, more particularly, the differential capacity to provide protection.

From a migrant vulnerability perspective, arguments for protection (as well as aid and assistance) are made in terms of humanitarianism in which regional, national and local responses best serve the needs of affected populations. As a result, the development, extension or adaptation of norms and legal instruments for the rights-based protection of environmental migrants also resides at these levels. This perspective reinforces the current position of much of the rights-protection apparatus for migrants, whereby the duty resides largely at state level.

From the broader perspective of climate change, the locus of responsibility for protection lies with developed countries and, therefore, in indirect restorative measures. Essential though these measures are, not just to mediate migration outcomes but to avert the impacts of climate change at a global level, they do not negate the responsibility of developed countries to support regional and state-level development of rights-based norms and policies for countries most likely to be impacted by environmentally induced migration. By agreeing restorative measures, developed countries cannot derogate their responsibility for 'humanitarian' justice, burden sharing and capacity building – not least because of the differential capacities of countries likely to be most affected, in terms of resources and governmental, civil society and institutional capabilities.

The interplay between national and international frameworks, issues of state sovereignty in applying protection instruments, and thus the responsibility to protect takes on unique meanings in the context of environmentally induced migration. They constitute a major challenge for the UN climate change conference in Copenhagen in December 2009.

Protecting those who remain

Individuals, groups or communities will respond differently to changes in their environments, which will lead to a variety of livelihood strategies and, in many cases, may not result in migration at all. Thus, a preoccupation with the conceptual and normative challenges of protecting the rights of people who will be displaced by climate-induced changes to their environments detracts from considering the rights of the much larger majority who will not, or cannot, migrate.

In this context, protecting the rights of those who remain presents two challenges:

i. In relation to slow-onset environmental degradation, the rights of those who remain may be equally affected by the same sorts of factors – declining availability of water, declining productivity of land, loss of access to land and other resources, loss of 'voice' – that will encourage or impel others to migrate. From this perspective, supporting resilience and adaptation highlights the need to consider ways in which their economic and social rights might be protected.

   Moreover, since environmental stress generally depletes household capital, and since certain levels of household capital are necessary for migration, it may well be that those most impacted by environmental stress, and most vulnerable to future shocks and stresses, are least able to migrate. Any protection regime focusing on migration could well overlook those remaining groups that are more vulnerable to human rights abuses and could, therefore, be self-defeating.

ii. Strategies for adaptation challenge deterministic notions of presumably vulnerable groups as passive victims, underscoring the importance of ensuring that rights-based protection recognizes people's agency. Beyond a reductionist, vulnerability perspective, an emphasis on adaptation and resilience embraces components such as well-being, livelihoods, assets, access to resources, self-protection and social capital (Cannon, 2000) – characteristics that are explicitly or implicitly the subjects of rights protection. In the same context, education, gender, race and ethnicity (Pelling & Hight, 2005; Paavola & Adger, 2006) should all be viewed through a rights-based lens. Community-based adaptation is crucial to promoting appropriate development policies and practices. Thus, political rights also need to be safeguarded and enhanced in governance and civil society institutions, as well as in terms of access to education and training, and enhanced institutional capacity and accountability (Adger, 2000: 754).

Inevitably, those countries and regions most impacted by climate-induced environmental migration have weak governance and civil society structures and are least able, or willing, to protect human rights and security. While the need for protection is clear, the capacity for providing it remains a challenge, as discussed in the previous section.

Conclusion

Given the inevitability of climate change, society is considered to have a special responsibility towards those people experiencing the severest impacts. When such change has the potential to generate migration, leaving affected groups highly vulnerable, institutionalized protection for these groups needs to be developed and coordinated. Consequently, recent efforts to determine how to use existing and new legal apparatuses to provide protection for affected groups represent an important conceptual and policy-related endeavour.

While drawing attention to the difficulties of adopting special responsibilities, this part of the chapter has explored the conceptual challenges involved in developing protection norms and instruments. In identifying the problematic, it has argued for a rethinking of the notion of protection and has suggested ways in which groups made vulnerable by climate change might be protected. The challenges reside in: 1) the manner in which we conceive of the relationship between climate change, the environment and migration; 2) the implied notions of force; 3) the terminology of environmental migrants; and 4) the moral grounds on which protection might be afforded and responsibility attributed. At the same time, this discussion has highlighted the equally important challenge of rights protection for those who do not migrate.

Implicitly, the discussion has argued for an adaptive approach rather than the invention of new norms and instruments. In this respect, it concurs with the current stance of governments and international agencies, which demonstrates a reluctance to negotiate new conventions or to create a new institutional architecture to address the migratory consequences.

It has also argued for the principle of retaining the existing locus of most norms and instruments of rights protection that apply in other fields. For those impacted by climate-induced environmental change, rights protection must similarly reside in national norms and legal instruments. However, this is not to say that developing this capacity and providing resources to enact effective protection is not a global responsibility. The interplay between national and international frameworks, issues of state sovereignty in applying protection instruments, and thus the responsibility to protect all take on unique meanings in this context.

Next, it is important to emphasize that the sedentary bias of much policy making, which views migration as the failure of households to adapt to climate change, must be reconsidered. Regardless of the fact that the majority will not migrate, migration is sometimes a positive strategy that demonstrates the determination of households, individuals and sometimes whole communities to improve their lives and to diversify risk and reduce vulnerability (Hussein & Nelson, 1998; Berkes & Jolly, 2001; Henry et al., 2003).

Finally, regarding the propensity to migrate, the protection of the rights of people whose environments will be severely impacted by climate change must strike a balance. While there is international concern for those who will be displaced, norms and legal instruments to protect the rights of the much larger number who will not migrate must also be implemented. The livelihoods of those who remain may be increasingly jeopardized as climate change accelerates the depletion of environmental resources, but we should not forget the evidence of human resilience and adaptation to environmental stressors. Conversely, migration is sometimes a positive household strategy rather than the last resort in times of extreme livelihood vulnerability. Accordingly, rights protection must thus strike a balance between the needs of both migrants and those who remain.

3. The role of existing legal and normative frameworks in affording effective protection to environmental migrants and the scope for enhancing these frameworks

The right to a level of environment adequate to permit a "life of dignity and well-being"[9], when considered to be a basic human right, raises many issues relating to protection for those subject to environmental displacement. The lack of a clear definition of the content and scope of such a right, as discussed, constitutes a severe problem. Current norms and international legal regimes for protection do not, at present, directly offer any coherent or concrete protection machinery for the environmentally displaced. Moreover, the opportunities for creating an entirely new set of international protection instruments are limited, if not non-existent, although exploratory proposals have been made (Prieur et al., 2008).

However, the potential for deploying existing norms and legal frameworks offers the most promising avenue for tackling the rights of those impacted by climate-induced changes to their environments. Thus, this part of the chapter draws on existing categories of protection to consider the scope for adapting international and national legal protection and obligations in human rights law and environmental law to these new demands.

In many respects, the Guiding Principles on Internal Displacement offers a model of how 'complementary' protection for the environmentally displaced might be constructed. Indeed, the Principles already afford protection to those displaced by natural disasters. Thus, the aim here is to outline the potential for further developing these Guidelines by enhancing and adapting existing norms and frameworks to fully address the different modalities of environmental displacement. Such an approach has the advantage of not proposing that people forcibly displaced because of environmental factors, and those who remain, constitute a wholly new category of persons without

[9] Stockholm Declaration, Principle 1; and Rio Declaration, Principle 1.

protection and international recognition. There is little international support for developing a new normative category (Kolmannskrog 2009a). Progressive adaptation of extant norms and instruments has the additional advantage of permitting the expansion of existing institutional competencies and capacities without necessarily requiring new organizational structures.

Human rights law

Climate change potentially impinges upon enjoyment of the full range of internationally protected human rights. Thus, a human rights approach offers a foundation point and significant possibilities for the development of proactive principles and guidelines to protect environmentally displaced persons. With its emphasis on need rather than causal conditions, human rights law potentially offers a more powerful way forward than recourse to refugee law, although it is notions of force, with respect to changing environmental conditions, that has prompted the latter approach.

The 1948 *Universal Declaration of Human Rights* (UDHR) is the principal instrument defining protection norms, including freedom of movement (important in the context of environmental displacement) and other social, cultural and economic rights that can be enjoyed under international human rights law and international humanitarian law.

While climate change cannot be construed as a violator of human rights, just as it cannot be conceived as a persecutory agent under the 1951 *UN Convention Relating to the Status of Refugees*, the UDHR lays out states' obligations to protect rights. The question then is the extent to which these obligations apply to rights that are affected by climate change – a question being explored by the Office of the United Nations High Commissioner for Human Rights (OHCHR).

According to McAdam and Saul (2008), there are three reasons why human rights law is of particular importance in addressing the rights of people susceptible to climate-induced displacement:

- It sets out minimum standards of treatment that states must afford to individuals, providing both a means of assessing which rights are compromised (in this case, by climate change), and which national authorities have primary responsibility for responding to those at risk.

- It can guarantee 'complementary protection': when rights are compromised by climate change, human rights law may provide a legal basis whereby protection may be sought (and granted) in another state.

- In cases of relocation, human rights law requires minimum standards of treatment to be observed in the host state.

Drawing on Ardiles-Martinez et al. (2008), the following outline suggests ways in which these rights might be applicable to those who are forced to migrate because of environmental change. It offers a basic reference kit:

- *The right to life*[10]*:* The quality of the environment affects people's ability to enjoy the universally held right to life. Direct impacts include the increased incidence of natural disasters, while indirect impacts include poorer standards of health, nutrition, access to clean drinking water, susceptibility to disease, and diminishing livelihood capacity as a result of desertification.

- *The right to development*[11]*:* The attainment of the right to development in developing countries will be severely impaired by impacts on food and water security, decreases in the Earth's landmass, traumatic weather patterns and ecosystem destruction.

[10] Universal Declaration of Human Rights (UDHR), 10 December 1948, Article 3. Available at: http://www.un.org/en/documents/udhr/index.shtml#a3.
International Covenant on Civil and Political Rights (ICCPR), 23 March 1976, Article 6. Available at: http://www.unhchr.ch/html/menu3/b/a_ccpr.htm.
International Covenant on Economic, Social and Cultural Rights (ICESCR), 16 December 1966. Available at: http://www.unhchr.ch/html/menu3/b/a_cescr.htm.
European Convention on the Protection of Human Rights. Available at: http://conventions.coe.int/Treaty/Commun/QueVoulezVous.asp?NT=005&CL=ENG.
American Convention on Human Rights. Available at: http://www.cidh.org/Basicos/English/Basic3.American%20Convention.htm.
UN News Centre, *United Nations adopts Declaration on Rights of Indigenous Peoples*. Available at: http://www.un.org/apps/news/story.asp?NewsID=23794&Cr=indigenous&Crl.
Convention on the Rights of the Child (CRC)http://www.unhchr.ch/html/menu3/b/k2crc.htm
Convention on the Elimination of All Forms of Discrimination against Women (CEDAW) http://www.un.org/womenwatch/daw/cedaw/
Economic and Social Council (ESC) General Comment on the Right to Water. Available at: http://www.unhchr.ch/tbs/doc.nsf/0/a5458d1d1bbd713fc1256cc400389e94?Opendocument

[11] Declaration on the Right to Development, 4 December 1986, Article 1, clause 1. Available at: http://www.unhchr.ch/html/menu3/b/74.htm.

- *The right to property[12]:* Climate change may result in the deprivation of property without compensation, particularly in coastal areas subject to flooding and permanent inundation, and may also have an effect on land uses as a result of changing weather and climate patterns.

- *The rights of indigenous peoples[13]:* As noted in the first section, indigenous people are particularly vulnerable to the impacts of climate change because of their close relationship with the environment and their reliance on the land for their livelihood.

- *The right to health[14]:* Climate change is likely to increase the number of worldwide deaths from malnutrition, heat stress and infectious diseases. It will also alter traditional sources of, and access to, clean water.

- *The right to food[15]:* Climate change will have potentially severe impacts on food security by reducing the availability of food, changing access to food, worsening the stability of food supply and affecting the utilization of food.

- *The right to water[16]:* Climate change will result in changes to the components of the hydrological cycle and hydrological systems, such as changing rainfall patterns, intensity and extremes.

A significant concern is that climate change threatens to disrupt civil and political rights, which may, in turn, provoke adverse changes in government and civil society institutions. Climate change will also have impacts on emergency response and disaster recovery, particularly for vulnerable peoples. McAdam and Saul (2008) have drawn attention to this inter-linkage between rights, their relationship to the physical

---

[12] Article 17 of the UDHR, European Convention for the Protection of Human Rights and Fundamental Freedoms, Protocol No. 1. *Article 1 Protection of property*, American Convention on Human Rights: *Article 21. Right to Property.*

[13] *Declaration on the Rights of Indigenous Peoples http://www.iwgia.org/sw248.asp Article 27 of the International Covenant on Civil and Political Rights 1966* http://www.unhchr.ch/html/menu3/b/a_cescr.htm, Convention (No 169) concerning Indigenous and Tribal Peoples in Independent Countries

[14] Article 25(1) UDHR. Article 12(1) of the ICESCR, the *Convention on the Rights of the Child* (CRC) http://www.unhchr.ch/html/menu3/b/k2crc.htm and the *Convention on the Elimination of All Forms of Discrimination against Women* (CEDAW) http://www.un.org/womenwatch/daw/cedaw/

[15] The right to food is most comprehensively addressed in Article 11 of the *International Covenant on Economic, Social and Cultural Rights* (ICESCR) http://www.unhchr.ch/html/menu3/b/a_cescr.htm

[16] The right to water has not been expressly mentioned in the original UN human rights documents; however, the Economic and Social Council (ESC), in General Comment 15 on the Right to Water, makes clear that they consider that the right to water is inherent in many of the other explicit rights. They set out that the right to water should be considered as part of Article 11 of ICESCR on the right to an adequate standard of living.

environment, and the way the accumulation of negative effects due to climate change may seriously affect the capacity of governments to guarantee life in some areas. It is not that climate change itself is responsible for rights violations; rather, it is the effects of climate change that weaken states' capacities and hinder them from fulfilling their obligation to protect people's rights.

This outline of the rights that may be impacted by climate change alerts states to their protection obligations under international and national laws. It indicates where adaptation of this machinery may ensure that conditions leading to displacement can be averted or, if displacement occurs, where the key rights of those who are displaced may be protected.

Environmental law

Environmental law offers a broader approach to the subject, which may, albeit indirectly, offer ways of protecting human rights in response to climate-induced displacement. Essentially, it requires that states implement programmes for the control and reduction of pollution of the atmosphere and the marine environment, and to conserve biodiversity.

Although these measures do not seem to directly relate to forced displacement issues, displacement can be due to a loss of livelihood or resource depletion caused by climate change. Drawing principally on the work of McAdam and Saul (2008), the following points can be made.

In general, every state has obligations not to knowingly allow acts on its territory that are contrary to other states' rights and to refrain from using their territory in ways that produce harm to the environment beyond their borders. These principles of transboundary environmental impacts are well established (McAdam & Saul, 2008).

Furthermore, the concept of sustainable development, as noted by McAdam and Saul, specifically limits the realization of some 'rights' to development. For example, Principle 2 of the 1992 *Rio Declaration*[17]

[17] Declaration of the United Nations Conference on Environment and Development (Rio Declaration), UN Doc. A/CONF/151/26/Rev.1 (1992).

notes the responsibility of states to ensure that the sovereign right to resource exploitation does "not cause damage to the environment of other states or of areas beyond the limits of national jurisdiction." This implies limitation on carbon emissions and the damage it potentially might cause by displacing vulnerable populations.

Principle 3[18] indicates that the right to development "must be fulfilled so as to equitably meet developmental and environmental needs of future generations." Again, this requires that there be limits – for example, on emissions – that might constrain the livelihoods of future generations: this could readily be construed to include those who are compelled to migrate, and certainly those who remain but whose livelihoods are depleted by the impacts of climate change on their environments (McAdam & Saul, 2008).

Environmental law does, therefore, have the merit of providing an additional and broader basis for responding to climate change damage and the potential migratory effects. But it requires upholding a duty to act collaboratively to ensure that an international system is sufficiently strong to protect human rights. To this extent, it complements human rights law, which acts principally at the national level. Thus, there is enormous scope, coupled with considerable challenges, in adapting international environment law to accommodate rights-based needs with respect to displacement.

The principal challenges of deploying environmental law in this context are as follows. As we have seen, there is considerable difficulty in quantifying the harm caused by the carbon emissions of any particular state, in quantifying this culpability, and in identifying causation between emissions and detrimental effects, when all states have contributed to emissions, at some point. It is also difficult to establish accountability of corporate institutions for carbon emissions in a legal system in which states remain the primary duty-bearers (ibid). Finally, at the international level, environmental law primarily regulates responsibility between states, while individuals enjoy lesser legal capacity than under human rights law.

This means that, where protection cannot be achieved through international collaboration, domestic measures must be taken to ensure

[18] Declaration of the United Nations Conference on Environment and Development (Rio Declaration), UN Doc. A/CONF/151/26/Rev.1 (1992).

that such rights are protected (ibid). This conclusion resonates with the fact that most environmentally induced migration will probably be internal and regional. Crucially, therefore, responses to environmental displacement, including resettlement and mitigation, confirm the importance of burden-sharing and the need for international financial obligations for mitigation and protection to be met by richer, carbon-producing nations. In this regard, proposals have been made for a new international environmental migration fund that could provide the financial basis for measures to deal with the impacts of migration[19], which, in some respects, extends the poorly implemented Global Environmental Facility for funding adaptation measures under Article 4 of the UN Framework Convention on Climate Change (UNFCCC).

In addition to the 'ability-to-pay' principle, burden-sharing mechanisms of this kind could be based on the 'polluter pays' principle (Principle 7 of the 1992 Rio Declaration), linking contributions to the level of country-specific greenhouse gas emissions as well as other variables such as gross national product. Again, this line of argument reinforces the need for rich countries to meet the 'costs' of migration caused by the environmental effects of climate change, as part of the Article 4 adaptation costs of the UNFCCC.

Subsidiary and complementary protection

Alongside human rights and environmental law, there are several subsidiary and complementary norms and instruments that afford different forms of human rights protection for migrant groups, either directly or indirectly – often applying to specific population categories. Their application to the needs of the environmentally displaced requires detailed consideration, but they hold considerable potential.

Most significant among these, as noted above, is the 1948 *Universal Declaration of Human Rights*. Other relevant international norms and instruments include the 1996 *International Convention on Civil and Political Right*s, the 1966 *International Covenant on Economic, Social and Cultural Rights (ICESCR)*, and the 1996 *International Convention on Civil and Political Rights*, as well as a range of international conventions dealing with specific social groups, such as the 1990 *International Convention on the Protection of the Rights of All Migrant Workers*, the

[19] *Climate Change as a Security Risk*, German Advisory Council on Global Change (WBGU), London, 2007.

1989 *Convention on the Rights of the Child 1989*, the 1981 *Convention on the Elimination of All Forms of Discrimination against Women*, the 1991 *ILO Convention 169 on the Rights of Indigenous People* and the 1954 *Convention Relating to Stateless Persons*, together with the 1991 *Convention on the Reduction of Statelessness*.

These conventions and covenants cover many rights-based norms that, by virtue of their general application, could also be construed as pertaining to the specific conditions of the environmentally displaced as well as those who remain. The following are some examples of the kinds of protection relevant to these conditions:

- rights such as freedom of movement, to be found in Article 12 of the 1996 *International Convention on Civil and Political Rights* and, especially pertinent to conditions of environmental degradation, freedom from hunger and the measures needed to ensure the improvement of methods of production, conservation and distribution of food, found in Article 11 of the 1966 *Covenant on Economic, Social and Cultural Rights.*[20]

- protection from collective expulsion (Article 22) and violence and intimidation (Article 16) to be found in the 1990 *International Convention on the Protection of the Rights of All Migrant Workers.*[21]

- significant social, economic and political rights of equality afforded by the 1981 *Convention on the Elimination of All Forms of Discrimination against Women*[22] (notably Article 14), applicable to the changing environmental conditions and the potential displacement consequences – for example, the particular problems faced by rural women – and ensuring equality of participation, given the significant roles that rural women play in the economic survival of their families.

- protection of stateless people is a critical challenge under the 1954 *Convention Relating to Stateless Persons*, the

[20] International Covenant on Economic, Social and Cultural Rights, G.A. res. 2200A (XXI), UN Doc. A/6316 (1966).

[21] International Convention on the Protection of the Rights of All Migrant Workers and Members of Their Families, adopted by General Assembly resolution 45/158 of 18 December 1990.

[22] Convention on the Elimination of All Forms of Discrimination against Women, adopted and opened for signature, ratification and accession by General Assembly resolution 34/180 of 18 December 1979; entry into force 3 September 1981, in accordance with article 27(1).

1991 *Convention on the Reduction of Statelessness* and the protection mandate of the UNHCR for stateless people, given that statelessness is the likely condition for citizens of small island states that will be submerged by rising sea levels. Here, provisions to enjoy equal rights as citizens of the host country in areas such as Housing (Article 21), Labour legislation and social security (Article 24), Freedom of movement (Article 26) and Expulsion (Article 31) are of notable significance.

With regard to guidelines and the development of good practice in relation to protecting the rights, and providing for the needs, of displaced people, there has been significant progress in recent years. In this regard, the *IFRC Code of Conduct in Disaster Response Programmes*[23] as well as the *Sphere Handbook and the Humanitarian Charter and Minimum Standards in Disaster Response*[24] provide useful reference points in situations of mass displacement and provide essential features of protection regimes, including IDPs. Both these frameworks are buttressed by a strong rights-based approach. More recently, the IASC, in 2007, also coordinated the production of protection guidelines for IDPs; these include extensive recommendations for mitigating protection risks and vulnerabilities.[25]

Regional Instruments

Since the majority of those displaced by environmental change will remain in country or at least within region, we should turn to regional instruments to consider the scope they offer for protection. Given that the two principal instruments – the Organization of African Unity (OAU) Convention[26] and the Cartagena Declaration – both offer a much more liberal interpretation of the definitions of people who are forcibly displaced, it would seem likely that they could be used to develop regional norms of protection. However, at present, there appears to be little scope for regional instruments to be brought into play in the context of migration induced by changing environmental conditions. Indeed, it was not designed for this purpose.

[23] International Federation of Red Cross and Red Crescent Societies (1994) *Code of Conduct*. Geneva: IFRC.

[24] Sphere Project (2004) *Humanitarian Charter and Minimum Standards in Disaster Response*. Geneva: Sphere Project. Available at: http://www.sphereproject.org/.

[25] Inter-Agency Standing Committee (2007) *Handbook for the Protection of Internally Displaced Persons*, Global Protection Cluster Working Group.

[26] Organization of African Unity Convention Governing the Specific Aspects of Refugees Problems in Africa (adopted 10 September 1969, entered into force 20 June 1974) 1001 UNTS 45.

In the African context, the OAU Convention contains a broader refugee definition than the Geneva Convention. Theoretically, it may be possible to afford protection to encompass environmental catastrophes as "events seriously disturbing the public order". Where people are fleeing a natural disaster, regional practice allows international borders to be crossed and temporary residence afforded. But this is not considered a convention obligation under the OAU Convention but, rather, a gesture of humanitarian goodwill (McAdam & Saul, 2008). However, the preparation of the Africa Union Convention on internal displacement, envisaged for adoption in 2009, will be a significant step forward.

The mere provision of temporary protection improves the situation of the environmentally displaced in Africa. It does not, however, remedy their vulnerability and raises fundamental questions about long-term impacts on, and burden sharing in, receiving countries.

Similarly, in 1984, the Organization of American States (OAS) adopted the *Cartagena Declaration on Refugees*, which provides the following definition in its Article 3: "[the definition] includes among refugees persons who have fled their country because their lives, safety or freedom have been threatened by generalized violence, foreign aggression, internal conflicts, massive violations of human rights or other circumstances which have seriously disturbed the public order."

Like the OAU Convention, the Cartagena Declaration was not designed to deal with environmentally displaced persons. Arguably, it could include victims of environmental crisis if such events could be considered serious enough to disturb the public order. However, the International Conference on Central American Refugees (CIREFCA)[27] report distinguishes between natural disasters victims, who do not qualify as refugees, and other events "seriously disturbing the public order." People displaced by human-made events might qualify for protection but here the challenge would be to determine the causal link between human activities and environmental displacement, which, as we have seen in the first section, remains methodologically and conceptually problematic. Moreover, since the Declaration is not binding on states parties to the OAS, the practical utility in the disputed circumstances of environmental displacement is doubtful.

[27] United Nations High Commissioner for Refugees, Document CIREFCA/89/9 (1989) (English Version)

Ad hoc disaster protection regulations

There are examples of ad hoc migration concessions for victims of natural disasters, which also offer some potential for application to environmentally displaced persons. At the same time, they offer a starting point for developed countries to address their rising concern to find ways of managing the potential impacts of the global movement of environmentally displaced people (see also Martin Chapter 7 in this volume).

Temporary protection measures seem to offer the most promising way forward. For example, in 2003, the US Immigration Service extended for two more years the Temporary Protection Status it granted to 80,000 Hondurans who had fled to the United States after 1998 Hurricane Mitch devastated large parts of Central America.

Similarly, after the 2004 tsunami, the Canadian, Malaysian and Swiss governments temporarily suspended involuntary returns of failed asylum seekers to affected areas of India, Indonesia, Sri Lanka and Thailand. Likewise, Australia fast-tracked the processing of temporary visas for asylum claimants and existing applications from these countries. On a case-by-case basis, rather than 'group determination', the Canadian Ministry of Citizenship and Immigration fast-tracked asylum claims for applicants related to resident families already in Canada (Kolmannskog, 2009b).

Other European countries have legal apparatus to provide subsidiary protection and temporary humanitarian responses to environmental catastrophes and victims of major natural disasters. For example, both Sweden and Finland extend forms of temporary protection to disaster-displaced persons. Both the Finnish and Swedish Aliens Acts (2004 and 2005, respectively) specifically cite environmental disasters (one of the three categories of large-scale population displacement) as grounds for providing temporary protection, although in-country responses and international humanitarian support are proposed as the first line of response (Kolmannskog, 2009b). Such temporary protection may also apply to longer-term conditions, such as the submergence of island states. In such cases, international burden sharing and durable solutions need to be established.

In a similar vein, it is often thought that New Zealand has undertaken to resettle people from Tuvalu who will be displaced by rising sea levels. In fact what exists is a temporary seasonal employment programme which is very limited in its terms. This is far from a model of international cooperation, global environmental responsibility and sharing of the burden of climate change-induced relocation which some authors suggest (Gemenne, 2006). Indeed, New Zealand refers to this agreement as a "migration programme", keeping the programme as low-key as possible, demonstrating the sensitivity of migration responses, even in life-threatening conditions.

In France, specific protection does not exist, but exceptions have been made for victims of disasters. Thus, in principle, Special Decree powers[28] could allow environmentally displaced persons to be accepted on a discretionary basis in exceptional circumstances, rather than on a rights basis, and with some protection from expulsion (Cournil and Mezzaga, 2006).

These ad hoc and special measures are far from offering an effective protection regime for those forced to migrate because of sudden catastrophic weather events, let alone those impacted by slow-onset changes to their environments. Moreover, whether or not this adds up to an evolving soft law norm is highly debatable. It does show some malleability at the edges of immigration policy (Brown, 2007), while, at the same time, highlighting how the environmental migration policy/ protection is mediated by the more pressing needs of immigration control (Cournil & Mezzega, 2006).

Internally displaced persons and the Guiding Principles

Large numbers of the migrants propelled by changing environmental conditions will differ from refugees for an important reason: that they will not have crossed an international border but will be internally displaced persons (IDPs). The 1998 *Guiding Principles on Internal Displacement*[29] cover an important protection gap in relation to the 1951 Geneva Convention and offer both the scope for extending protection to those who are forcibly displaced by environmental conditions, and a model for developing similar guidelines for these involuntary migrants.

The 1998 Guiding Principles are a synthesis of human rights law, parts of refugee law and international humanitarian law – arguably, the synthesis that might equally apply to people displaced because

[28] Translated from the French 'acceuil régalien/protection régalienne'.

[29] United Nations, *Guiding Principles on Internal Displacement*, presented by the UN Secretary-General Francis M. Deng to the United Nations Commission on Human Rights, UN Doc. E/CN.4/1998/53/Add.2.

of environmental conditions. Although not binding, the 1998 Guiding Principles reflect existing, binding international law and are incorporated into an increasing number of national legal frameworks. This constitutes a significant innovation in international normative development (NRC, 2008; see also World Summit[30]).

The *Guiding Principles on Internal Displacement* expressly encompass people who have fled their homes due to natural or human-made disaster. Recognizing that people displaced by disasters have protection needs requiring international attention[31] is a very significant extension to the precepts of refugee protection, providing a crucial entry point to encompass the rights of those forcibly displaced by environmental factors. The incorporation of natural or human-made disasters could certainly be construed to include population displacement caused by extreme weather events due to climate change and, more controversially, might also include displacement precipitated by progressive changes to environmental conditions.

Where the Guiding Principles are invoked, they provide the right for displaced people to request and receive protection and humanitarian assistance from national authorities or, if these are unable or unwilling to do so, from international assistance. In line with human rights law and international humanitarian law, the Guiding Principles require that assistance not discriminate and that IDPs have the right to enjoy the same freedoms as other persons within their country – notably, the right to an adequate standard of living, including the provision of food and water, basic shelter, clothing, essential medical services and sanitation, and protection of property. There is special attention to women, children, persons with disabilities, and the elderly, during periods of displacement. IDPs have the right to seek safety in another part of the country, to leave their country, or to seek asylum in another country; they are also protected against forcible resettlement where they might be at risk. IDPs should also be allowed to return voluntarily and safely to their home, if they so wish, and to enjoy equal access to public services (Ardiles-Martinez et al., 2008).

---

[30] World Summit Outcome, A/RES/60/1, UNGA, 60 sess, para.132 (2005).

[31] The Internal Displacement Monitoring Centre (IDMC) report on non-conflict-induced displacement claims that many individuals and communities displaced by natural disasters confront similar problems and present similar challenges to those displaced by conflicts (McDowell and Morrell, 2007).

Can these Guiding Principles be applied to people who migrate because of more gradual environmental degradation and disasters? Certainly, there are remarkable parallels between the application of the principles to IDPs and the situation of those who are forcibly displaced by environmental conditions, not to mention the express conditions of natural disasters. For example, it could be argued that the environmentally displaced have a right not to be returned to a region were their lives may be at risk due to the frequent occurrence of climate-related natural disasters and where they are incapable of sustaining the most basic human rights. Drought and environmental degradation can possibly be included as well, although slow-onset change is more disputed and would require a definition of the impacts of environmental degradation to be open to an evolutionary interpretation of displacement and protection.

An important consideration here, in extending the protection norms of the Guiding Principles to environmental displacement (or their use as a model for developing similar norms) is that, like the refugee conventions, they assume that displacement is involuntary. Here, the issue is one of categories. As discussed earlier, those moving due to gradual environmental degradation may be more analogous to voluntary/economic migrants rather than those who may be considered forcibly displaced. People may migrate voluntarily because the traditional forms of income or employment are insufficient (NRC, 2008), or as a positive adaptation strategy. These rights and needs might be better addressed by development agencies and strategies. But as environmental conditions worsen, the concern may not be for economic opportunities but for basic survival and, thus, the concept of force comes into play: in this respect, rights protection would more easily fall within protection norms more analogous to IDPs.

There is also the 'visibility challenge' in situations of gradual environmental degradation, in contrast to more dramatic, sudden disasters. Both situations can generate internally or environmentally displaced persons, from the point of view of norms and rights embodied in the Guiding Principles. But Birkeland (2003) claims that environmental factors are seldom given due attention, in this context. She argues that an inclusive meaning of IDPs is preferable to creating discrete and separate categories such as environmental or climate refugees in order to secure rights and assistance.

A protection gap – transborder environmental migrants

Clearly, extant protection norms and instruments provide considerable scope for enhancing the protection needs of people forcibly displaced by environmental factors. However, there is a significant gap regarding transborder environmental migrants. They fall in the gap between the Guiding Principles, human rights norms that place obligations on national governments primarily for their own citizens, and statelessness conventions, since, with few predicted exceptions, such people will not be stateless. Thus, those forced to migrate across borders because of depleting environmental conditions have little, if any, recourse to protection. To some extent, the ad hoc measures, outlined above, represent an approach to the protection for transborder migrants, but these provisions are directed towards temporary protection for migrants from sudden extreme events, such as disasters. The needs of those who are forced to migrate permanently across national borders have yet to be addressed.

Accordingly, it would seem well worth investigating how the various legal instruments, norms and frameworks outlined in this section of the chapter might be constructed into a preliminary framework of rights protection guidelines on environmental displacement. Such a framework would give added impetus to the development of a more coherent set of international guidelines while, at the same time, supporting the development of national-level apparatus.

Conclusions: Developing guidelines for people forcibly displaced by environmental factors

Having outlined the parameters of extant international law and norms of rights protection for displaced people, we now consider their potential to meet the challenges of rights protection for the environmentally displaced. The evidence suggests that, in principle, there is considerable scope for adapting or building on existing norms and instruments to develop a framework of guidelines for a protection regime for those forcibly displaced by depleting environmental conditions and to support the rights of those who do not migrate. As already demonstrated, many of those who will be forcibly displaced due to environmental factors could probably fit into already existing categories of protected persons, although the way they fit within these categories needs to be much more clearly identified. Even so, such an approach is far from offering a comprehensive framework and begs the question of whether it would ever be possible to agree a systematic application of norms and practices to which all states would adhere.

Nonetheless, this is a promising starting point.

What might be the best ways forward? How may the various strands of law and norms discussed above, and especially the 1998 Guiding Principles (as model and substantive framework) be developed to afford protection to those displaced by climate-induced environmental change?

Kolmannskog (2009a and 2009b) suggests that there are two essential steps. First, it is important to clarify the different scenarios and typologies of displacement, since these will invoke different, though overlapping, protection norms.[32] In this respect, he identifies five different scenarios: disaster displacement caused by sudden disasters but where displacement may not be permanent; gradual environmental degradation leading to progressive displacement; displacement resulting in statelessness due to either or both of the previous two phenomena; conflict-induced displacement as a result of environmental degradation and extreme competition over resource depletion; and the impoverishment and vulnerability of those who do not migrate.

Second, Kolmannskog argues that, in addition to considering the different typologies of displacement, protection guidelines must also address the different stages of displacement. Here, he draws on recent work in developing national-level guidelines for domestic law and policy on internal displacement (Brookings-Bern Project, 2008):

- Prevention from displacement – through adaptation measures, as outlined in the first part of this chapter dealing with protecting those who remain, and minimizing relocation, except in high-risk areas where there are imminent threats to lives and livelihoods.
- Protection during displacement – where displacement is organized with due process and deployment of rights-protection machinery, where the displaced fully participate, and where temporary or complementary/subsidiary protection is provided.
- Durable solutions for those displaced – through local integration or resettlement.

[32] See also IASC, *Climate Change, Migration and Displacement: Who will be affected?* Working paper submitted by the informal group on Migration/Displacement and Climate Change of the IASC, 31 October 2008.

To sum up, a combination of extant norms and frameworks of humanitarian, refugee, human rights, environment law and regional and ad hoc measures offers the most promising path to developing a protection regime in respect of environmental migration. An exception to this general proposition is the rights of transborder migrants, both temporary and, more problematically, those who are permanently, forcibly displaced by changing environmental conditions. Short of binding international obligations, but mirroring the process and formulation of the 1998 Guiding Principles on Internal Displacement, these complementary bodies of law set out the principles and the practice for protecting and safeguarding the rights of people who may be displaced by environmental change. Overall, a soft law approach seems to offer the most potential.

4. The extent to which legal and normative frameworks can support the capacity of local and regional governance and civil society structures to implement adaptation and resilience strategies

Migration is sometimes a positive strategy that households, individuals and sometimes whole communities adopt to improve their lives and to reduce risk and vulnerability. Nevertheless, despite the preoccupation of national governments and international organizations with the spectre of mass migration induced by climate change, a much larger number of people will not migrate. A key concept in mitigating migration as an outcome should be the reduction of vulnerability and the promotion of adaptation, resilience and sustainability. This involves recognizing people's agency while bearing in mind the vulnerability of certain social groups, such as the elderly.

There are many options and levels of response for mitigating or adapting to climate change, discussed in the first section of this chapter (the institutional level, technological developments, community development, education and training initiatives) (Boano et al., 2008:18–19).

Just as the issue of migration raises a host of rights-based protection issues, policies aimed at supporting adaptation, resilience and mitigation also have a bearing on the protection discourse. Mirroring national and international policy making, in this context, the rights-based discourse is similarly disjointed and fragmented.

A desk-based study cannot provide a clear overview of the extent to which legal and normative frameworks can support the capacity of local and regional governance and civil society structures to implement adaptation and resilience strategies. Major research is therefore recommended, in the next section, to address this need.

We may conclude the following:

a. Many of the rights-protection norms and legislative frameworks outlined in the previous section apply as much to those who remain as to the displaced. As argued in the conclusion of the previous section, it is important to clarify the different scenarios and typologies of environmental change and, thus, the rights that might be impacted among populations that remain.

b. Adaptation and resilience measures are generally local, while mitigation is predominantly (inter)national-orientated (Klein et al., 2007; Wilbanks et al., 2007). Assessing the role and scope of mitigation and adaptation confirms that "the more localised the scale, the more attractive is adaptation, since many beneficial effects of mitigation are external to the region" (Wilbanks et al., 2007: 724). The conclusion from this is that rights-based protection must similarly be rooted in a bottom-up approach. This further suggests that local governments, local stakeholders and civil society organizations must play a crucial role in developing and protecting the rights-based needs of their populations – within, of course, the framework of national and international norms and legislation.

c. Reinforcing this conclusion, there is increasing awareness, in the climate change discourse, of the need to shift from government to governance and civil society. This has implications for the displacement agenda and, specifically, rights-based protection. It confirms that, whilst governments have a duty to protect rights, responsibility for the advocacy of people's right to be protected from displacement is likely to rest largely with civil society organizations. This shift makes institutional accountability for rights protection more complex.

d. The countries most likely to be affected by the impacts of climate-induced environmental change and the need to develop adaptation strategies displacement are those that already have fragile rights-protection machinery and, often, limited rights-based civil society structures. Environmental degradation will accentuate pressures on this protection machinery and the role of protection in sustaining access to land and water as well as in ensuring that civil and political rights, such as participation in the preparation of development and mitigation strategies, are also safeguarded. In this regard, Bogardi et al. (2007) argue that strengthening institutional capacity needs to be a central element in responses. Indeed, the need to strengthen institutional capacity applies as much to the development of protection machinery as to the wider framework of social and economic development policies to support adaptation and resilience.

5. Research needs and priorities

This chapter has highlighted some of the challenges and opportunities involved in developing norms and legal frameworks for the protection of people whose livelihoods and social and institutional structures are likely to be irreversibly impacted by changes to their environments induced by climate change. Many issues remain unresolved, and more information is needed before an effective framework of rights protection can be put in place for those affected. The following are among the most significant areas of research required to address these issues.

Climate change discourse and human protection

There is a need for further research on how current thinking about adaptation to climate change, and the literature on climate change, more generally, is influencing the way we see development and human protection. The current focus on the environment as a separate and special driver of migration may, paradoxically, have limited the effectiveness with which we have considered protection needs – an argument made in the first section of this chapter. To perceive the environmental impacts of climate change as the drivers of migration promotes the impression that there is a separate category of claims for protection. As we have seen, this position is hard to sustain, given the challenge of establishing the causal chain of climate change–environmental degradation–migration – particularly in the case of incremental rather than acute migratory movements. Moreover, it favours migrants over the larger numbers who will not or cannot migrate. Above all, the putative demand for a separate category of protection needs for environmental migrants precludes the much more fruitful avenue of adapting and reconfiguring exiting apparatus of protection norms and legislative frameworks.

Climate change responsibility, burden sharing and the development of rights-based protection

Another key area requiring investigation is the relationship between the discourse of responsibility for climate change and how this is shaping funding for protection agendas and institutional capacity building. As the interlinked issues of protection and responsibility rise up the agenda and seem certain to be advanced at the UN Climate Change Conference in Copenhagen, more work on this relationship is vital to ensuring that protection is provided as efficiently and effectively as possible. As the third section of this chapter has emphasized, protection needs are becoming more pressing. However, those countries most likely to be impacted by the migratory consequences of climate-induced changes to their environments have the least capacity to resource, defend and promote rights-based agendas across a spectrum of social, economic, political and cultural needs, including the impact of changing environmental conditions. Both in relation to government capacity and civil society organizations in developing countries, we need to develop a much greater understanding of the principles, issues of sovereignty, institutional architecture, and modalities for burden sharing.

Developing the capacity for rights-based protection in developing countries

The issue of burden sharing is linked to a third area of research – developing capacity for rights-based protection in the most impacted developing countries. Given that the migration impacts of climate change will be predominantly on national governments, research in this area is a particularly pressing need.

As the third section of this chapter has emphasized, lack of detailed empirical understanding of how environmental change will precipitate migration or adaptation is mirrored by a lack of understanding of the availability and, more importantly, the scope and efficacy of rights-based frameworks to protect people in those countries most susceptible to the impact of environmental change.

The proposal here is that a spectrum of 'environmentally stressed' country case studies be explored, identifying different typologies of climate change and the contrasting range of changing environmental conditions in developing countries: sudden disasters; gradual environmental degradation; statelessness; conflict-induced displacement as a result of environmental degradation; and the impoverishment and vulnerability of those who do not migrate. Given that the majority of those who are, or will be, displaced will add to the already significant process of rapid urbanization in sub-Saharan Africa and South-East Asia, a focus on rural and urban sectors is needed.

The case studies would investigate, in more detail than can be achieved by desk-based study, the current practices, initiatives and portfolios of legal and normative protection and rights-based frameworks, including procedural and institutional elements, and how these are implemented at national and local levels. The adoption of recently developed national-level instruments and norms under the 1998 *Guiding Principles on Internal Displacement* (Brookings-Bern Project, 2008), and the adaptation of these to the needs of those forcibly displaced by environmental change, would form part of the study. The research would include both governmental and civil society organizations whose remits relate to rights-based advocacy and/or to environmental issues.

The aim would be to establish existing rights-based protection capacities and the scope for adapting and extending these to the needs of those who may be susceptible to environmentally induced forced displacement (both temporary and permanent), as well as those who are similarly impacted but do not migrate. This would provide a baseline understanding of the kinds of protection apparatus that is available and would thus form the basis for developing more robust and comprehensive guidelines and norms. It would also highlight the 'gaps in protection', which are likely to be substantial, and indicate the potential scope for overcoming these gaps.

This study should also include an assessment of the existing and potential institutional capacity of government and civil society, the resource and professional needs and, of course, the modalities of funding. The political sensitivity of this proposal should not be underestimated, given the fragility of rights protection in developing countries.

One significant gap relates to issues of urban land rights and access to land for those forcibly displaced by the impacts of climate change. Even the lowest estimates of forced migration due to climate change indicate that tens of millions of permanent migrants will head predominantly for the cities. They will add to the already huge challenges of managing spontaneous urbanization in developing countries, generated by rural-to-urban economic migration in recent decades. Thus, developing strategies and capacity for urban land delivery and urban management and development are a key priority, as are resettlement programmes.

Baseline mapping would not only provide practical support for the case study countries. Ideally, it should also result in generic findings, principles and practical strategies that could be adapted to assist other countries in confronting the same pressures.

This proposal sits alongside growing international calls to establish national observatories/laboratories to monitor and respond to climate change and environmental displacement in impacted countries. The baseline studies of protection capacity could be one of the functions of these observatories.

The issue of rising sea levels – statelessness and landlessness

The fourth research agenda pertains to the specific case of those whose island states will disappear with rising sea levels.[33] While the incidence, timing and scale of displacement generated by slow-onset environmental change and by extreme weather events remain tentative, where rising sea levels are concerned, the evidence is more compelling than for many other areas of climate change impacts.

This issue raises two specific concerns compared to other climate-change-induced migration outcomes: the greater predictability of this outcome and the fact that people affected will not be able to migrate to higher ground in their own or in adjoining countries. Of course, a key issue here is the fact that, although they will be 'landless' in a literal sense, they may not be stateless in the sense that they will still be citizens of a state, albeit one that no longer exists.

How will the protection needs of this category of people be addressed? Among other key areas, research is needed on: the effectiveness and potential of current initiatives, notably in the Pacific region, which were discussed earlier (ad hoc Protection measures); the extent to which resettlement rights in host countries can be provided and how these

[33] A related issue is whether such people are technically stateless, nationless or merely landless. Such people will still have a nationality, despite being landless and stateless; but their conditions of statelessness are different from those normally covered by the statelessness conventions.

might be protected, both in transit and at the stage of permanent resettlement; the scope for extending legal and normative protection and rights-based frameworks of the Statelessness Conventions to this new category; how claims might be addressed under these conventions.

Transborder migration

Section 2 drew attention to the significant protection gap for the needs of transborder environmental migrants. The scope for extending and adapting current norms and instruments for those forcibly displaced by changing environmental conditions is most limited with respect to this group. Some ad hoc measures were outlined. Research and action are needed (at an intergovernmental level, in the first instance), in order to assess the potential for developing these ad hoc measures, so as to establish more secure doctrines and norms for this group. Addressing the needs of permanent transborder environmental migrants is particularly problematic.

The challenge of protection in developed countries

A sixth research area concerns rights protection for climate change migrants in a global setting.

In the last decade or so, developed countries have struggled to manage the so-called 'new international migration'. Reconciling the conflicting agendas of sustaining international obligations to protect refugees while managing domestic labour market needs and open borders in a restructured global economy, has resulted in contradictory outcomes and the high political saliency of migration. Protection norms for refugees have been severely undermined by increasingly restrictive legislation and border controls, while the need for cheap unskilled labour has encouraged south-north migration, although tighter border controls have accentuated irregular migration such as trafficking and people smuggling – often with dire consequences.

The spectre of large-scale global migration as a result of climate change, although lacking robust methodological or empirical foundations, has alerted developed countries to the possibility of a new era of mixed and complex international migration. As we have seen in section two, some countries – the Nordics and New Zealand – have already anticipated these new dynamics of international migration.

There is currently little understanding of these new dynamics and a lack of coherent responses to them, although the responsibilities of states are implied in the Hyogo Framework for Action[34], are fully articulated in the International Council on Human Rights report (ICHRP, 2008), and are certain to feature at the UN Climate Change Conference in Copenhagen in 2009.

This complex and fluid situation points to a significant research agenda, if global migration patterns resulting from climate change are to be fully addressed. These demands go well beyond rights protection, but this element raises particular research issues. To a large extent, it could mirror the research aims proposed above for developing the capacity for rights-based protection in developing countries. The agenda would address issues such as current policy discourses and initiatives in relation to legal and normative protection and rights-based frameworks for environmentally displaced migrants; the status and protection needs of environmental migrants; the claims that they might make and under what norms and legislative frameworks; the extent to which they might be considered 'forced' migrants, with perhaps special claims for protection; and gaps in protection. These and other issues pose severe challenges to extant norms and legislative frameworks for supporting the rights-based needs of environmental migrants and the obligations of receiving countries to afford such migrants temporary or permanent protection.

The research should encompass a range of countries as well as intergovernmental organizations, such as the European Union.

Developing normative guidelines for environmentally displaced persons

Identifying whether there is an overarching structure to these specific areas of proposed research represents a possible seventh area of study. Although still contentious and disputed territory, an academic and policy discourse is developing for the preparation of international guidelines on climate change and displacement (see, for example, NRC, 2008; ICHRP, 2008; Boano et al., 2008: 24, 30–31; Zetter, 2008;

[34] Hyogo Framework for Action 2005–2015: Building the resilience of nations and communities to disasters (HFA). Available at: www.unisdr.org/eng/hfa/hfa.htm.

Kolmannskog, 2009b). To some extent, this mirrors the preliminary stages of the process in the early 1990s that led to international agreement on the 1998 Guiding Principles on Internal Displacement.

Accordingly, it would potentially be worth researching how the various legal instruments, norms and frameworks outlined in section two of this chapter might be constructed into a preliminary framework of rights protection guidelines on environmental displacement. Such a framework would give added impetus to the development of a more coherent set of international guidelines while, at the same time, supporting the development of national-level apparatus.

An initiative of this kind has particular saliency in light of growing international concern to 'manage' environmental migration, and the crucial window of opportunity presented by the UN Copenhagen Climate Change Conference in December 2009.

An important coda to research priorities

Migration – and forced migration – due to climate change is one possible response. Moreover, concepts and practices of resilience and adaptation by those who remain, as well as the role of migration as a proactive response to changing environmental conditions, all challenge the deterministic and reductionist notion of climate-affected people as vulnerable groups and passive victims. By contrast, these responses highlight people's skills, strategic responses and agency – necessarily built on enhanced institutional capacity and reform to governance and civil society.

In all the research proposals, therefore, especially with regard to the country case studies, it is essential that research into rights protection and human security norms and instruments focus as much on how rights may enable or constrain adaptation and resilience strategies as on those who migrate. For those who remain in increasingly pressurized environments, rights-protection issues include protection of access to land and water rights; the scope, rights and empowerment of community-based and civil society organizations to engage in strategies and decision making; and the effectiveness of these governance structures.

6. Conclusions: the supremacy of a rights-based approach

The scenario of millions of people on the move (erroneously labelled 'climate refugees' and based on weak empirical and methodological evidence) has generated alarm at the potential international scale of impact, the threats to human security and the risk of ensuing conflict.

In light of these concerns, what can be concluded from this chapter is that upholding and developing international and national human rights obligations is a vital element in the global response to the migratory impacts of climate change. A rights-based approach provides the means of both averting some of these migratory outcomes of climate change and addressing some of the challenges that such migration will create, regardless of the degree of 'force'. Although it has evolved and developed in the context of specific categories of forced migration, such as refugees, the normative basis of human rights protection provides a conceptually and operationally valuable framework for addressing the forced migration consequences of climate change.

It provides the framework to redress situations where rights are threatened or breached. It offers the means of addressing the needs of migrants who are forced to leave their familiar surroundings and enter new environments with new rights entitlement and protection needs. At the same time, climate change, as argued above, will not necessarily lead to displacement; in fact, the majority of affected people will probably not migrate. Similarly, therefore, a rights-based approach offers protection and support for the adaptation and resilience strategies for those who remain, thereby also reducing the impacts of climate change and the propensity to migrate.

Protection and human security instruments and norms do not have the immediacy of physical, spatial and developmental strategies and policies needed to respond to climate-induced displacement. Nonetheless, providing and enhancing protection capacity remains an essential component of a comprehensive approach to the challenges of displacement and adaptation at local, national and international levels.

Linking the protection discourse to environmental displacement opens up significant policy-relevant potential and outputs. For intergovernmental organizations and agencies, the investigation would promote international dialogue and facilitate development of international norms, instruments and guidelines for protection and human security.

At national and local levels, strengthening protection norms and instruments constitutes an essential foundation for supporting vulnerable populations in impacted countries. At the same time, promoting a rights-based perspective of protection and an analysis based on entitlements, can also be used as a tool to indicate the parameters for other 'hard' and 'soft' policies and strategies for the environmentally displaced – for example, rights of access to land, housing, free movement, participation and empowerment in decision making on resettlement.

Environmental change does not, of course, undermine rights and security in isolation from the broader conditions of poverty, weak development and poor governance. It is important, therefore, to locate and promote the inclusion of environmental displacement impacts in this wider policy discourse on protection, rights and security. Thus, the growing significance of the protection discourse in the field of environmental displacement indicates the need for enhancing governance and civil society structures to ensure that affected populations are fully involved in developing response strategies, and that advocacy tools and processes are in place to promote their rights. A concern with people's rights highlights community-focused needs and approaches.

Given the inevitability of ecosystem degradation and the resulting increase in the numbers of those who will be displaced, as well as the vulnerability of those who remain, there is a compelling case for ensuring that the protection machinery embraces this new challenge of climate change.

7. References

Adger, W.N.
2000 "Institutional adaptation to environmental risk under the transition in Vietnam", *Annals of the Association of American Geographer*, 90(4): 738–758.

Adger, W.N.
2006 "Vulnerability", *Global Environmental Change,* 16(3): 268–281.

Ardiles-Martinez, D. et al.
2008 "Human rights and climate change study", OHCHR and New South Wales Young Lawyers, Sidney. Available at: http://www.lawsociety.com.au/uploads/files/1221178952462_0.992712916169245.pdf

Baechler, G.
1999 "Environmental degradation in the south as a cause of armed conflict", in: *Environmental Change and Security: A European Perspective*, Carius, A., Lietzmann, K. (eds), Springer, Berlin: 107–130.

Belcher, J. & Bates, F.
1983 "Aftermath of natural disasters: coping through residential mobility", *Disasters,* 7: 118–128.

Bell, D.
2004 "Environmental refugees: what rights? Which duties?" Deborah Charles Publications, Liverpool, UK.

Berkes, F. & Jolly, D.
2001 "Adapting to Climate Change: Social-Ecological Resilience in a Canadian Western Arctic Community", *Conservation Ecology,* 5(2). Available at: www.ecologyandsociety.org/vol5/iss2/art18/print.pdf

Birkeland, N.
2003 "Last time I fled because of war, this time because of hunger" – Environmental Change and Internal Displacement in the Huambo Province, Angola", in *In the Maze of Displacement*, Høyskoleforlaget, Kristiansand, Norway.

Boano, C. T. et al.
2008 *Environmentally displaced people: understanding linkages between environmental change, livelihoods and forced migration*, Forced Migration Policy Brief No.1, Refugee Studies Centre, University of Oxford. Available at: www.rsc.ox.ac.uk/PDFs/RSCPB1-Environment.pdf

Bogardi, J. et al.
2007 *Environmental Degradation and Migration*. Available at: http://www.berlin-institut.org/online-handbookdemography/environmental-migration.html

Brookings-Bern Project
2008 *Protecting Internally Displaced Persons: A Manual for Law and Policymakers*, Brookings Institution, Washington, DC.

Brown, O.
2008 "Climate change and forced migration; observations, projections and implications", *UNDP, Human Development Report 2007/2008*, Occasional Paper: Fighting climate change: human solidarity in a divided world, UNDP, Geneva. Available at: www.hdr.undp.org/en/reports/global/hdr2007-2008/papers/brown_oli.pdf

Campbell, K.M.
2008 "Climatic Cataclysm: The Foreign Policy and National Security Implications of Climate Change", Brookings Institution, Washington, DC.

Cannon, T.
2000 "Vulnerability analysis and disasters", in *Floods*, Parker, D.J. (ed.), Routledge, London.

Castles, S.
2002 *Environmental change and forced migration: making sense of the debate*, UNHCR Issues in Refugee Research, Working Paper No. 70, UNHCR, Geneva. Available at: www.unhcr.org/research/RESEARCH/3de344fd9.pdf

Christian Aid
2007 *Human Tide: the Real Migration Crisis,* Christian Aid, London. Available at: www.christianaid.org.uk/Images/human_tide3__tcm15-23335.pdf

Conisbee, M. & Simms, A.
2003 "Environmental refugees: The case for recognition", *New Economics Foundation*, London.

Cournil, C. & Mazzega P.
2006 "*Catastrophes écologiques et flux migratoires: Comment protéger les 'réfugiés écologiques'?*", *Revue Européenne de Droit de l'Environnement*, 4: 417–427. Available at : www.flautre.net/IMG/pdf/cournil_mazzega_REDE.pdf

De Bruijn, M. & van Dijk, H.
2003 "Changing population mobility in West Africa: Fulbe pastoralists in Central and South Mali", *African Affairs,* 102(407): 285–307.

Findley, S.E.
1994 "Does drought increase migration? A study of migration from rural Mali during the 1983–1985 drought", *International Migration Review,* 28:539–553.

Gemenne, F.
2006 *Climate Change and Forced Displacements: Towards a Global Environmental Responsibility? The Case of the Small Island Developing States (SIDS) in the South Pacific Ocean,* Centre for Ethnic and Migration Studies (CEDEM), University of Liège, Belgium. Available at: www.cedem.ulg.ac.be/m/cdc/12.pdf

Gorlick, B.
2007 *Environmentally Displaced Persons: a UNHCR Perspective*, a presentation [TO BE CONFIRMED/ADDED] http://www.equatorinitiative.org/images/stories/events/2009events/brian_gorlick_environmentally_displaced_persons_unhcr_perspective.pdf

Hampshire,K. & Randall, S.
1999 Seasonal Labour Migration Strategies in the Sahel: Coping with Poverty or Optimising Security? *International Journal of Population Geography,* **5**, pp. 367-385.

Henry, S. et al.
2003 "Impacts of environmental conditions on migration in Burkina Faso", *Population and Environment*, 25(5): 423–460, Population Association of America, Minneapolis.

Hussein, K. & Nelson, J.
1998 *Sustainable Livelihoods and Livelihood Diversification,* Working Paper 69, Institute of Development Studies, Brighton. Available at: www.ids.ac.uk/ids/bookshop/wp/wp69.pdf

ICHRP
2008 *Climate Change and Human Rights: A Rough Guide*, International Council on Human Rights Policy (ICHRP), Geneva.

IOM
2007 *Expert Seminar: Migration and the environment, International Dialogue on Migration*, No. 10. IOM, Geneva.

Keane, D.
2004 "Environmental causes and consequences of migration: A search for the meaning of 'environmental refugees'", *Georgetown International Environmental Law Review*. Available at: http://findarticles.com/p/articles/mi_qa3970/is_200401/ai_n9353848

Kibreab, G.
1997 "Environmental causes and impact of refugee movements: a critique of the current debate", *Disasters* 21(1): 20–38.

Klein, R. J. T. et al.
2007 "Inter-relationships between adaptation and mitigation", in *Climate change 2007: Impacts, adaptation and vulnerability. Contribution of Working Group II to the Fourth Assessment Report of the Intergovernmental Panel on Climate Change,* Parry, M.L. et al. (eds.), Cambridge University Press, Cambridge: 745–777.

Kolmannskog, K.
2009a "To what extent can existing forms of legal and operational protection apply to climate change-related displacement in general and cross border displacement in particular". Paper presented at the RSC Workshop on Environmental Change and Migration, University of Oxford, 8–9 January 2009.

Kolmannskrog, K.
2009b *Guidelines for Climate Change, Migration and Displacement*, unpublished communication, February 2009.

Lee, S.
2001 *Environment matters: Conflict, Refugees and International Relations*. World Human Development Institute Press, Tokyo.

Lonergan, S.
1998 "The role of environmental degradation in population displacement", *Environmental change and security program report*, 4:5–15, Woodrow Wilson International Centre for Scholars, Washington, DC.

Lopez, A.
2007 "The protection of environmentally-displaced persons in international law", *Environmental Law Review 365,* Clark Law School. Available at: www.lclark.edu/org/envtl/objects/37-2_Lopez.pdf

McAdam, J. & Saul B.
2008 *An insecure climate for human security? Climate-induced displacement and international law,* Sydney Centre for International Law, Working Paper No. 4, Faculty of Law, University of Sydney. Available at: www.law.usyd.edu.au/scil/pdf/SCIL%20WP%204%20Final.pdf

McDowell, C. & Morrell, G.
2007 *Non-conflict Displacement: A Thematic Literature and Organisational Review*, Internal Displacement Monitoring Centre (IDMC) of the Norwegian Refugee Council (NRC), Geneva.

McGregor, J.A.
1993 "Refugees and the environment", in *Geography and refugees: Patterns and Processes of Change*, Black, R. & Robinson, V. (Eds), Belhaven Press, London: 159–70.

Myers, N.
2005 "Environmental refugees and emergent security issue", *13th Economic Forum,* 23–27 May 2005, Prague. Available at: www.osce.org/documents/eea/2005/05/14488_en.pdf

NRC
2008 *Future Flood of Refugees: A Comment on Climate Change, Conflict and Forced Migration*, Norwegian Refugee Council (NRC), Oslo.

Oreskes, N
2004 "Beyond the ivory tower: The scientific consensus on climate change", *Science,* 306(5702): 1686.

Paavola, J. & Adger, W.N.
2006 "Fair adaptation to climate change", *Ecological Economics*, 56(3): 594–609.

Pelling, M. & Hight, S.
2005 "Understanding adaptation: What can social capital offer assessments of adaptive capacity?" *Global Environmental Change*, 15(3): 308–319.

Prieur, M. et al.
2008 "Projet de convention relative au statut international; des déplacés environnementaux", *Revue Européenne de Droit de l'Environnement*, 4: 381–393.

Renaud, F. et al.
2007 "Control, adapt or flee: How to face environmental migration", *InterSecTions,* UNU–EHS, No. 5/2007. Available at: http://www.ehs.unu.edu/article:345

Reuveny, R.
2007 "Climate change-induced migration and violent conflict", *Political Geography,* 26: 656–673.

Reuveny, R.
2008 "Ecomigration and violent conflict: Case studies and public policy implications". *Human Ecology,* 36(1): 1–13.

See, M.
2007 *Draft protocol on environmental refugees: A report on the original meeting and the proposed amendments to the 1951 convention and 1967 protocol.* Ministry of Environment, Energy and Water, Republic of Maldives.

Suhrke, A.
1994 "Environmental degradation and population flows", *Journal of International Affairs*, 47(2): 437–496.

Wilbanks, T. J. et al.
2007 "Toward an integrated analysis of mitigation and adaptation: some preliminary findings", *Mitigation and Adaptation Strategies for Global Change*, 12(5):713–725.

Zetter, R.
2008 "Legal and normative frameworks", *Forced Migration Review*, 31:62–63, Special issue on Climate Change and Displacement, Refugee Studies Centre, University of Oxford. Available at: www.fmreview.org/FMRpdfs/FMR31/62-63.pdf